A CULTURAL HISTORY OF THE HUMAN BODY

VOLUME 6

A Cultural History of the Human Body
General Editors: Linda Kalof and William Bynum

Volume 1
A Cultural History of the Human Body in Antiquity
Edited by Daniel H. Garrison

Volume 2
A Cultural History of the Human Body in the Medieval Age
Edited by Linda Kalof

Volume 3
A Cultural History of the Human Body in the Renaissance
Edited by Linda Kalof and William Bynum

Volume 4
A Cultural History of the Human Body in the Age of Enlightenment
Edited by Carole Reeves

Volume 5
A Cultural History of the Human Body in the Age of Empire
Edited by Michael Sappol and Stephen P. Rice

Volume 6
A Cultural History of the Human Body in the Modern Age
Edited by Ivan Crozier

A CULTURAL HISTORY OF THE HUMAN BODY

IN THE MODERN AGE

Edited by Ivan Crozier

BLOOMSBURY

LONDON • NEW DELHI • NEW YORK • SYDNEY

Bloomsbury Academic
An imprint of Bloomsbury Publishing Plc

50 Bedford Square
London
WC1B 3DP
UK

1385 Broadway
New York
NY 10018
USA

www.bloomsbury.com

Hardback edition first published in 2010 by Berg Publishers, an imprint of Bloomsbury Academic
Paperback edition first published by Bloomsbury Academic 2014

British Library Cataloguing-in-Publication Data
A catalogue record for this book is available from the British Library.

ISBN: HB: 978-1-84788-793-1
PB: 978-1-4725-5467-3
HB Set: 978-1-84520-495-2
PB Set: 978-1-4725-5468-0

Library of Congress Cataloging-in-Publication Data
A catalog record for this book is available from the Library of Congress.

Typeset by Apex CoVantage, LLC, Madison, WI, USA
Printed and bound in Great Britain

CONTENTS

List of Illustrations vii

Acknowledgments xi

Series Preface xiii

Introduction: Bodies in History—The Task of the Historian 1
Ivan Crozier

1 Death and Birth 23
Malcolm Nicolson

2 Performing the Western Sexual Body after 1920 43
Ivan Crozier

3 The Technological Fix and the Modern Body: Surgery as a Paradigmatic Case 71
Thomas Schlich

4 Diseased Bodies in the Modern World 93
Anna Crozier

5 Popular Beliefs 109
Dan O'Connor

6 Beauty and Concepts of the Ideal 127
Christopher E. Forth

7 Re-markable Bodies 147
Anna Cole and Anna Haebich

8 Body Marks (Bestial/Divine/Natural): An Essay into the Social and Biotechnological Imaginaries, 1920–2005 and Bodies to Come 165
Michael M. J. Fischer

9 Dissolution, Reconstruction, and Reaction in Visual Art, 1920 to the Present 201
Ana Carden-Coyne

10 The History of the Body: Self and Society, 1920–2000 221
D. M. Vyleta

Notes 249

Bibliography 305

Contributors 339

Index 341

ILLUSTRATIONS

INTRODUCTION

Figure 0.1: Eric Avery, *Blood Test.* 15

Figure 0.2: Eric Avery, *LIVER DIE.* 16

Figure 0.3: Eric Avery, *Arriverderci.* 17

Figure 0.4: Eric Avery, *SARS Ward (Now).* 18

CHAPTER 1

Figure 1.1: The typhus louse shaking hands with death. 25

Figure 1.2: Voluntary Euthanasia Society poster. 28

Figure 1.3: Ultrasound image used in advertising. 38

Figure 1.4: Ultrasound image of the chambers of the heart of a dead fetus. 39

Figure 1.5: Rectangularization of Survivorship Curve. 40

CHAPTER 2

Figure 2.1: Eric Avery, *How to Use a Male and Female Condom.* 52

Figure 2.2: A Princess Diana and a vertical clitoral hood piercing. 59

Figure 2.3: Prince Albert piercing. 60

CHAPTER 3

Figure 3.1: Physician Walter Freedman and neurosurgeon James Watt examine a lobotomy patient. 77

Figure 3.2: Portrait of Dr. Carrel, *Chanteclair*, no. 138, Paris, 1906. 84

Figure 3.3: *Undreamt of Perspectives, or: Surgery in the Year 2000.* 85

CHAPTER 4

Figure 4.1: A handkerchief being blown away by a sneeze. 97

Figure 4.2: Face of a man suffering from smallpox; vaccination against smallpox. 98

Figure 4.3: The malaria mosquito. 103

Figure 4.4: *He Trusts You.* 105

CHAPTER 5

Figure 5.1: Runner in action (D.G.A. Lowe). 111

Figure 5.2: Male athlete having cardiovascular fitness tested. 115

Figure 5.3: Dianne Harris, *Man or machine?* 119

CHAPTER 6

Figure 6.1: Leaflet and coupon for Oatine Cream and Oatine Snow facial cream. 130

Figure 6.2: Photographs of noses. 132

Figure 6.3: Two paths to physical perfection. 136

Figure 6.4: Speedos advertisement. 137

Figure 6.5: Oliver Burston, *Illustration of a Man Running on a Treadmill.* 145

CHAPTER 7

Figure 7.1: Ali Ismaeel Abbas after U.K./U.S. bombing raid on Iraq, April 2005. 149

Figure 7.2: Australian Aboriginal man from Montgomery Island. 152

Figure 7.3: An elder Mbakwa-Manja woman of Ubango Shari, Central African Republic, holding knife used for female genital excision. 158

Figure 7.4: Six arms showing different tattoo designs. 163

CHAPTER 8

Figure 8.1: Eric Avery, "Hands Healing: A Photographic Essay" 167

Figure 8.2: Shirin Neshat, *Offered Eyes*, 1993. 171

Figure 8.3: Guillermo Gómez-Peña, *La Geisha Apocaliptica.* 180

Figure 8.4: Guillermo Gómez-Peña, *El Indio Amazonico.* 181

CHAPTER 9

Figure 9.1: Carolee Schneemann, *Interior Scroll*, 1975. 213

Figure 9.2: Guy Feldman, photo of Marc Quinn's statue of Alison Lapper pregnant. 218

CHAPTER 10

Figure 10.1: Soldier and peasant woman. 225

Figure 10.2: "In Praise of Technology"—scientists working for the benefit of the family. 229

Figure 10.3: Idealized portrait of the "Peasant and Worker State." 239

ACKNOWLEDGMENTS

The editor wishes to thank the British Academy for a Small Research Grant used to visit Eric Avery in Texas and Washington; the Wellcome Trust and Eric Avery for allowing the reproduction of so many images; and Anna Crozier for her help in the editing process. For her brilliance at indexing, thanks is also due to Amanda Sordes.

SERIES PREFACE

A Cultural History of the Human Body is a six-volume series reviewing the changing cultural construction of the human body throughout history. Each volume follows the same basic structure and begins with an outline account of the human body in the period under consideration. Next, specialists examine major aspects of the human body under seven key headings: birth/death, health/disease, sex, medical knowledge/technology, popular beliefs, beauty/concepts of the ideal, marked bodies of gender/race/class, marked bodies of the bestial/divine, cultural representations and self and society. Thus, readers can choose a synchronic or a diachronic approach to the material—a single volume can be read to obtain a thorough knowledge of the body in a given period, or one of the seven themes can be followed through time by reading the relevant chapters of all six volumes, thus providing a thematic understanding of changes and developments over the long term. The six volumes divide the history of the body as follows:

Volume 1: A Cultural History of the Human Body in Antiquity (750 B.C.E.–1000 C.E.)
Volume 2: A Cultural History of the Human Body in the Medieval Age (500–1500)
Volume 3: A Cultural History of the Human Body in the Renaissance (1400–1650)
Volume 4: A Cultural History of the Human Body in the Age of Enlightenment (1650–1800)
Volume 5: A Cultural History of the Human Body in the Age of Empire (1800–1920)
Volume 6: A Cultural History of the Human Body in the Modern Age (1920–21st century)

General Editors, Linda Kalof and William Bynum

Introduction

Bodies in History—The Task of the Historian

IVAN CROZIER

Forty years ago, a cultural history of the body would have been essentially unthinkable. Histories were written of events, of people, of ideas, of politics, of societies, but bodies had somehow slipped through the historiographical net. And yet, when a number of historians, enriched by various anthropological and sociological writings, began to focus on cultural history, the body became a central site of reference—as something on the cusp of the prediscursive that demonstrated the direct action of power.[1] Early historians who explored such interests, such as Norbert Elias and Michel Foucault, did much to encourage an interest in the body.[2] So, too, did medical historians, who looked up from their detailed studies of the development of medical theory and practice and started to take account of the bodies that were persistently being spoken of in their sources.[3] Before long, the body was as much of a historiographical object as the French Revolution or World War I—and, indeed, the body has been prominently placed in these two historiographies.[4] This tectonic shift was also in part brought about by the rising status of theoretical speculation, mingling aspects of psychoanalysis and Marxism,[5] and the broadening of historiography in this abstract arena, to include a wider range of interpretations of culture. The body became an important testing ground for numerous theories about gender, sexuality, identity, psychoanalysis, and postcolonial discourses about race. Now, as this series shows, body history is well and truly situated within academic historical studies. Even "traditional" historical topics such as war,[6] medicine,[7] class,[8] and education[9] have taken a corporeal turn and emphasize the body as a mediating point between ideas and social reality: as something at once "real" and imagined, objective and subjective.

Accounts of the body in the twentieth century have profited from the increase in self-conscious reflection that accompanied developments in the plastic arts, literature, the sciences, and popular culture. All of these fields have in turn benefited in various ways from the increasing uncoupling of sexuality from "traditional" (or religiously induced) morality and from reproduction, which has consequently put the uses of bodies up for negotiation at a point where there was previously more sanctioning. The increasing articulation concerning what bodies do (their actions and pleasures, their destruction in war, the ways they can be depicted or described) and how they differ (across gender, class, ethnicity, aesthetics, etc.) gives historians of the modern period much material to draw on, as these issues were very much at the heart of the birth of modernism. The corporeal turn of the twentieth century has involved new conceptions of the self that are visited in many of the chapters in this book. Such self-conceptions cannot be disassociated from the changes being wrought in this context. This period covers, in no particular order: world wars; the rise of feminism; a growth in class consciousness and the introduction of increasingly automated workplaces; interactions with bodies from other cultures, made possible by imperialism and the formulation of sciences of ethnography and anthropology; reflections on issues of the self, the state, and sexuality in terms of psychology as well as eugenics; and developments in medicine that afforded major changes in the understanding and treatment of disease. The resulting knowledges and experiences—of bodies damaged and destroyed by the horrors of war; the recapitulation of the "rights" over women's bodies and the uses to which they could and should be put; the effects of factories on working bodies; the reconceptualization of racial bodies; and the changing status of illness (and the possibilities of treatments)—brought about drastic new circumstances in which to think of the body. These changes, when drawn together with advances in artistic representation, including photography, and the wide distribution of body images through magazines, cinema, and eventually the Internet, have meant that bodies can be represented in ways that were impossible in earlier periods. This increased access to bodies in the twentieth century has had both a cementing and a destabilizing effect—by the end of the century, television shows (*Extreme Makeover*, *Ten Years Younger*, etc.) had put the body in such a position that it became the final arbiter of social acceptance, above intellectualism, social position, or talent, while at the same time becoming one of the key sites for anxiety (in terms of weight, fashion, aesthetics, etc.). In order to map these monumental social changes onto the twentieth-century body, it is profitable to focus on separate problems, although it will soon be recognized that these categories meld into one another.

One of the key problems historians face when approaching the body concerns evidence and interpretation. This problem is initially a psychological one—rather than abjection, there is a Cartesian sense that all bodies are experienced as our own. Bodies are both internalized and cast out in our comprehension; they maintain something of the prediscursive, despite the fact that historians have learned to think of bodies in terms of systems of discourse. When we read about pain or pleasure, it is appreciated as our own—or it remains unsympathetically unimaginable. Studies in the transmission of tacit knowledge highlight formal aspects of this problem well.[10] If knowledge has a tacit component, and this transmission has to involve learning how to perform successfully,[11] then we are left with an assumption that equivalent actions are based on identical experiences. But if bodies differ, then knowledge and action must also be "personal," as Hungarian philosopher Michael Polanyi has it.[12] We see this problem particularly when the body is situated in various ways.

If bodies are not to be treated at this personal level by the historian, if we are to move away from the conception that all bodies are our own, then bodies must be cast into some schema for interpretation: We have to become reflexively aware of the discursive parameters through which we access and interpret bodies. This approach has largely been adopted in this series, with bodies placed alongside one another in terms of general themes, including the medical, sexual, social, cultural, artistic, aesthetic, and religious. In the chapters in this volume, bodies are shown to be negotiated through specific regimes and conceptualizations. They are not treated as purely natural entities but as the product of various context-specific interventions. Only by placing bodies in their (discursive and historical) contexts can they be understood.

This standpoint further relates to the problem of representation. As we cannot access past bodies directly, we require them to be represented, either visually or through other forms of conceptualization—written sources being the most common form of historical material. Accessing past bodies requires a skill at reading through sources, as "the body" is not always engaged with directly in the text. Rather, hints and suggestions must be scraped together and interpreted. It is for this reason that bodies lay under the historiographical radar for so long, failing to register as we focused on other events than those through which the body was inscribed directly.[13] Much of what is found in this book follows this reliance on fragmentary evidence rather than offering an overview of *the* body in history and culture. Nevertheless, some general historiographical themes emerge, which are tackled in the following.

DISCIPLINED BODIES

Historians have, on the whole, been rather comfortable with a sociocultural appreciation of disciplined bodies since Foucault's genealogical writings, especially

Discipline and Punish (1975). In this work, Foucault traced the way that bodies at the end of the late eighteenth and early nineteenth century came under particular regimes of power within specific institutions that situated them not in opposition to the king or the state but in terms of new organizations—the prison, the school, the military barracks, the hospital, and so on—that exerted control over the ways in which subjects lived and related. He wrote,

> What was then being formed was a policy of coercions that act upon the body, a calculated manipulation of its elements, its gestures, its behaviour. The human body was entering a machinery of power that explores it, breaks it down and rearranges it. A "political anatomy", which was also a "mechanics of power", was being born; it defined how one may have a hold over others' bodies, not only so that they may do what one wishes, but so that they may operate as one wishes, with the techniques, the speed and the efficiency that one determines. Thus discipline produces subjected and practiced bodies, "docile" bodies.[14]

Such corporeal techniques for control—which typified the particular institutions in the modern age that Foucault described—are seen in a modified form in any number of practices that are adopted as "normal" in society: hygiene, sexual relations, diet, health, appearance, manners, clothing, poise, and so on. Bodies are situated inextricably within these particular power relations. In each of these regimes, different aspects of the disciplined body are focused on in sui generis specificity to the point that there is no "ontological independence of the body outside of any one of those specific regimes."[15] To situate these bodies, I concentrate in this introduction on health, gender, race, and sexuality before turning to the role played by modern biomedicine in the preceding processes. These different fields are often interlinked, so that different disciplinary orders play off one another in the control of modern bodies. Some of the possibilities for such intertwining are explored in the works of physician-artist Eric Avery.

HEALTHY BODIES

Healthiness, and particularly the conceptions of health that have been widely disseminated throughout the twentieth century (diet, regimes of exercise, lifestyle, medical checkups, etc.), can be considered a paradigmatic form of discipline, although in ways that are not identical to the specific processes described by Foucault when focusing on post-Enlightenment institutions or on classical Greek sexuality. Conceptions of health beg the question of why there is such an imperative for living a healthy lifestyle. Of what moral value does healthiness consist, other than the long-term avoidance of suffering? Answers could include a political economy of lifestyle and could take issue with the state's role in constructing health to serve its own economic ends. They could also take into account issues such as class difference—which is demonstrated clearly when one considers eating habits and the price of fresh vegetables in contemporary Britain, making healthy eating a marker of economic and social status, evidenced by the organic chic of the farmers' market. But such answers would skirt the embodiment of health at an individual level—the action of power directly on the body and the associated exercising of agency in order to fit the sanctioning of the particular health regime. In this case, healthiness is a point of accessibility in the discussion of disciplined bodies, as it is an embodiment of external values. Foucault had much to say about this embodiment:

> Mastery and awareness of one's own body can be acquired only through the effect of an investment of power in the body: gymnastics, exercises, muscle-building, nudism, glorification of the body beautiful. All of this belongs to the pathway leading to the desire of one's own body, by way of the insistent, persistent, meticulous work of power on the bodies of children or soldiers, the healthy bodies.[16]

Such ideas hang together well with Robert Baden-Powell's *Scouting for Boys: A Handbook for Instruction in Good Citizenship* (1908), which extolled to young boys the message "you must train yourself to be strong, healthy, and active as a lad."[17] Healthiness, more than simply avoiding long-term disease, is allied to fortitude and to manliness.[18] It signifies much more in culture than the moral terms in which it has been wrapped. This aspect can be better examined through attention to the growth of health movements throughout the twentieth century, which have received much attention by historians in recent years.[19] Particularly interesting work has been done on the way modern subjects adopted a variety of bodily practices, such as naturism (and the *Freikörperkultur* movement in particular), vegetarianism, yoga, calisthenics, and so on, in order to illustrate individual agency and the embracement of new corporeal possibilities through health practices—essentially attempts to reconstruct the self through health.[20] These subjects are not merely cultural dopes following trends but are actively involved in the establishment and maintenance of social institutions of health in their actions—while at the same time they are able to be situated in specific, and historically contingent, power relations beyond the scope of their bodily desires.[21] Healthy acts are thus both consecutively determined and the products of (limited) agency.

In opposition to the healthy, fit, and often masculine body was the slovenly, unhealthy, fat body. This fat body not only opposed ideals of health; it also was increasingly gendered as feminine. Christopher Forth has shown specifically how the gut came to signify a lack of fortitude, and to be gendered feminine even on the most masculine of men, such as Émile Zola.[22] But, more usually, fat has been taken up as a feminist issue.[23] This movement was kicked off by Judy Freespirit and Sara Aldebaran, who wrote the *Fat Liberation Manifesto* in 1973, soon followed by Susie Orbach's *Fat Is a Feminist Issue* (1978). In these texts the case was made for "size acceptance." Their political heirs are now members of organizations such as Largesse, whose mission is "to create personal awareness and social change which creates a positive image, health and equal rights for people of size."[24] Such efforts consist of renegotiating the politics of fat; they struggle to effect a shift in power relations, moving the body from a regime based on slim aesthetics and health to a heavier situation in which the espoused feminist ideals come to the fore. Such a vignette shows that bodies and their interpretations are not fixed; they can be anchored to a particular system, but they can slip their moorings just as easily.

GENDERED BODIES

To backpedal a little, it is highly important to consider the broader development of feminism in relation to the body in the modern period; indeed, it is perhaps one of the defining social movements of our century. Feminism drastically changed the status of the body in the twentieth century. Topics such as birth control, marriage rights, education (and physical ability to undertake such), work, and sexual pleasure dominated much feminist writing in the early part of the century and helped to reconfigure the body as a political site. This attention not only changed the way that individual women experienced their bodies (via childbirth, sexual pleasure, medical innovation, etc.) but became something of a social-reform agenda in terms of eugenics.[25]

Not all feminist concerns were as socially demanding or overtly political. One aspect of women's changing position in society was the altering of female dress codes during this period in ways that displayed the body in a new light—trousers became one option, and short-length skirts another. Newer (post-1970s) feminist concerns maintain a fashion-conscious edge. Issues such as eating disorders, weight, physical appearance, and so on have all been discussed in terms of the pressures put on modern women to conform to attractive ideals.[26] These issues escalated after World War I, when women entered "public" society in terms of work and other roles, in much more conspicuous ways. The coincidence with the rise of mass media such as magazines and cinema also played a significant role in shaping the way that women were perceived, and widely distributed ideals of beauty.[27] One of the ramifications of this has been the resort to body-modification surgery—from minor procedures that can be performed at a "beauty" salon, such as Botox injections to remove wrinkles and laser treatments to remove unruly pubic and facial hair, to major surgery, such as liposuction, nose reshaping, breast augmentation and reduction, and even reconstructing the genitalia to conform to aesthetic ideals. Such efforts to reshape the body are not new—they have been around since at least the 1920s.[28] But these practices have become widely dispersed, to the point where they are used as entertainment on British and American television (see shows such as *Ten Years Younger* and *The Swan*, in which women are subjected to a variety of procedures to give them an "ideal" appearance, superficially at least).[29] These body modifications are an attempt to stave off time, even death, as the aging body has a somewhat decreased value in modern society.[30] Although men also participate in cosmetic plastic surgery, they are not (yet) targeted to the same level as women. Such practices can be seen as examples of the "technical fix" provided by modern medicine.

Cosmetic plastic surgery is often directed to the bodily performance of a gendered ideal. This is clear in the case of breast-enhancement surgery but is also present in the removal of fat from thighs or hair from women's lips. But surgery is, of course, not always necessary to adopt a feminine mien. Rather, while gendered performances of bodies are situated in past performances, these do not determine future sanctioned gender performances in any hard sense.[31] Individuals adapt the accepted norms of femininity (and masculinity as well) to new instances, although not without the sanctioning of the society to which they belong.

This notion of performing gender (rather than "being" a gender) is most forcefully promulgated by theorist Judith Butler, the current last word in gender theory, which has had an immense impact on contemporary feminist theorization of the body.[32] For Butler, gender is not part of an essential identity. It is not something fixed on an individual. It is not the last point of difference between bodies (male and female). Rather, it is something that is performed; it is acted out in specific, ritualized, repeated performances of the gendered body. To quote Butler:

> Performativity cannot be understood outside of a process of iterability, a regularized and constrained repetition of norms. And this repetition is not performed *by* a subject; this repetition is what enables a subject and constitutes the temporal condition for the subject. This iterability implies that "performance" is not a singular "act" or event, but a ritualized production, a ritual reiterated under and through constraint, under and through the force of prohibition and taboo, with the threat of ostracism and even death controlling and compelling the shape of the production, but not, I will insist, determining it fully in advance.[33]

As such, bodies should be considered to be underdetermined; they have to be thought of as social institutions as much as parts of the natural world.[34] Butler is not saying that there is no materiality of the body, but she is problematizing the idea that subjectivity can be thought of as deriving from the natural body directly, in terms of naturalized binary gender forms. This idea is held dearly by many poststructuralist philosophers and historians of the body, and the many examples in this series can be thought of as demonstrating the fluid position that bodies hold in history. In this sense, gender, like other bodily aspects, is historically contingent on power relations that ebb and flow over time.

RACIAL BODIES

Crucial for understanding bodies in the twentieth century is a racialized reading.[35] Race and ethnicity—and the difference they entail—have been central to a number of social trends and theoretical ventures—colonialism, exoticism, postcolonialism, and so on—at the heart of which is the body. These stances are, of course, not solely products of the twentieth century; no doubt a number of other chapters in this series will document clearly how racial ideas were inherited from earlier periods as far back as the Greeks.[36] But the changes in value of "racial" difference—and in particular racial bodily difference—do seem especially pronounced in this century, not least because of the drastic changes in sociopolitical status that formerly subaltern groups faced after the collapse of apartheid in South Africa; political changes in the United States that gave equal humanitarian and political rights to blacks; the acknowledgment of British citizenship for South Asians, Caribbeans, and Africans from former colonies; and the recognition of the political status of Australian Aboriginals, to use some well-known examples. In other words, the world changed very much in regard to ethnicity in the twentieth century as it was indelibly marked on the political and social landscape of the West. Some of the ramifications for the body with regard to race have involved moving away from models of difference that were predicated on racialist conceptions of the inferiority of the non-European body. These precepts underlay many of the discourses produced in the colonial period and can be seen in the medical and scientific writing of the period.[37]

A flip side to these ideologies about racial bodies can be demonstrated by the susceptibility of the white body (and mind) in "hot climates." Ideas about the tender frame of the white man in the tropics have existed for centuries, although again these notions were exacerbated in the colonial period when colonials were overcome by the heat or succumbed to the evils of drink or opium as a way of coping with the hostile environment.[38] Tropical neurasthenia, as the concomitant mental debilitation with attendant physical symptoms came to be known, was a constant concern of those far from their European homelands, which they nevertheless did their utmost to reconstruct in the cooler climates of the higher altitudes, such as the Ngong Hills of Kenya or the hill stations of Shimla in India.[39] This is not to undermine the very real effects of tropical illnesses such as malaria or a host of other indigenous diseases that burdened the colonialists—another issue that raises the specter of racialization in terms of local health problems.[40] It seems that early twentieth-century colonialists spent much time preoccupied with bodily difference along racial lines as a means of rationalizing the imperial enterprise.[41] One of the chief axes through which this ideology cuts is the body, which recently has attained a more prominent position in colonial historiography.

These ideas of racially different bodies are perpetuated in a more positive light with regard to the black athletic body, generally regarded as a superior manifestation of physical prowess. Sport provides for the objectification of bodies in significant ways, some of which are aesthetic. But as John Hoberman has argued in the case of African Americans, athletic prowess has become a significant feature of the community, as sport became a dominant defining characteristic of black success, with other role models from the social and cultural fields lacking.[42] Of course, not all sports have been so dominated. For example, the rather colonial sport of cricket, played throughout many of England's former colonies, is currently dominated by the Australian team.

There is nevertheless a racial dimension to cricket that is worth exploring in terms of the objectification of particular players. In the summer of 1976, ethnic tensions were running high in London. The Notting Hill Carnival (an annual celebration of Caribbean culture held in August) ended in a riot against the police. In the same month, the West Indian cricket team was touring England for a test series. Regarding the tour, South African–born English captain Tony Greig said of the West Indies in a television interview, "We'll make them grovel." The groveling was all England's, with a 3–0 series win to the West Indies, capped by Michael Holding's marvelous 8/92 and 6/57 at The Oval and Viv Richards's score of 291 in the first inning, not needing to bat in the second. Particular attention should be called to the way that Holding, who earned the sobriquet "Whispering Death" because of his graceful, silent bowling action, was portrayed in various commentaries—a combination of brutal and beautiful emerges, with raptures over his long, hypnotic run-up.[43] Holding was a key figure in the next so-called "Blackwashes" of English cricket in 1984 and 1985–1986, when the West Indies won all ten tests, placing them at the top of the world cricket stage. Various discussions of Holding's action ensued; currently, it is described as follows on cricinfo.com (one of the premier cricket Web sites): "It began intimidatingly [*sic*] far away. He turned, and began the most elegant long-striding run of them all, feet kissing the turf silently, his head turning gently and ever so slightly from side to side, rhythmically, like that of a cobra hypnotising its prey."[44] Such lyrical descriptions help to underline Holding's aesthetic dimensions. Other contemporary white fast bowlers, such as Dennis Lillee, Richard Hadlee, Ian Bowtham, and others, are not described in such aesthetic terms; their bodies are not so central to the depiction of their style, despite being equally formidable.

The various intersections between the body and race, which are only touched on here, show the ways in which the body must be situated in terms of other social institutions in order to be understood. Such social institutions both constrain and support possible actions; they do so in a way that appears natural but is heavily imbued with culture.[45] This is not to suggest that all instances of the body are reducible to race but rather to emphasize that this is one aspect that is present in any representation of the body, whether it be based on difference or on homogeneity that ignores the racial element in a totalizing way.

SEXUAL BODIES

As I argue in the chapter on sexuality, focusing on the sexual body allows us to bring together many of the developments in medicine, feminism, and self-expression that can be found more generally in the twentieth century. Fundamental issues of sexuality, such as masturbation and homosexuality, became acceptable to many, a position that these did not hold in earlier centuries. Some of the more recent details concerning developments in the corporeal performance of sexuality in the twentieth century are discussed in my chapter in this volume,[46] so I limit my discussion here to some notes on the way that sexuality has changed both the body and the conceptions of the sexual self in the past century.

The politics of sexuality and identity has been a major aspect of twentieth-century thought, going so far as to lie at the center of the work of two of the century's brightest philosophical stars, Michel Foucault and Judith Butler. This situation developed largely as an application of rational Enlightenment ideals—particularly skepticism—to notions that sex was a natural phenomenon. The growth of ethnographic techniques also allowed sex researchers to consider what people were actually *doing* for sexual pleasures, and the findings, which showed all manner of polymorphous behaviors, came to undermine seriously some of the conceptions of "natural" sexual behavior.[47] This undermining opened the way for new theories of sexuality based on fluidity of identity (after some of the fixed notions of psychoanalysis came under criticism in the later part of the century) and also encouraged a celebration of sexual identities in such a way that labels and fixed concepts of a "sexual self" opened up to less rigid interpretations.[48] Changing conceptions (and practices) of sexual pleasure are pivotal in this development.

One of the most significant changes in sexual practices in the twentieth century has been the widespread effects of HIV/AIDS. Following a brief window of sexual "liberation" in the post-pill 1970s, in which sexual mores were forever changed in the liberal West, the specter of AIDS again linked sex to death in ways reminiscent of syphilis in the nineteenth century. In the post-AIDS world, sexual propaganda emphasizes "safe sex," or lower-risk sex, such as using condoms. This message has seen its critics from the religious right as well as in other cultures, where sexual expressions are hampered by other dogmatic ideologies. The impact of epidemiological factors on the uses of the sexual body, as well as the cultural representations of the sick body that closely followed the AIDS society of the 1980s, both impacted on sexuality in significant ways, with a major response coming in the form of a more open discussion of sexuality, its varieties, and associated risks. Sexual education now no longer pertains only to getting pregnant but encompasses a wide range of other sexual health issues in which the body is key.

Despite its relative unpopularity in the discourse of the contemporary history of sexuality, it is worth remembering that reproduction is allied to sexuality. Although it has been widely accepted since at least the work of the nineteenth-century sexologists that reproduction does not exhaust the possibilities of sexuality, it is nevertheless important to consider the way that sex has been medicalized so as to allow for biomedical interventions into what was hitherto thought of as the "natural" purpose of sex (i.e., reproduction). The reproduction of bodies has been a major site for contest and development throughout the twentieth century—more so than at any other time. At the beginning of the century, the struggle by radical political reformers and feminists to have the spreading of birth control knowledge be accepted continued the work of their nineteenth-century colleagues.[49] An increase in eugenic discourse across Europe soon followed.[50] These efforts to control reproduction on the basis of the "perfection" of the race—an issue in which the body is a key signifier—were taken to extreme lengths under the National Socialists in Germany, as discussed by Paul Weindling among others.[51]

Apart from separating reproduction from sexual intercourse, efforts to engage with reproduction included a number of biomedical interventions. In the past thirty years, a great number of techniques have developed that allow intervention into the reproductive process. Sexual intercourse—although still the main way to get pregnant—is not the only essential criterion for reproducing in an age of test-tube babies and assisted births. "Producing parents" has become a new aspect of biomedical practice that has changed the place of the body in reproduction.[52] A large amount of work addresses the "new reproductive sciences," from genomics to invasive technologies that enable pregnancy, although the extent to which the body is problematized in this work is not always ideal.[53]

The body plays a fundamental—but not the only—role in all of these aspects of sex. Understanding sexuality is key to understanding the culture of the twentieth-century West, and to understand that, it is essential to address the many places of the body in sexual acts of all kinds. The sexual body marks the twentieth century as different from other epochs in its unsubtle display of sexual acts in a number of locations, from popular culture to

Internet pornography. As the prohibitions of censorship lifted throughout the West over the course of the second half of the twentieth century, sexual bodies were ever more widely exhibited. These issues are related to specific examples in the chapter on sexuality later in this volume.

BIOMEDICAL BODIES

The dominance of biomedical understandings of the body is clear in the preceding brief investigations into aspects of the disciplined body, especially in relation to health and sex. We live in an age dominated by modern medical concerns—be they panics about avian influenza or methicillin-resistant *Staphylococcus aureus* outbreaks, or moral and ethical consternation about biomedical advances such as stem cell research, organ transplants, and genomic practices—not to mention the vast interest in reproductive technologies, from so-called designer babies through to women's right to choose to terminate unwanted pregnancies (both of which have seen considerable fallout from the right in the spurious name of morality and religion). In all of these cases, the biomedical body is not just the domain of specialists concerned with the production of knowledge about particular therapeutic or physiological issues. Politics has a great impact on the development of scientific knowledge in these areas. And modern politics, ever out to please the voters, takes a particular interest in the voices of the lay public. This lay interest in modern biomedicine, in addition to its political status, is significant, as it shows the depth to which biomedical ideas have been almost universally adopted in modern Western society.[54] It is now widely recognized that medical encounters are not simply "top-down" practices in which medical ideas are forced on unwilling participants.[55] Rather, patients are active in their engagement with biomedical issues, from low-level decisions about eating a healthy diet and drinking moderate levels of alcohol through to decisions that are made in consultation with a health-care professional. Informed consent is the norm for medical intervention in the current period, now that doctors' control over passive patients has been largely relinquished in a litigious culture mindful of patient rights.

That patients are taking an active role in the care of their selves underlines the idea that a modern biomedical framework has been widely adopted, not just by the medical elite but by the wider society. Issues such as pain relief, contraception, and childbearing, to give a few examples, are all framed in terms of a body of medical knowledge and practice. This is partially the product of regulation that consolidated medical power (as well as systems of health care, from the state in Europe and Australasia to private insurance in North America); it also shows a wide acceptance of the place of medicine in society—not unjustifiably, as we now live in an age in which major diseases have either been eradicated (smallpox) or can be held at bay (HIV), provided the proper health-care arrangements are in place. The impact of this on people's lives is incalculably significant. But resistance is still possible; to use the examples I cited earlier, people can administer their own pain relief ("medical" use of cannabis or alternative health practices such as herbalism or acupuncture), they can adopt nonmedical contraception techniques (e.g., condoms or relying on body temperature to calculate safe periods), and they can forgo the usual hospitalization during childbirth by using a midwife rather than an obstetrician, preferring a home birth to one in a hospital. These decisions are not taken lightly and are done in terms of an awareness of biomedicine—they are not a wacky return to nature but are informed choices that involve individuals taking control of their bodies in medical situations.

Despite these alternative paths, biomedicine is the dominant way of conceiving of the body in the West. The work of medical anthropologists and sociologists has done a lot to show the ways in which the body is both mediated and experienced through these biomedical means, both in the contemporary West and in non-Western cultures. The results of this wide body of knowledge have shown that biomedicine is a locally performed institution, not a universal, homogeneous concept or practice. While regulative practices, as well as the globalization of pharmaceutical companies, have done much to assimilate a variety of biomedical traditions, to properly understand biomedical encounters with the body, local interactions must be studied.

ART, MEDICINE, AND THE BODY: AN INTERLUDE WITH ERIC AVERY

A mediation between biomedical bodies and artistic bodily representations is found in the work of American artist and physician Eric Avery.[56] Focusing on Avery's work allows many of the strands of this introduction to be pulled together, as issues of health, illness, sexuality, gender, race, and the body are all in some way present in his oeuvre. Indeed, much of Avery's work has been directed toward politics—from the ravages of drought and starvation the artist saw when he worked as a medical superintendent at the Las Dhure refugee camp in Somalia in 1980, to other human-rights issues, such as the plight of refugees fleeing the wars of Central America, as addressed in *View across the Rio Grande, the River of Death, from San Ygnacio, Texas* (1983) and *The Sleep of Reason from Behind* (1986). While these political works have depicted suffering in one way or another, the bulk of Avery's artistic attention has been directed toward medical issues—primarily AIDS, but more recently hepatitis C and general health issues, themes in which the body is central.

Avery's AIDS works began with *Blood Test* (1983), an early reaction to the HIV epidemic that was beginning to be noticed in the United States and Europe.[57] Avery's first HIV test was taken in Houston in 1983 after he moved there from New York City. In the notes that accompany the image on his Web site, he recollects, "My pre-test counselling indicated that I was at risk for getting a positive result. During the stressful two weeks while I waited for my results (HIV-), I drew my arm and cut the image."[58] This engagement with HIV/AIDS led to a number of other powerful works that explore medical knowledge through visual imagery. These medical encounters with art have evolved in Avery's work into "art/medicine actions," efforts to disrupt the boundaries between medical practice and artistic representation, as well as to alter the perceived function of the gallery space. In these actions, he makes the artworks into medical acts—passing on medical knowledge, depicting the results of poor health practices, raising the viewer's consciousness regarding the state of the world's health, incorporating blood tests for various viruses (HCV, HIV)—while also making medical practice into an artistic act by transposing it from the clinic to the gallery, by breaking the private/public divide that is maintained in regular medical activity, by placing medical encounters in an aesthetic space. Avery goes well beyond the idea that prints are to be hung on the wall for decoration. His actions are political acts; they challenge the viewers to reevaluate their relationships with biomedicine, disease, art, galleries, and the role of these in society. Such actions are best displayed in a recent example of Avery's work, *LIVER DIE* (2005).

LIVER DIE was an art/medicine action that intervened into debates surrounding the representation of the (sick) body and into medical practices, public-health campaigns, and the role of galleries in society. It choreographed all of these roles in a space created within the gallery. The clinical art space for *LIVER DIE* was a ten-by-ten-foot structure with three paper walls made from six-by-nine-foot linoleum block prints of the *BIG SICK LIVER* (2005). The print is based on an anatomical image showing the stages of liver damage caused by the hepatitis C virus. The front of the clinic is open, allowing viewers to observe the clinic in action.

Avery describes *LIVER DIE* as follows:

> At scheduled times during the conference, various clinical activities will occur in the art space. A blood drive is scheduled so that conference participants can donate blood and in the process find out if they have been exposed to the Hepatitis C virus. A physician and a nurse will offer clinical care to patients with Hepatitis who have agreed to participate. A Hepatitis Support Group meeting will be held in the clinic.
>
> Hepatitis C is the most common blood-borne disease in the U.S., affecting about 4 million Americans. It is more common than HIV. Because Hepatitis C often has no symptoms, many people do not know they have the disease and may be infecting others. Effective treatment is available.

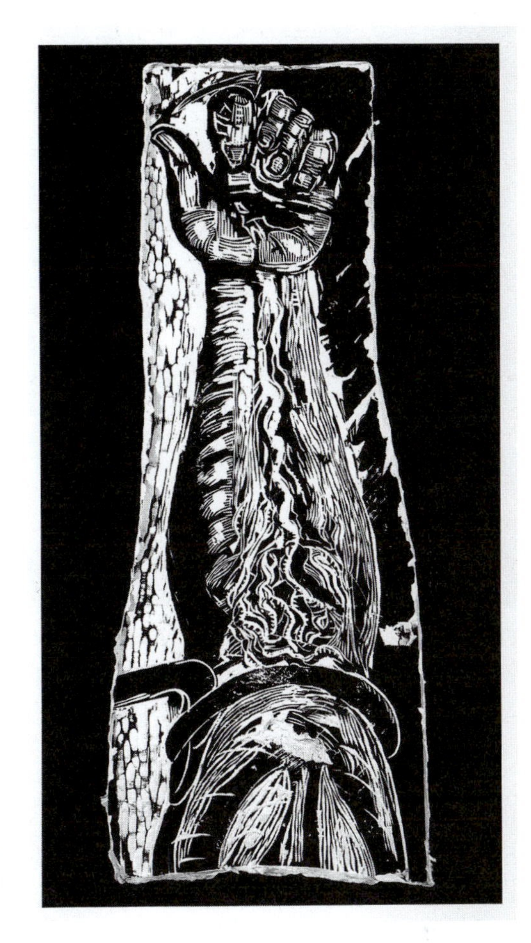

FIGURE 0.1: Eric Avery, *Blood Test*, 1983. Reproduced with permission of the artist.

> A clinic in an art museum is disorienting because it disrupts the protective shield constructed by museums against the existence of the transitory and debilitating facts of human physical life. When medicine, usually confined to hermetic spaces in clinics and hospitals, is presented in the unexpected context of beauty and human creativity, we are reminded of the ancient connections between healing and art.[59]

The impact of *LIVER DIE* was more than an artistic exploration of the body in a gallery; money was made from the sale of *LIVER DIE: A Print Action for Health* (2005), a pamphlet that incorporated a reduced image of *BIG SICK LIVER* and medical commentary on the illness.[60] This money was donated to a HCV clinic in Washington, D.C. Further, blood that was collected on April 1, 2005, was distributed through the American Red Cross, taking the transition from art to medicine a step further in that sick people benefited directly from Avery's action.

FIGURE 0.2: Eric Avery, *LIVER DIE*, Corcoran Gallery of Art, Washington, D.C., March 31–April 3, 2005. Reproduced by permission of the artist.

Other important medical prints in Avery's work include representations of his patients' bodies. In a series of portraits of his HIV patients, Avery refocuses his work on the sufferers of AIDS. This added dimension of his biomedical art interventions allows Avery to consider the human side of illness and health and to explore the individual suffering rather than the more abstract conceptions of disease or contagion. The bodies represented are fragile, mediated through various medical techniques. They demonstrate other aspects of AIDS than simply a lifestyle—such as ethnicity, poverty, and so on. They are a far cry from the vigorous bodies seen in depictions of the sexually active homosexual community, such as *Blue Bath* (1987) and *View of Galveston from the Gay Beach* (2001), having more in common with the suffering represented in Avery's human-rights prints of the 1980s.

More recently, Avery has focused on representing other health epidemics, such as a print of a patient with severe acute respiratory syndrome (SARS), isolated from his caregivers, who keep their distance, hands behind their backs or making notes, without any physical interactions with the patient, who is attached to various medical technologies, a distant prospect from the *Pieta*, which the work echoes. In this image, bodies are biomedically mediated even to the point of death—a bleak future awaits the sick man at the foreground of the print. The context of Avery's work—half of his time is spent as a hospital psychiatrist at the University of Texas Medical Branch in Galveston—becomes clear in the sanitized space in which the SARS encounter takes place.

FIGURE 0.3: Eric Avery, *Arriverderci*, 1999. Reproduced with permission of the artist.

Focusing on Avery's work allows many of the themes germane to the body in the twentieth century (sexuality, race, health, illness, difference, gender, etc.) to be drawn together in novel ways. Avery's interpretation of bodies, informed by his medical work but also placed in the context of Western art, is brimming with a real concern for suffering that is powerful in its directness. Through Avery's work we are shown the repercussions of our actions—they work as talismans against unsafe sex and other unhealthy lifestyle choices. We are also shown worlds of suffering that are not usually presented—refugee camps and human-rights abuses, as well as illness and medical intervention. The message

FIGURE 0.4: Eric Avery, *SARS Ward (Now)*, 2006. Woodcut on Japanese handmade Houshoshi paper, 21 × 30 in. Edition: Artist's proof. Reproduced with permission of the artist.

that Avery gives, however, is wholly one of hope. By raising these issues for the viewer, Avery strives to provide the necessary means of survival. He does this in a corporeal-centric way by focusing on illness, sexual pleasure, death, pain, and biopolitics, confronting viewers with their own mortality and the decisions they can make to extend their lives.

IDENTITY AND THE BODY

Following Julia Kristeva's *Powers of Horror*, *abjection* can be seen as letting go of something one would still like to keep—casting out that which is dear.[61] The abject is removed from our life while remaining as fragments in our selves. This process is central to the construction of our self (and our identity) and is perpetual. An example commonly used in relation to the body is that of a corpse.[62] As Kristeva sees it, the abject is outside the symbolic order of the self—in this example, death is outside of life but comes from it. As such, being forced to come to terms with the dead creates trauma, not only because of the loss of life but also because we are seeing ourselves after ceasing to live. On finding a corpse, one is (usually) repulsed as one is forced to face an object outside of one's world. The corpse is, to quote Mary Douglas's famous words pertaining to dirt, "matter out of place": death in the midst of life.[63] But abjection here also refers to the object that is a part of ourselves—namely, the body as a corpse that is animated. This seeing of the self outside of the self, seeing the living self in the corpse, characterizes much of the way we relate to bodies. We construct our identities—our inner selves—through relating to the "outside" world, to other bodies, internalizing other projections. To see other bodies is to see our self; we construct our bodies according to sanctioning practices that allow them to be cast out into society in a manner that causes least trauma to the self. It is in this sense that I wish to use Kristeva's idea—to address the way that bodily identity is cast out as a projected self-image.

Despite portraying the body as having been subjected to a multitude of disciplinary regimes that operated more or less through techniques of authority such as biomedicine, the twentieth century has also allowed for the use of the body as a method of self-expression. These two need not be seen as necessarily contradictory—some of the possibilities for expressing the self through the body have been consciously and positively adopted. Others have taken the opportunity to extend the body—and thus the self—through a variety of techniques designed to mark the body as individual. One of the key ways this happens is through the adoption of health practices, which act as a bonding agent in a number of ways, from pumped-up gay men enacting hypermasculinities, to the state of belonging to a sports team.[64] In these two examples, bodily identity is enacted so as to convey belonging and therefore identity. Other examples of bodily identity might include goths dressing in black, dyeing their hair various colors, and piercing their bodies as a form of belonging to a group—and actively disassociating themselves from others through their bodies. Other examples found wandering the streets of modern life include those with heavy piercings and tattoos (not simply the bourgeois enactments of fashion, such as a safely hidden tattoo on the small of the back that is exposed out of office hours, but those for whom bodily modification is a form of life—an inked reconstruction of the self). These corporeal projects clearly position the person's identity in a specific cultural milieu.[65]

Bodily identity is not simply reducible to group identity; some more extreme examples demonstrate the way that the body *becomes* a work of art itself—here I am thinking not only of fantastic tattoos but also of performance artists such as ORLAN and Stelarc, to use two famous examples discussed in chapters in this volume, who use the body as a tool for reconstructing the possibilities of the self.[66] This reconstruction renders identity fluid and also cuts it away from fixed identities. If bodies are perpetually able to be re-performed (as I argue here), identities follow in a similarly fluid way. The ability to adapt the body, while central to the construction of identity, should also make us wary of those who try to permanently ground such essences in the body itself; the body is a site for action and interpretation, not a final arbiter in the construction of the self, despite the prominent role it plays.

The ability of the body to be morphed, convoluted, constructed, and construed by the person does not imply, however, that one has rights over the body. At an obvious level, there are various legal restrictions on the kinds of pleasure that people may take with their bodies—prohibitions of drug use or of varieties of sexual practice (remember that until the late 1960s, many countries prohibited sodomy, and many still do so); until recently, sanctions included body piercings and tattooing in some jurisdictions. I address a final example in slight detail to illustrate that bodies are not solely controlled by the owner. The case of an English homosexual sadomasochistic trial—the so-called Spanner Trial—is considered in order to argue that there are still limits to the use of bodies, and therefore to the construction of identity, despite the recession of other bodily prohibitions over the course of the twentieth century. Sadomasochism makes a particularly suitable example, as it involves physically manipulating the body through a variety of games and tortures, in relation to power relations that are established in the act, in order to follow a specific desire. It enacts what Foucault has called a "limit experience."[67]

In 1987, a videotape showing a number of English men engaging in heavy sadomasochistic sexual activities, including beatings and genital torture, was sent anonymously to the police. The police instigated a murder investigation, although after interviewing numerous gay men, they learned that none of the men in the video had been killed nor had any required medical attention for their injuries. A number of men were tried, and sixteen were found guilty on December 20, 1990, of a number of offences, including causing actual bodily harm. They were either sent to jail, given suspended jail sentences, or fined. The men had all pleaded not guilty based on the fact that they had all consented to the activities. Rant J, who presided over the case, did not accept consensual sadomasochistic sex as constituting a just exception to the law of assault. Some of the defendants appealed their convictions, unsuccessfully—although they were awarded some reductions in sentence, as it was felt that the men might have been unaware that their activities were illegal. Further petition against the Court of Appeal's ruling took the case to the House of Lords, where the Law Lords heard the case in 1992, delivering their judgment in January 1993. The prosecutions were upheld by a majority of three to two. Further, the Law Lords ruled that sadomasochistic activity provided no exception to the rule that consent is no defense to charges of assault occasioning actual bodily harm or causing grievous bodily harm.

The Spanner case introduced some new definitions as to lawful and unlawful bodily harm. Rant J noted that bodily harm applied or received during sexual activities was lawful if the pain it caused was "just momentary" and "so slight that it can be discounted." His judgment applied also to bodily marks such as those produced by whipping or chaining up the bottom in a rough manner. As such, any injury, pain, or mark that was more than trifling and momentary was illegal and would be considered an assault under the law. This case shows that prohibitions on sexual practices are still in place. Clearly, childbirth as a result of sexual activity would not meet any of the criteria for a lawful sexual encounter, being neither momentary nor slight and often leaving a variety of scars and marks on the woman's body.[68]

In this example, the use of the body for personal ends is prohibited, regardless of consent or desire. The body in law is not simply the property of the individual, but as the case made clear, what we do with our bodies is significantly regulated.[69] The same argument could be directed toward a number of other activities, from the pleasurable (e.g., drug prohibitions and their impact on individual experience)[70] to the extreme (consider cases of amputation of healthy limbs).[71] In appreciating the body, in experiencing it, it is important not to lose sight of the sanctions that are placed on it, not just socially but legally. We may live in a time when the body is able to be performed in an ever-widening array of forms, but it is (we are) not free.

CONCLUSION

In the preceding introduction, a number of themes were raised that aimed at a few key points. The central issue is that bodies are performed social institutions. Individuals have agency, they have volition, but this agency is not universal. It is constrained through various techniques of training, practice, and sanctioning. Bodies become apparent through use; if we want to study them (outside of the anatomical tradition of the sixteenth century),[72] we have to consider them in action. This is not, however, a reductionist argument about culture. My second concluding point is that bodies are not simply natural kinds[73] or the last arbiter of corporeal truth. They have to be understood in a sociocultural and historical context, as the product of a set of specific individual, societal, and historical contingencies. Such contingencies derive from my third point: Bodies are mediated through a variety of discourses and arrangements of power. These power arrangements might appear through practices of health, punishment, aesthetics, and so on. The analyst needs to be aware of such contingent regimes and to place the body according to their action. Again, this is not a suggestion that leads us toward totalization. Another aspect of considering bodies in space and time is an awareness that the "same" body differs according to locale. The body is not used the same way when it is sick, during sex, as it ages, for pleasure, for work, for sport, or when it is represented (according to art or fashion). Bodies may appear to be the one stable aspect of our identity, but when we look at their uses, and the experiences we have through the body, we find that they are malleable, fluid, adaptable entities that are the artificial stratum of life.[74] Only by appreciating this underdetermined character of the corporeal, by taking on board the ramifications of such a position, can we properly treat the body in history.

If we accept the preceding points, then it must also be accepted that the proper way to study bodies is in their sociohistorical context. Bodies do not make sense in the abstract. They are deeply imbedded in culture and cannot be prized out like an oyster from a shell for inspection before devouring. What follows in this *Cultural History of the Human Body* in the twentieth century is an excursion into the fragmentation of the body into a myriad of discursive schemes. Rather than remaining whole, a common denominator for experiences or representations, the body in this volume is scattered among a variety of approaches, an ontological entity at the limit of (historical) discourse. It is seen to be instrumental in reconfiguring the self via any number of intellectual or cultural practices, from art to medicine to techniques of pleasure and sanctioning. The body is shown in this series to be highly unstable, in terms of changes over time and in terms of direct attempts at modifying the body, either through "official" interventions or personal efforts to adapt the body to new aesthetically, athletically, culturally, or sexually desired forms. The body has become a project for both participants and would-be eugenic masters of the human race. This project of making the body is enacted through technologies that enhance the techniques for control, modification, and representation. The body cannot be looked on outside these parameters. It has to be situated in order to be understood and experienced, as the following chapters clearly show.

CHAPTER ONE

Death and Birth

MALCOLM NICOLSON

Until the middle of the nineteenth century, Death was something of a free spirit. He (Death is usually, but not always, as we shall see, personified as male) swung his scythe about quite haphazardly. It was well known, of course, that he displayed a strong preference for gathering in the old and the very young. Anthony Trollope captured a common experience when, in *Barchester Towers*, he described the elders of Saint Ewold's assembling in the churchyard surrounded by the "graves of their children and their forefathers."[1] But apart from those two favorite targets, Death was fickle and indiscriminate, arbitrarily striking first in one direction and then the other. This capricious behavior was encapsulated in the popular image of the "Dance of Death."[2]

In the second half of the century, however, the rationalizing spirit of industrial capitalism, having already successfully imposed new disciplines on manufacturing production, time, and the laboring body, turned its attention to the activities of the Grim Reaper. British epidemiologist William Farr, working in the Office of the Registrar-General, developed his "statistical nosology," standardizing fatal-disease categories into 27 precise definitions, later increased to 180.[3] Deaths were tabulated by disease, parish, age, sex, and occupation. Consistent patterns of mortality began to emerge. Dying was no longer quite such a random event as it had previously appeared to be.

Karl Pearson brought this statistical disciplining of Death to fruition in his pioneering essay, "The Chances of Death," published in 1897.[4] Human beings died, it was revealed, in strict formation, along a sinuous S-curve of mortality. The initial section of the curve sloped sharply downward, then it leveled off as mortality rates fell in late childhood and adulthood, only to plunge again as old age arrived. What was particularly interesting to Pearson was that the exact shape and slope of the survivorship curve were specific to the standards of civilization prevailing in any given population. It is telling that Pearson suggested a new metaphor for Death. He was now neither a dancer nor a scythe man, but a "marksman," whose aim was "perfectly regular in the mass, if unpredictable in the individual instance." When, in 1920, Raymond Pearl, a student of Pearson and professor of biometry at Johns Hopkins University, gave his seminal Lowell Institute Lectures on "The Biology of Death," the scientific study of Death had begun in earnest.[5]

In the early decades of the twentieth century, biologists developed the idea, derived originally from August Weismann, that the life span of each animal species was a product of natural selection, as indeed was the fact that animals died at all.[6] (It had been noted by this time that unicellular organisms appeared not to senesce.) In other words, the timing of the final plunge of the S-curve was fixed genetically, for the benefit of the species. By implication, Weismann's arguments about the selective nature of the life span applied to human beings as well as animals. Humankind's "three score years and ten" was no longer seen as a divine dispensation but rather as an aspect of our evolutionary inheritance. The theorists of the New Darwinian Synthesis, which came to dominate evolutionary biology in the 1940s, rejected the group selection that was implicit in Weismann's original thesis but still sought to explain life span in terms of selective advantage.[7] Our deaths were timed, so it seemed, to maximize the transmission of our genes to following generations. Death had, of course, long been thought of as subject to powers greater than his own—"Death, thou shalt die," as John Donne put it.[8] But, to the evolutionary biologists of the twentieth century, the forces that controlled Death were mundane and secular rather than divine.

As well as having to submit to the indignity of scientific scrutiny and secular control, Death suffered a few further vicissitudes in the twentieth century. By 1920, western Europe and parts of North America had seen a decline in overall mortality rates that had already been maintained for forty years and that would continue for the rest of the century.[9] This was a sustained reverse for the Grim Reaper that was unprecedented in human history. Admittedly, from his perspective, the prospect was not wholly bleak. World War I had seen the methods of industrial production applied to the destruction of human beings with remarkable success. The 1919 influenza pandemic, aided by the rapidity of modern transportation, had killed even more people.[10] But the populations of Europe and North America, better nourished and better housed than ever before, did not decline. Any deficit was quickly made up. In a generation or so, numbers were up to what they would have been had the twin disasters never happened.

To a large extent, the key to the resilience of twentieth-century population levels, in the face of huge mortality pressure, was the steady decline of

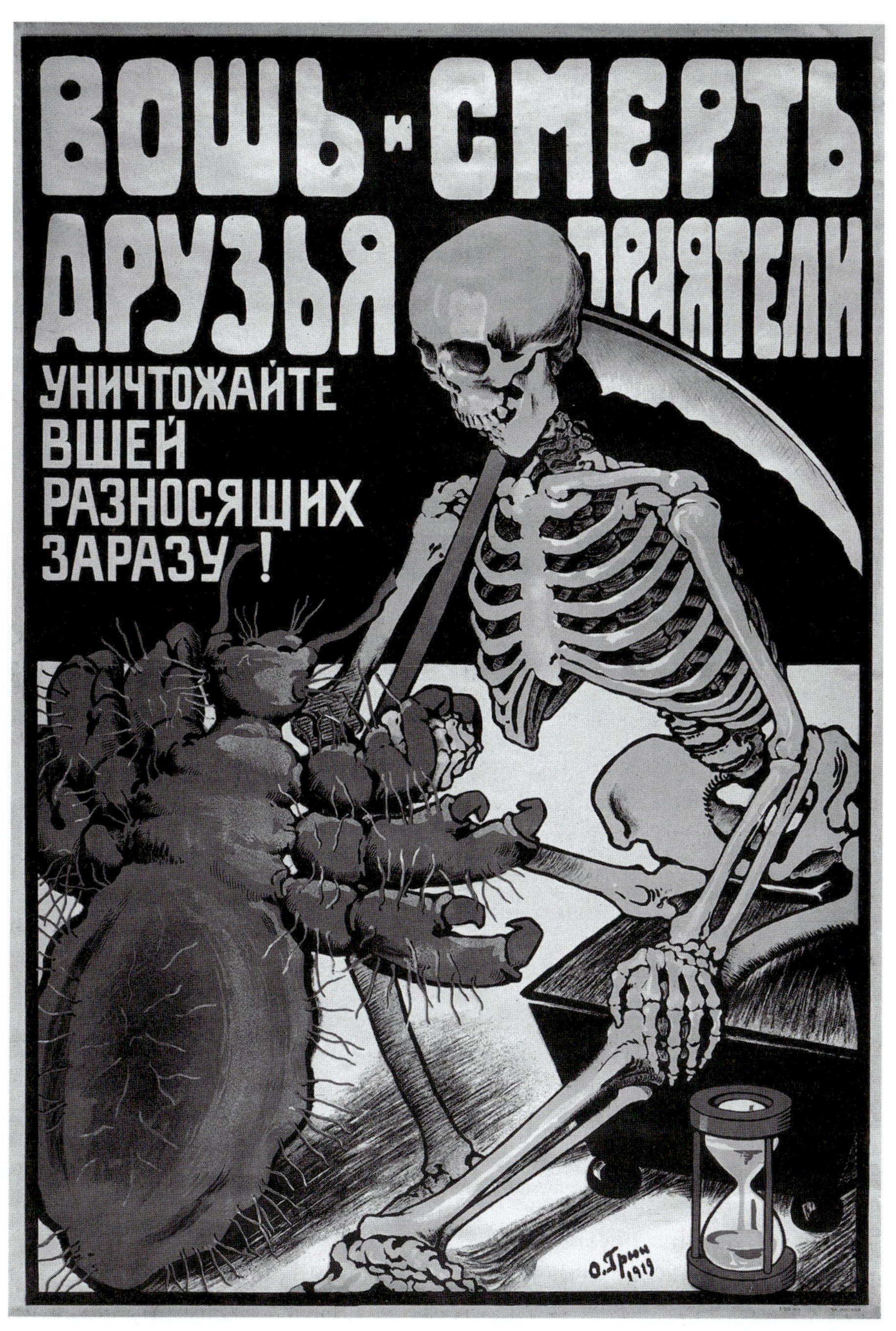

FIGURE 1.1: The louse and death are friends and comrades. Kill all lice carrying infection! The typhus louse shaking hands with death, 1919. Wellcome Library, London.

the childhood death rate, once the Grim Reaper's strongest suit.[11] Mortality in the first year of life is the most sensitive indicator of the health of a population, and, in the nineteenth century, the infant mortality rates of the large industrial cities were appallingly high. But even in the notoriously unhealthy city of Glasgow, the rate fell from above 180 per thousand in 1860 to about 120 per thousand by 1920.[12] Some of the rural areas of Britain were down to below 50 per thousand by this time. This downward trend continued, more or less steadily, for the remainder of the twentieth century. The decline was, however, considerably faster among the affluent middle classes than in traditional working-class areas. The infant mortality statistics are indicative of the sort of company that Death regularly keeps—poverty and inequality. Both absolute and relative poverty are bad for one's health.[13] In the 1990s, as the gap between the rich and the poor in Britain widened, infant mortalities actually began to rise again, in a few areas of particular urban deprivation, for the first time since World War II.[14]

Birth and Death are intimately linked, and not merely by being the opposite ends of the life course. As evolutionary theory developed in the twentieth century, birth and death began to be conceptualized as the necessary biological conditions for sexual reproduction. Sex is advantageous, evolutionarily speaking, because the mixing of genetic material in the production of gametes and the fertilization of eggs produces new, genetically unique individuals. Genetic mixing creates variation, which is the raw material for natural selection. Death, thus, makes room for new answers to the fresh challenges posed by changing environments. Death, as the feminist author Charlotte Perkins Gilman recognized, is the "essential condition of life."[15]

Positive images of Death are not new, of course. "Come, sweet Death," implores the text of one of J. S. Bach's most melodic chorales, composed in 1736 but given renewed popularity in the twentieth century by Leopold Stokowski's lushly Romantic orchestration. The same title was used, if somewhat ironically, in 1998 by the novelist Wolf Hass.[16] Similar themes have been developed by many other artists in the intervening period. Even if not everyone can now share Christianity's "sure and certain hope" of resurrection in a better place, the idea of Death as the sufferer's friend, bringing an end to pain in this world, remains an attractive one.

If Death can bring good things, or at least end bad things, then dying should be nothing to be afraid of. It might even be actively sought, under the right circumstances. In 1932, Gilman was diagnosed with breast cancer. She had campaigned all her adult life for women's right to personal autonomy and had, as we have already noted, a robustly positive attitude toward death. Three years later, she killed herself. Her suicide note stated that she preferred "chloroform to cancer."[17] The manner of her death was an inspiration to the right-to-die movement in the United States. To attempt suicide was still to commit a criminal act in the majority of Western countries into the last third of the twentieth century, but most of these laws have now either been repealed or lie unenforced on the statute book. However, assisting a person to commit suicide remains illegal in many jurisdictions. Nevertheless, campaigns to decriminalize assisted suicide have gathered strength throughout the twentieth century. In recent years, Belgium and the Netherlands have gone a step further by offering legal protection to physicians who actively hasten the death of a terminally ill patient, provided the initial request has come from the patient and proper procedure is adhered to.[18]

Several attempts to change the British law concerning euthanasia have failed. In 2002, the British Medical Association (BMA) narrowly voted down a motion calling for "the 1991 Suicide Act to be amended so that mentally competent individuals who are physically incapable of ending their own life could have assistance to do so."[19] Nevertheless, Michael Wilks, chair of the BMA Ethics Commission, has recently asserted that there is widespread unease within the medical profession concerning the current legal position.[20] A considerable number of doctors admit to having helped, to some extent, a patient to die, at his or her request. It is likely, however, that many of the cases in which death has been accelerated by medical intervention would not meet the criteria in the legislation proposed by the BMA (and long advocated by the Voluntary Euthanasia Society, now renamed Dignity in Dying), since they involve patients who are not mentally competent, by legal standards, to give their consent. Cases involving children or in which the patient was unconscious would not be covered, for instance.

Euthanasia means, literally, "good death." The notion of a good death is an old one. But in earlier centuries, a good death was generally a "heroic" one. "And how can a man die better than facing fearful odds, for the ashes of his fathers and the temples of his gods?" as Thomas Macaulay praised Horatius's self-sacrificial defense of Rome against the Etruscans.[21] Despite Wilfred Owen's condemnation of such jingoist attitudes as "the old Lie,"[22] heroic deaths were still celebrated at the end of the twentieth century, but perhaps less so than previously. If asked to suggest a model death for the postimperial era, many people might prefer to nominate Bing Crosby's. On the afternoon of October 14, 1977, at the age of seventy-four, Crosby played a round of golf at a course near Madrid. He completed eighteen holes in a very creditable eighty-five shots, and he and his partner won their match. After his last putt, Crosby bowed to applause and said, "It was a great game." Walking back to the clubhouse, he collapsed and died, instantly, from a massive heart attack.[23] "What a way to go!" many people said, even those who were not very fond of golf. It seemed admirable to have been active and alert enough to enjoy life into one's eighth decade and then to die suddenly, with only the briefest interlude of pain or fear. By the end of the century, however, with increasing life expectancies and many

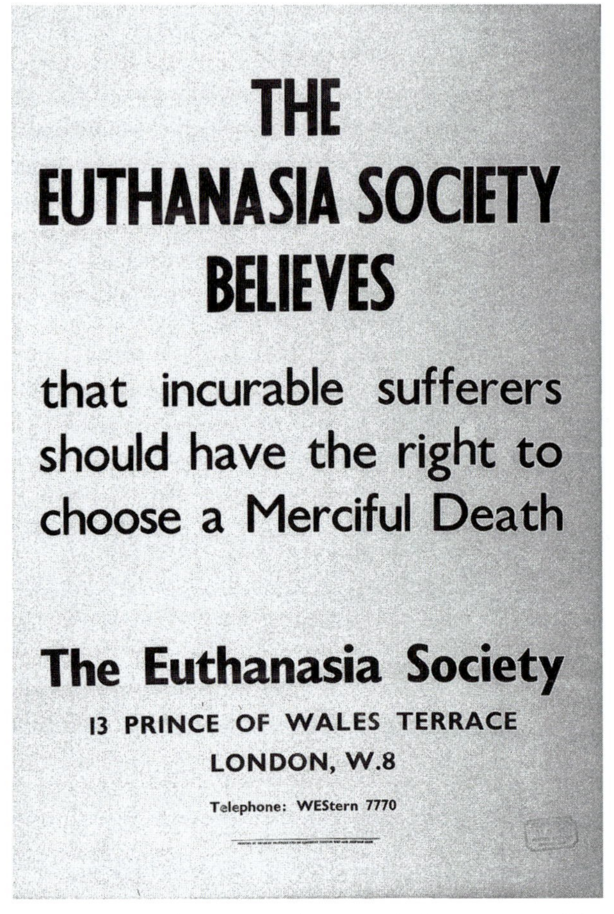

FIGURE 1.2: Voluntary Euthanasia Society poster. "The Euthanasia Society believes that incurable sufferers should have the right to choose a Merciful Death." Wellcome Library, London.

retired people enjoying improved health and greater affluence, seventy-four no longer seemed very old. Considerable numbers of golfers, many of them assisted by artificial hips and cardiac pacemakers, now enjoy the game well into their eighties. The Grim Reaper has to spend a little more time patiently hovering around the eighteenth green than he had been accustomed to.

Death used to be a simple matter. The dead stopped breathing, their hearts ceased to beat, their blood vessels no longer pulsated, they became cold, gradually stiff, and eventually malodorous. Dying was the process by which life lost the struggle against the (always threatening) forces of entropy and disorder. The dying person slipped, sometimes quickly, sometimes slowly, but always irrevocably, from one state to another. One was either alive or dead. In fact, death was not always quite as neatly demarcated from life as this dichotomy would suggest. Or perhaps it would be more accurate to say that this apparent lack of ambiguity surrounding the condition of being dead was a relatively modern cultural achievement. In earlier centuries, the arterial pulse and the cardiac apex beat could be difficult to detect in comatose or moribund patients. There was, in any case, a widespread belief that the arteries of some people did not noticeably pulsate even in health, except during vigorous exercise. Respiration might likewise be virtually imperceptible. Well into the nineteenth century, fear of premature burial, as a consequence of an erroneous pronouncement of death, was widespread.

Developments in medical technology and diagnostic acumen sorted the matter out. From the invention of the stethoscope in 1817 to that of the electrocardiograph and the electroencephalograph in the early twentieth century, the objectifying gaze of modern medicine subjected the dead body, as it had the living body, to ever more detailed scrutiny.[24] Life and its absence were defined more precisely. If one lacked detectable mechanical or electrical activity in the heart, one was dead. If there were any remaining doubt, the absence of electrical activity in the brain could confirm the fact.[25] Diagnosis by clinical examination became more reliable, too, as the semiology of death became better understood. Even embalmers were taught to verify death by looking carefully into the eyes of the corpse, to confirm the absence of pulsation in the retinal blood vessels and to note the first, reassuring indication that decomposition had begun.

This modernist certainty surrounding Death did not last long, however. In the final quarter of the twentieth century, the Grim Reaper, like everything and everyone else, went postmodern. Death was no longer a monolithic, unitary condition. As Rosemary Rhodes put it, dying came to be regarded as merely "a rough marker" for a complex event.[26] There were now several different Deaths, each surrounded by its own nexus of medical, technological, and social circumstances. For example, as the technology of intensive life support improved, brain death came to be distinguished from cardiorespiratory death.[27] With respiratory and circulatory functions maintained by heart-lung machines, patients could be regarded as alive even if their hearts had stopped beating. Or, alternatively, they could be regarded as dead even if their hearts still beat. The dead could be in the paradoxical situation of being sustained by a "life-support" system.

In 1968, a committee of Harvard Medical School proposed that, if the entire brain was not functioning and this state was judged permanent, as evidenced by a flat electroencephalogram and the absence of reflex responses, then the patient in question could, and should, be regarded as dead.[28] The adoption of these brain-death criteria largely resulted from the successful development of organ-transplantation surgery and the subsequent creation of a demand for fresh organs to transplant. A definition of death based on brain function provided the legal and social sanctions that allowed surgeons to remove transplantable organs from warm, well-oxygenated bodies so as to maximize the chances of the organ functioning properly in its recipient. This change in definition was not accepted universally, however. In Japan, for example, the removal of organs from "beating heart cadavers" remains culturally unacceptable.[29] And even in cultures where the practice is permissible, it is surrounded with some anxiety. As W. R. Albury has noted, the fear of premature dismemberment has replaced the fear of premature burial in the popular imagination.[30]

More recently, some physicians and bioethicists have advocated a further change in the definition of death.[31] They maintain that the death of the cerebral cortex, rather than the death of the whole brain, should constitute the standard for declaring an individual dead. It is argued that it is the absence of cortical activity that is crucially significant since a human being impaired in this way has lost not merely the powers of movement and sensation but his or her psychological and social identity. He or she has lost the ability to function as a person, lost indeed everything that makes continued life worthwhile or meaningful. That person no longer effectively exists, and his or her body could be regarded, from a moral and ethical standpoint, as having died. The use of a cortical death standard would permit individuals who are in a permanent coma yet who retain brain-stem function, that is, those in the so-called persistent vegetative state (PVS), to be declared dead. It is argued that expanding the criteria of death in this manner is a rational response to technological advances in critical-care medicine that, when not fully successful, can leave the patient in a condition that no reasonable person would wish for. Maintaining PVS patients involves, moreover, so the argument goes, inappropriate use of scarce resources. It is certainly the case that if the cortical definition of death were widely adopted, a considerable number of intensive-care beds would be freed and many more organs made available for transplantation.

Several religious traditions find the use of the cortical death definition objectionable, however. If life has transcendent value as a gift from God that we must cherish and preserve, then, it is asserted, even the life of an individual without a functioning cerebral cortex is worth maintaining. At this point, changing biological or medical definitions of death necessarily impinge on legal and social ones. If we have a social right to health care, as Western cultures assume, then at what stage is that right extinguished? A PVS patient may have long been dead by the cortical definition, have long ceased to be a social being, but his or her loved ones may not deem their relationship to have ended. A woman may indeed yet seek to become pregnant by her partner after his cortical death, if the practical and legal difficulties of collecting semen by electroejaculation stimulation can be surmounted.[32] Even a PVS woman may become a parent. In one case in the United States, a long-term coma patient who had been raped by a nursing attendant became pregnant and was duly delivered of a healthy baby.[33] If one applies the cortical death criterion in her case, she became both the victim of a crime and a mother posthumously. But if she were deemed legally dead, the offense against her body was, presumably, not rape. But if not, what? One may presume that the legal definition of death might be made flexible enough to protect the PVS body from sexual assault but to facilitate access, nevertheless, for the surgeon's knife. But this demonstrates that sociolegal death, the point at which recognition of legal standing by the courts ceases, property is transferred, life insurance pays out, husbands become widowers, wives widows, children orphans, and so forth, has become as fractured and postmodern in character as its medical equivalents.

Despite all these uncertainties surrounding death and dying, eventually the issue of what to do with an unequivocally dead body has to be faced. Here again we are presented with a greater choice than ever before. Many people still opt for a traditional burial, in a wooden casket, six feet down in consecrated ground. Within some faiths, interment is the only acceptable means of disposal. Others take comfort in the notion that, while in the earth, the substance of their bodies will again become part of the great circle of life, feeding the grass and the trees and the animals that live on them. The popularity of this view, albeit among an environmentally minded minority, has encouraged the development of biodegradable coffins and the setting aside of areas of woodland for "green" burials.[34] But the most remarkable development in the disposal of human remains in the twentieth century has been the increased popularity of cremation.[35] Cremation remained illegal in Britain until 1902 and was regarded with widespread suspicion for many years afterward. As late as the early 1960s, one of my primary-school teachers, in the God-fearing Western Isles, would regularly fulminate against the practice, which she saw as an attempt, futile she assured us, to foil the resurrection of the body. By the 1980s, however, cremation had become widely socially acceptable and was indeed chosen by a majority of bereaved families. It is relatively cheap and has the advantages of seeming to be hygienic—many people are not sanguine about the processes of bodily decay and decomposition.

The increased use of cremation as a means for the disposal of human remains has changed the relationship between the living and the dead in interesting ways. As L. Kellaher and D. Prendergast point out, ashes represent a highly ambiguous form of object or material.[36] They are of the person and of the corpse yet resemble neither of them in sensory terms. Ashes are socially acceptable in places where a dead body would not be. Cremation, thus, allows a postmodern, socially fragmented form of bereavement. While many families choose to place the caskets containing the ashes of their loved ones in the gardens of remembrance that are adjacent to most crematoria, an increasing number do not. Reduction of the body to ashes has encouraged a more individualized and differentiated relationship with the dead. We can, if we wish, keep them closer to us, on our mantelpieces, in our bedroom cupboards, or wherever. Cremation has also rendered the dead mobile, like the rest of society. That mobility fractures the collective community identification with the places of the dead, such as that represented by the churchyard of Saint Ewold's, alluded to earlier. Instead, the dead can revisit the places of the living. They can even climb mountains. I have sat on the summit of Haystacks in the Lake District with the ashes of some late mountaineer, unknown to me in life, swirling around my ankles.

As noted earlier, when personified, Death is usually portrayed as male and hideous. Since Death began to make a regular appearance in visual art in the fifteenth century, he has been routinely drawn as a skeleton, or even a rotting corpse, hooded, cloaked in black, menacing. Sometimes, however, an alternative personification is used, the "Angel of Death," who is sometimes represented as being female and beautiful, if in an aloof and inaccessible style. But the angel makes few appearances in twentieth-century art, at least in English-speaking countries, as far as I have been able to discern. *The Death of the Gravedigger*, by the symbolist painter Carlos Schwabe, is an interesting late nineteenth-century example.[37] She retains her popularity, however, in Latin America, and especially in Mexico, where there is a cult of "Santa Muerte."

In the twentieth century, however, a number of quite positive images of Death have been produced, in both popular and elite art forms. In Terry Pratchett's best-selling *Discworld* series of fantasy novels, for example, Death retains his traditionally ominous appearance, complete with scythe.[38] Pratchett's Death is not, however, otherwise an unattractive character, being efficient and not gratuitously cruel, if occasionally impulsive. In George Balanchine's ballet *La Valse* (1951), Death is tall, handsome, and dangerously erotic. His seductive good looks persuade a young girl to leave the company of her friends. A lingering kiss seals her fate. Her friends are downcast but soon forget her absence and dance happily again. Life goes on.

In the 1998 Hollywood film *Meet Joe Black*, Death's cosmetic makeover is complete. He is accorded the ultimate compliment of being played by Brad Pitt. Inevitably he, too, attracts the pretty girl, but in a neat twist, her kiss persuades him of the value of life and he rewards her by sending her dead boyfriend back from the Great Unknown. In Neil Gaiman's graphic novels, we are presented with a most unusual Death. She is a pleasant, gamine young woman, albeit one who dresses as a goth.[39] A sympathetic presence who enjoys a cup of tea, she visits people as they are born as well as when they die, apparently to symbolize the link between birth and death. In J. K. Rowling's Harry Potter series of novels, Death does not make a personal appearance but gets a good write-up, nevertheless.[40] In *Harry Potter and the Philosopher's Stone*, two wizards have possession of the stone and might thus live forever.[41] However, they decide, admittedly after several centuries, to relinquish it. In a passage that is strikingly didactic for a fantasy novel, their sentiments echo Dylan Thomas's affirmation that "wise men at their end know dark is right."[42]

If death can be dangerously erotic, then eroticism can also be dangerous and deadly. Campaigns against sexually transmitted diseases, syphilis in the first half of the twentieth century and AIDS in its last decades, have often used images that conflate sex and death. In one famous anti–venereal disease poster, a prostitute's face morphs into a skull. In another, a skeletal Grim Reaper voyeuristically watches over a commercial sex act, grinning with evident anticipation. These are, of course, negative images; nevertheless, it would seem that the portrayals of death in the popular culture of the twentieth century have been surprisingly positive, at least when compared with those of earlier centuries. This trend may be an expression of a more secular and materialist view of Death or may spring from a widespread pessimism about the purpose of life and a concomitant skepticism about personal immortality.

The secularization of Death has certainly played its part in the creation of one distinctively twentieth-century form of cultural pessimism. The Danish theologian Søren Kierkegaard (1813–1855) is often regarded as the first existentialist, dwelling as he did on the intrinsic absurdity of human life.[43] There being no logical reason to believe in God, the only option left to the Christian is to make an irrational "leap of faith." Making such a commitment entails adherence to the moral teachings of one's religion and thereby imposes meaning and structure on one's life. It also brings with it the promise of an afterlife, the basis of which will, Kierkegaard hopes, be less absurd than the mundane one. Thus, because Death does not end our individual existences, life on earth can be ascribed some meaning, if only because, after death, the meaning of existence will become clear in heaven (or even more so, perhaps, in hell). However, the development of a materialist view of Death as the end, not merely of life but of personal existence, changed the nature of this calculation. To the twentieth-century atheistic existentialists, notably Jean-Paul Sartre, Kierkegaard's strategy was, literally, hopeless.[44] If Death ended not only our lives but also our individual existences, then a religious commitment could not rescue us from existential angst. Sartre took refuge in Stalinism, which might be said to be a fate worse than death.[45]

Oddly enough, although Sartre did not believe in life after death, he wrote a play about it, *No Exit*, first produced in 1944.[46] Here, we see an irony of the materialistic reconceptualization of Death. The repudiation of belief in an afterlife has abolished the hope of heaven but not, at least not wholly, the fear of hell. Rather, it turns out that hell is alive and well and living among us. Hell is here and now. The grim theme of Sartre's play is that hell is other people. But it is a hell from which, paradoxically, only Death can release us. It is remarkable how, even in these unbelieving times, artists continue to be inspired by the

Greco-Christian notions of an afterlife. The choreographer David Dawson, for instance, would seem to have located his recent ballet *The Grey Area* (2002) in Limbo.

Of course, regardless of the skepticism of secular intellectuals, many people (a majority, apparently, in the United States) continue to believe in some form of personal immortality. It is a tenet of most forms of Christianity that the body itself will be resurrected, some time shortly before the end of the world. Many Christians believe, indeed, that Death will be vanquished more directly and immediately than even John Donne imagined. In 1918, Joseph Rutherford, president of the Watch Tower Bible and Tract Society, gave a public lecture in Los Angeles, entitled "The World Has Ended; Millions Now Living May Never Die." Identifying World War I as Armageddon (or one of its preliminary skirmishes), Rutherford prophesied that the end of the world was imminent and that the true believers (Jehovah's Witnesses) would soon, like Elijah in the Old Testament, be gathered up into heaven without having first to undergo the inconvenience of a bodily demise. In 1920, Rutherford published the same material in a book, now even more confidently titled *Millions Now Living Will Never Die*, in which he proclaimed that the Last Trump would sound in 1925.[47] Rutherford's hopes were sadly disappointed, but the idea of direct ascension into heaven has proved an attractive one among the adherents of the increasingly popular millenarian varieties of Christianity in the United States. Many millions now eagerly await the rapture, as the longed-for event is termed.

It might be said that secular versions of the resurrection and the rapture also exist.[48] Cryonics is the practice of preserving bodies by storing them at temperatures low enough to prevent decay, in the hope that, some time in the future, medical science will allow their revivification.[49] Immediately after death, the blood vessels are injected with "cryoprotectant" substances, which cause water to vitrify rather than freeze. This procedure helps protect the tissues from the damage caused by the formation of ice crystals, if at the cost of toxicity from the cryoprotectants. Cryonics is viewed with skepticism by most expert commentators, but its advocates counter that the future progress of medical knowledge cannot be predicted and a slim chance of revival is better than none. Despite the fact that the cryonic procedures cannot begin until a legal pronouncement of death has been made, the providers of cryonic services do not regard their clients as dead and refer to them as "patients." The term *corpsicle*, a portmanteau of *corpse* and *popsicle*, has been suggested.[50]

A handful of optimists believe that this technology may not be necessary. With the continued advance of medical knowledge, it is sometimes argued that human beings may eventually be able to live for a thousand years or perhaps even extend their lifetimes indefinitely. As already noted, the inhabitants of the developed world are certainly living longer and more active lives than they did previously. In the affluent areas of western Europe, Japan, and North America, the S-curve of survivorship is becoming "rectangularized."[51] The increasing rarity of premature deaths has flattened the central section of the curve. Moreover, the final drop, representing the end of the life span, is being displaced along the x-axis as average life expectancy for adults extends into the ninth decade. Some of this reduction in mortality may reasonably be attributed to medical intervention, although it is debatable exactly how much. However, the fact that the survivorship curve does eventually plunge steeply downward suggests that human life span is indeed strongly biologically determined. Centenarians remain rare, and survival into a twelfth decade is truly exceptional. It seems unlikely that any conceivable advancement in medical knowledge could radically alter this situation—even if we are able to maintain, in the long term, our present economic affluence and our current levels of investment in health and medicine.

Affluence brings its own health problems, moreover. An epidemic of obesity, which began in North America and is now spreading to Europe and even parts of Asia, will inevitably bring in its wake a greater prevalence of diabetes, heart disease, and other degenerative disorders.[52] It is not impossible that these increased levels of morbidity will, in due course, become demographically significant. Meanwhile, the Grim Reaper is again mustering those forces that have traditionally been his most reliable—infectious diseases. AIDS is already having a devastating impact on life expectancy in large parts of sub-Saharan Africa.[53] Tuberculosis, long feared as the "Captain of the Men of Death," is on the march again, aided by drug resistance and relative economic deprivation. Malaria is resurgent. Meanwhile, the West anxiously watches Southeast Asia as its ducks and geese brood the next influenza pandemic.

One does not, however, have to make extravagant claims for medical science, or even to subscribe to the tenets of conventional religion, to deny the finality of death. Spiritualists and mediums promise a direct connection to the next world that is eagerly sought by many. In the late twentieth century, contacting the dead in this way has become mass entertainment, with television shows, indeed entire television channels, being devoted to the work of photogenic and charismatic mediums. While not wishing to deny the considerable comfort that such sessions evidently bring to their audiences, what the dear departed have to say, when given the opportunity to enlighten the living about the afterlife, often seems disappointingly trivial. Bereaved family members are often delighted to receive messages about ornaments, pet dogs, or the forgiving of minor slights. Having gone to the bother of turning up and paying my money, I think I would want revelations of somewhat more cosmic significance. But communication across the Great Divide is a tricky and imperfect business, or so the spiritualists assure us.

Another area in which attitudes toward death and dying have changed radically in the late twentieth century is medical education. Medical students used to be systematically desensitized to the presence of death. That was one of the purposes served by the ritual dissection of a cadaver, formerly a rite of passage for every would-be doctor.[54] Nowadays, however, many medical schools devote much less time to the teaching of anatomy, and, coincidentally or otherwise, students are encouraged to develop a caring appreciation of the special needs of the dying and the bereaved. Some medical faculties even have departments of thanatology, which is the scientific and academic study of Death. Thanatologists tend not, however, to be natural scientists in the quantitative tradition of Pearson and Pearl but, instead, humanistic scholars interested in the cultural, social, and psychological aspects of death and dying. The writings of the American psychiatrist Elisabeth Kübler-Ross have been very influential in these developments. In 1969, her seminal work *On Death and Dying* challenged the taboo that then surrounded the discussion of dying.[55] She deplored what she identified as a widespread reluctance on the part of hospital staff to interact with terminal patients, other than when administrating clinical treatment. As a result, she argued, the dying often became isolated and did not receive the social and psychological support they deserved. Kübler-Ross also proposed a five-stage model of the patient's trajectory to death, from "denial and isolation" to "acceptance." Although it is now recognized that arbitrary imposition of this model can itself be coercive, the value of Kübler-Ross's teaching in the shaping of more compassionate attitudes to dying cannot be gainsaid. A similar accolade may be paid to Cicely Saunders, the founder of the first modern hospice for the dying, St. Christopher's, in London in 1967.[56]

If one turns to the other end of the life cycle, one can recognize that many of the changes that Death underwent in the twentieth century have their counterpart in Birth. There now exist, for example, several different sorts of motherhood. Biological motherhood is distinguished from social motherhood. And, as reproductive medicine has enabled a wider range of interventions, further differentiations have become necessary.[57] The egg might be supplied by one woman, the so-called genetic mother; the uterus by another, the gestational or birth mother. Where there have been custody disputes between the genetic and gestational mothers (Ova and Birtha, as they were dubbed in one landmark American case), some legal authorities have found for the former, others for the latter.[58] Religious authorities have likewise been divided about who is to be regarded as the rightful mother.

The status of the fetus also changed in the second half of the twentieth century. Traditionally, the presence of a fetus was recognized upon "quickening," the moment at which its movements were first felt by the mother.[59] It was from this point, in several jurisdictions, that the fetus was accorded its own legal protection, against induced abortion, for example. From a medical perspective, however, in the period from quickening to birth, the fetus continued to be regarded as part of its mother. The testimony of the mother was, moreover, crucial in many aspects of the medical understanding of her pregnancy. Her recollection of the date of her last menstrual period might form the basis on which the gestational age of the fetus was estimated, for instance.

However, from the late 1950s on, the development of obstetric ultrasound, which allowed the imaging of the fetus in utero, radically changed the status of both mother and fetus within the practice of obstetrics. The ultrasound scanner could reveal fetal movements long before they became perceptible to the mother.[60] Ultrasonograms also enabled gestational age to be more accurately determined than was possible from menstrual history. The various modern means of imaging fetal anatomy, and monitoring fetal physiology and biochemistry, have had the effect of making the fetus a patient in its own right.[61] It became, in this sense, independent of its mother even before birth. Therapeutic interventions, such as blood transfusions and surgical procedures, began to be carried out on it. The ultrasound scanner has, moreover, given the fetus a public presence that it did not previously possess. It has appeared, for instance, in advertisements for employment opportunities and soft drinks. It might be said that, by unveiling its appearance and behavior, ultrasound has made the fetus into not merely a patient but a member of society.

The capacity of obstetric ultrasound to individualize the fetus has been enthusiastically utilized by the antiabortion lobby.[62] It is a legal requirement, in some U.S. states, that a woman seeking to terminate a pregnancy must view an ultrasound scan of her fetus. The vivid impression of fetal life conveyed by ultrasound images has undoubtedly deterred some women from proceeding to abortion. Images of the fetus can signify, however, not only life before birth but also death before birth. By far the major impact of the ultrasound image within obstetrics has been in intrauterine diagnosis, in the detection of fetal abnormalities. The individualization of the fetus, the delineation of its individual characteristics, has led to the ever more detailed scrutiny of fetal structure for pathology.[63] In this area of its use, ultrasound imaging and abortion are not in contention but are complementary. Indeed, they are mutually dependent on each other. Given that the possibilities of detection far outstrip those of therapy, the only option that can be offered to most of the parents who are presented with a diagnosis of fetal abnormality is termination of pregnancy. Research into the detection of fetal abnormality would not have proceeded at anything like the pace it has done, over the last thirty years, if that clinical option had not been readily available. Life and death were, one might say, revealed simultaneously.

The resting of the medical gaze on the fetus represented the end point of a trajectory that had been underway for a large part of the twentieth century. As noted earlier, overall mortality rates began to fall in Britain in the 1890s and have continued to decline steadily, more or less ever since. However, there was one category of mortality that bucked this general trend, that is, deaths

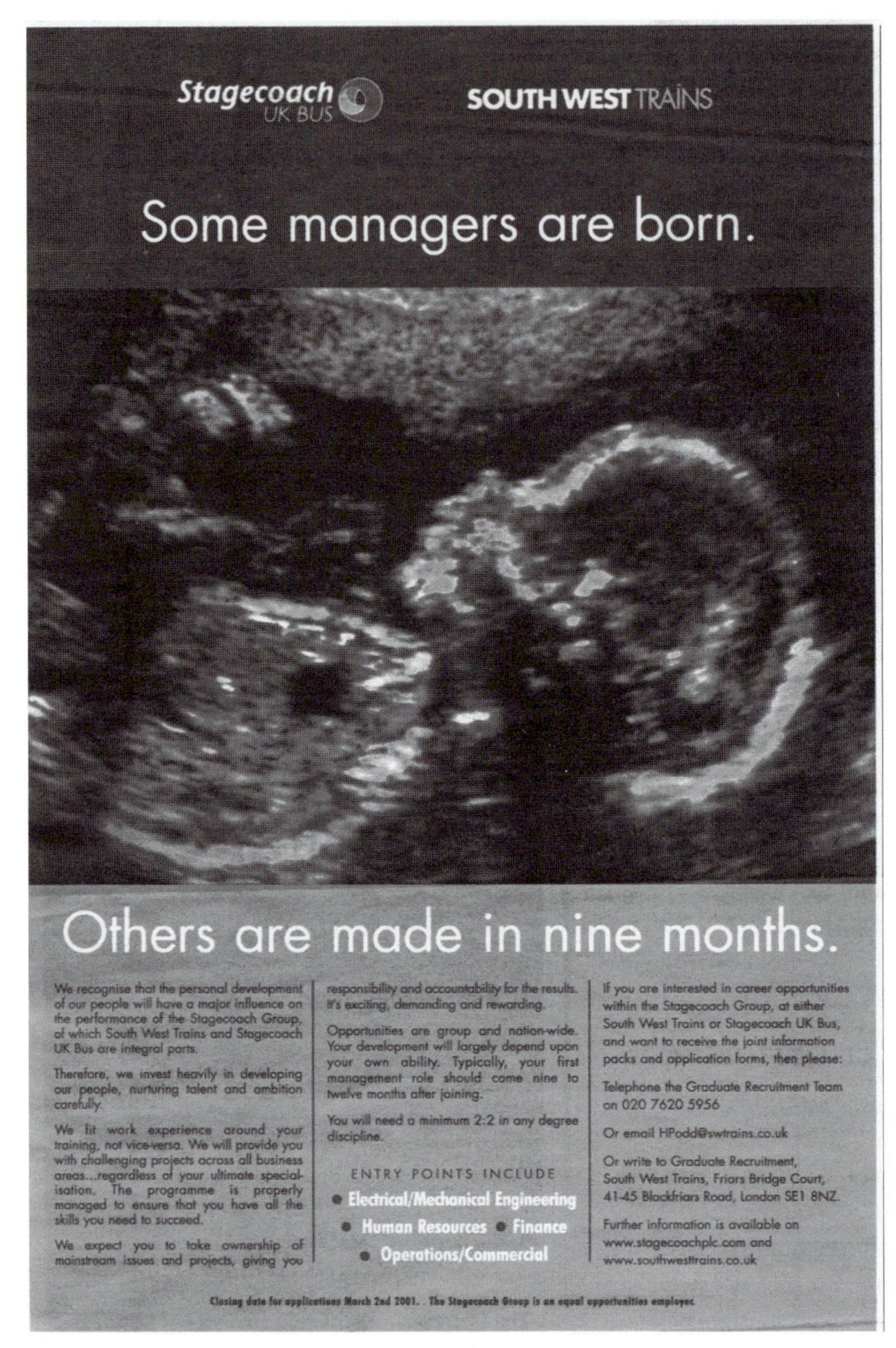

FIGURE 1.3: Ultrasound image used in advertising. Used by permission of SouthWest trains.

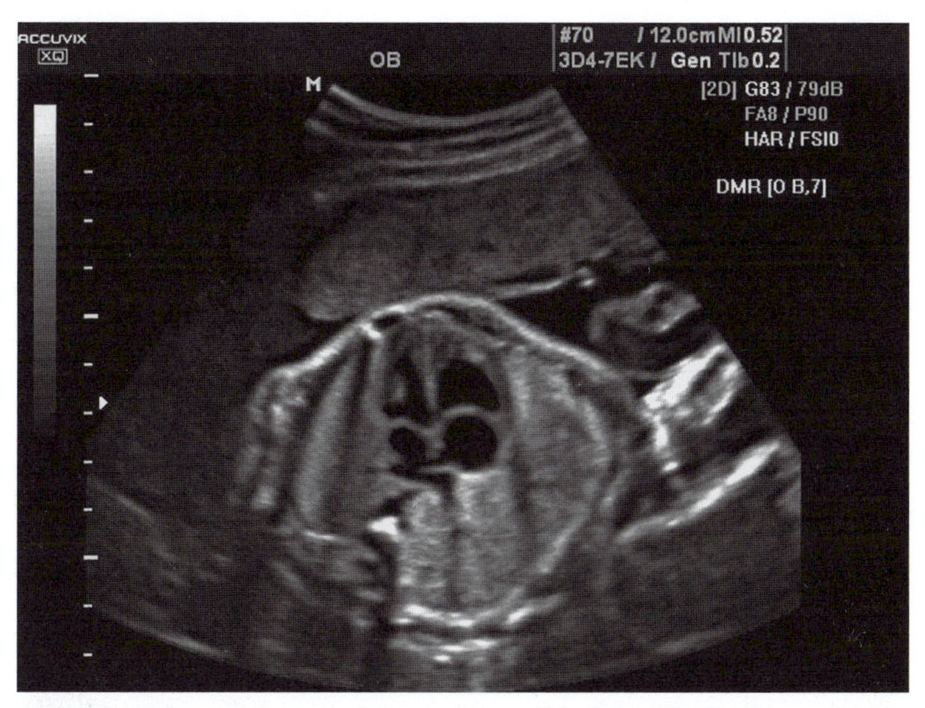

FIGURE 1.4: Ultrasound image of the chambers of the heart of a dead fetus. Used by permission of Diagnostic Sonar Ltd.

of women in childbirth. In the early twentieth century, far from declining, the number of mothers dying in labor, or its aftermath, actually rose. A woman was significantly more likely to die in labor in the early 1930s than she had been in 1880. In the decade following 1920, the annual maternal mortality rate in England and Wales averaged approximately 4.4 per 1,000 live births. The figures for Scotland were even worse. Irvine Loudon, in his magisterial survey *Death in Childbirth*, unreservedly lays the blame for this disgraceful situation on poor standards of obstetric care.[64] A general neglect of education, training, and innovation in obstetrics had had very serious implications for the quality of care being received by pregnant and parturient women. The official *Report on an Investigation into Maternal Mortality*, which was published in 1937, concluded that at least half of maternal deaths in childbirth were due to avoidable causes, the chief of these being medical mismanagement, notably unwarranted or unskilled use of obstetric instruments.[65] The principal concern, therefore, of the new generation of obstetricians who trained in the late 1920s and 1930s was reduction in the maternal death rate.

In the late 1930s, the maternal mortality rate finally began to improve. By 1950, the incidence of death in childbirth had fallen dramatically and was continuing to decline. Overall, maternal mortality was down to about 20 percent

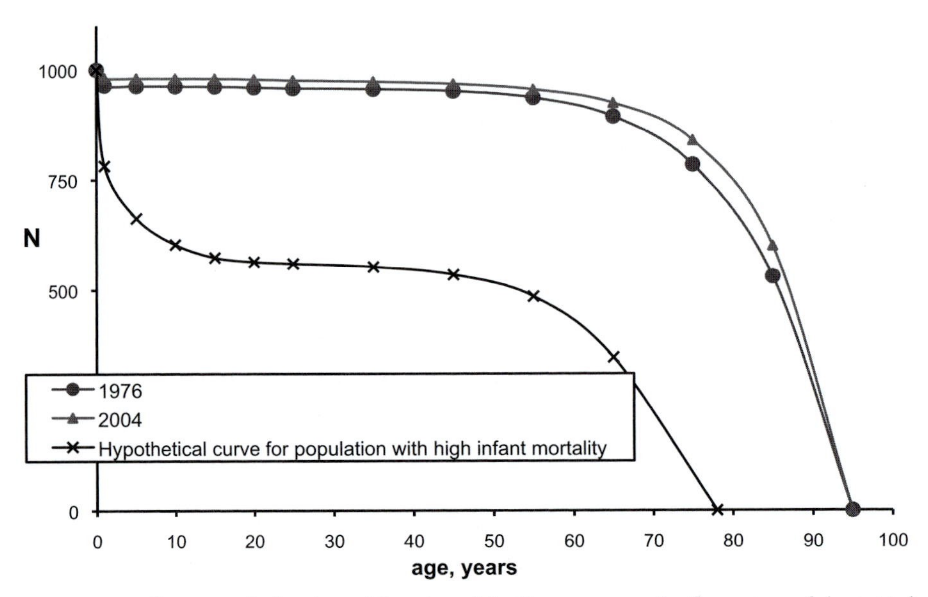

FIGURE 1.5: Rectangularization of Survivorship Curve—note the flattening of the initial and central section of the two upper curves, and the rightward displacement of their final drops. (Prepared by John Fleming)

of what it had been in 1935.[66] Standards of obstetric education and practice had improved considerably, and obstetricians had new clinical tools for the secure management of complicated labor, such as ergometrine to stimulate uterine contractions, a reliable blood-transfusion service, and improved anesthesia. First the sulphonamides and, later, antibiotics greatly diminished the dangers of puerperal infections. As confidence grew in the mother's safety, obstetricians were able to accord greater attention to the well-being of the baby, during and immediately after birth.

Here also there was much to do. As already noted, infant mortality (deaths in the first year of life) had declined more or less steadily in Britain since the beginning of the twentieth century. But the rate of neonatal mortality (deaths in the first four weeks of life) had fallen far less quickly.[67] In the 1940s and 1950s, therefore, neonatal health became a matter of considerable concern in both professional and lay circles. And concern for the life of the newborn led to a more intense focus on the life, and death, of the fetus.

Increased interest in the life and death of the fetus/neonate expressed itself in a change of attitude toward that most archetypal and poignant of death-in-life events, stillbirth. Stillbirths were not officially registered in England and Wales until 1927 and in Scotland until 1939.[68] The timing of this change in attitude toward death before or during birth may be seen as an expression of the gradual shifting of the medical gaze from the mother toward the neonate in this period. Obstetricians had, moreover, become interested in distinguishing between different kinds of stillbirth. On the one hand, a "fresh" stillbirth, in which the fetus had apparently died shortly before birth, raised questions about the quality of care that its mother had received while in labor. On the other hand, a "macerated" stillbirth, in which the fetus had died in the womb sometime before labor had commenced, could not be blamed on faulty management of the delivery.[69] Stillbirths were classified, in other words, according to how they impinged on obstetricians' practice and professional interest.

The medical gaze moved, thus, from maternal death to neonatal death and on to the death or life of the fetus. Concerns with perinatal mortality led increasingly to attempts to monitor intrauterine conditions and events. Ultrasound imaging proved to be particularly effective at many aspects of this task, culminating in the ever closer scrutiny of the fetus for abnormality, as described earlier. Moreover, just as ultrasonography could detect the life of the fetus before the mother felt its movements for the first time, the scanners could reveal whether the fetus was dead before the mother miscarried or even noticed any change in the course of her pregnancy. Indeed, improvements in fetal imaging during the 1960s and 1970s allowed the obstetrician not only to identify fetal death but to predict it, sometimes well in advance of the event. Death was revealed in a more direct fashion than ever before. Mothers whose fetuses were affected in this way would usually be offered the option of proceeding to a therapeutic termination of pregnancy, what might be termed euthanasia before birth. Thus, by the last decades of the twentieth century, even the death of a fetus, wrapped within its mother's body, had become subjected, like most other sorts of death, to the objectifying gaze and reductive discipline of modern medicine.

Death, one might conclude, changes as we change the way we are born and the way we live. Perhaps, indeed, Death, like God, is what we want it to be. As a young man, Richard Strauss wrote a somewhat untypical programmatic piece, *Death and Transfiguration*. Although its critical reception was mixed, it remained one of his favorite compositions. On his deathbed, he is said to have remarked to his daughter, "Funny thing, Alice, Death is just as I composed it."[70]

CHAPTER TWO

Performing the Western Sexual Body after 1920

IVAN CROZIER

The twentieth century was the time of the sexual body in the West. The combination of internalized disciplinary procedures that rendered docile bodies explicitly sexual with the possibility of actively embracing the sexual opportunities available in the twentieth century allowed new subjectivities to emerge.[1] A variety of new sexual forms of life came into being via a series of novel bodily procedures, some of which this chapter considers.[2] Sex took on new meanings and values in the twentieth century that altered conceptions of identity and of what was considered natural. Sex offered possibilities for extending the performances of the embodied self. New sexual acts, and new ways of modifying the sexual body, situate the body in time. If it is agreed that sexuality is more than the biological meeting of reproductive organs,[3] then cultural pressures have to be considered when addressing the ways in which individuals negotiate their bodies in search of sexual pleasures, performing sexual acts that are imbued with political, social, and personal meanings. In twentieth-century terms, these changes are far-reaching and in some cases involve modifications to the body itself for sexual purposes. The sexual body, in its wider understanding, is not simply the sweaty, thrusting, ecstatic body engaged in one of many forms of sex but a broader cultural phenomenon.

The late nineteenth-century emergence of sexology and Freudian psychoanalysis, and the concurrent rising profile of feminism and homosexual rights and identities, raised sexual consciousness of the body in a number of new ways.

Sexuality was problematized by medical, feminist, and homosexual groups in ways that led to a fundamental questioning of the meaning and purpose of sexuality.[4] It was soon realized that, Victorian rhetoric aside, men and women of a variety of sexual persuasions sought pleasure through a number of different sexual performances and valorized sexuality as something of a "secret of the self"—a key element of identity that had a specific value in the constitution of the person's identity.[5] This drastic increase in doctors' sexual knowledge, as well as that of various radical sex reformers, contributed to general changes in sexuality. Sex increasingly became something that was discussed explicitly in public. One of the reasons for the popularity of the work of Sigmund Freud in the early twentieth century was that it gave people a means and a reason for talking about sex.[6] Concurrently, there were a number of attempts to explicitly delimit sexual appetites and knowledges through various laws.[7] Tensions following the sexualization of society can be seen in various areas, from anxieties about women having the "right" kind of orgasm (i.e., vaginal, not clitoral), to concerns over sexual education, to general ideas about eugenics, to challenges to laws pertaining to homosexuality, through to large organizations, such as the World League for Sexual Reform, and to debates about censorship and so-called obscene literature.[8] Further, a number of sex surveys in the twentieth century, from Alfred Kinsey to Shere Hite, made it clear that sexual behavior was much more multifarious than dominant discourses had previously implied.[9] The so-called sexual revolution of the 1960s, with its associated technology, the female contraceptive pill, broadened the scope for sexual practices relative to what they had been hitherto. At the center of lots of these developments was the body, and in particular the sanctioning of the sexual body at numerous levels of social practice, from the intimate to the popular.[10]

One of the key aims of this chapter is to move from official discourses about the sexual body—sexological, medical, psychoanalytic, legal, political—in order to attend to sexual performances themselves. While many historians of earlier periods have relied through necessity on official sexual discourses (especially those with sexually explicit medical or legal content), the proliferation of publicly accessible discourses of sexuality and the rising visibility of sexual practices and representations have meant that attention can be given to these various other sources that redress this focus on official published sources. The twentieth century has given rise to a wide variety of sanctioned performances of sexuality that have extended the possibilities of the sexual body well beyond the scope of parenting or uxorious canoodling. These developments are archived in many popular representations, from magazines to literature to film, and now most widely the Internet, with a myriad of sexual performances captured in explicit detail for all to read, see, or watch. These representations and their interpretation are at the core of this chapter.

NATURAL KINDS, SOCIAL KINDS, AND THE "BOOTSTRAPPED" SEXUAL BODY

Centering on the body in such a way raises one of the perennial problems related to sexuality: the nature/nurture debate.[11] Here is not the place to rehearse the extensive arguments surrounding this issue—although it is central to understanding tensions in explanations of humanity in the twentieth century.[12] This debate, nevertheless, raises a tricky problem: How can the sexual body be both natural and cultural? How can we make connections between the physicality of sex—its very biology—and the way that it is manifest in individual lives, in particular in the cultural meanings that these manifestations signify in various social groups? These problems have been highlighted through standpoint debates between psychologists and sociobiologists and are still a topic of great anxiety to genetic determinists as well as garden-variety postmodernists, both of which largely despise the explanations of the other camp. In the social sciences, this problem has been the bane of constructivists—and indeed the backlash against the nonsophisticated versions of constructivism (such as naive deconstructionists who emphasize discourse above all else, and other species of cultural determinists who have sought to move away from any biologicization of sexuality) has been brought about largely by the way that such positions have trouble coming to terms with the physical reality of the body. Not until the 1990s did constructivist theorists take seriously the contention that bodies do matter in accounting for the self.[13] While sexualities are constructed, they are inextricably connected with the biological body, so care needs to be taken not to privilege either the social or the physical at the expense of the other.

As I focus on changes in the sexual body, these issues are very present. To avoid falling into one or the other extreme, biological or social determinism, I employ the performative theory of social institutions, derived from Barry Barnes's 1983 concept of "bootstrapped induction," which helps to understand how cultures relate to the biological and the cultural in ways that are naturalized.[14] Bringing this sociological theory to the cultural history of the body can help explain cultural changes surrounding the sexual body while also explaining the body as perceived as natural (as naturalness and artificiality are often used as means of explaining the relationship to the body and to sex—for example, complaints against condom use are sometimes framed in terms of the lack of natural sensation, as if sex is an unproblematically natural thing rather than a learned response). Such a model allows us to appreciate changes in conceptions of the sexual body and to address the all-important issue of sanctioning, and its relationship with identity, without rejecting the physicality of the body.

To think about how the sexual body is practiced, we need to drop to the level of microsociological interactions. Monolithic structures, or a lower-level habitus,[15] are of no real use in describing the relation between an individual and her body or the relationship between this body and others, as they operate at a reified level that does not take proper account of individual practices as the result of collective dynamics—such as the sanctioning processes that legitimate performances.[16] Rather than try to bridge the gap between determining structures and individual actions, it makes more sense to adopt a framework that can explain the generation of social institutions and can inform us as to how social sanctioning takes place from within or by the individual whose actions are being considered.[17]

The standpoint I am advocating advances an analytic framework that helps to understand the constitutive dynamics of individual practices. These, however, have to, crucially, be understood as bootstrapped loops. An individual's practices (verbal and otherwise) are learned in and through social interaction. It is in this interactive dynamics that individuals ostensively learn within a community of knowledge emerging from referring practices. Referring activity constitutes different types of reality. To analytically make sense of the social construction of knowledge, Barnes distinguishes between Natural kind terms and Social kind terms. Both are the product of referring activities within a collective of mutually susceptible interacting individuals that in the process generate the knowledge of a community.[18] Barnes notes that "in the course of social interaction much referring activity is self-referring, and much inference self-validating. This occurs to the extent that our inductive inferences become permeated with feedback-loops or 'bootstraps.'"[19] This social-constructionist account is presented in its full force in this quotation: Natural kind terms as well as Social kind terms contain a self-referring component, meaning that categories, knowledge, beliefs, and practices get established via the sanctioning processes within a collective. So it is bootstrapped in the sense that there is no external reality as such that guides the self-referential process. For my argument, this entails accepting the natural body as a social and cultural construction.

Barnes calls the product of the self-referential dynamics social institutions. The implication for this chapter is that the body is a social institution, one in which individual performances are continually sanctioned, and in the process get constituted, by the performances of others. This conception helps us to understand how the body is treated as natural, as something appearing stable, and yet is constantly adapting to social circumstances. The body is a product of the self-referential activity of a community of category users. The body emerges from a web of linguistic (and other) practices through which it becomes intelligible and thus comes to exist.[20] At this point it is important to briefly note that Barnes's analytic distinction between Natural kind terms and Social kind terms allows us to consider the fact that some of our categories—like the body—rely on the existence of external reality but one that is made intelligible (and, of course, unintelligible) via cultural categories and practices.[21]

In this way, then, uses of the body are self-referring; they refer to the lived experiences of the body but in relation to an existing cultural context. A successful performance, in this chapter a sexual performance, is so considered because it employs *valid* action, meets required ends, and does not overstep sanctioned boundaries—which are in flux and which are continually negotiated in individual practices. Individual performances are indeed constrained by culture. However, this is not to say that individual performances are totally determined. Self-referential processes are *performative*, by which J. L. Austin's notion that certain linguistic acts are constitutive is meant.[22] At this juncture a clarification regarding performance and performativity is required. It is a common misunderstanding to collapse these two concepts into one another. Performance, a concept directly inherited from a Goffmanian analysis of self-presentation, is an individual act. Performativity, in contrast, is a collective achievement that refers to the constitution of social life. Performativity is a conglomeration of performances of interconnected individuals within a collective.[23]

In the context of sexuality, the performative model is especially profitable. It helps to unravel the fact that sexual practices, bodies, beliefs, and so on are constructed as natural, as fixed by nature, and yet are subject to constant social and historical causality. In this chapter I consider how the natural sexual body is best conceived of as a performance that adapts to changing ideas about sexuality as represented in public space. It is an argument about the power of calling the body natural, and yet one that takes account of changes that construct that self-same nature.

Following the preceding methodological statements, it is important to emphasize that this conception of the body is temporally situated. A successful performance relies on other successful ventures in the same social group. But the maintenance of these sanctioned practices is not externally controlled; rather, performances are part of a finitist system, in which past performances, while *informing* or *resourcing* future performances, do not *determine* these future performances in any absolute sense. The rules for one successful performance are constantly—self-referentially—renegotiated each time a new action is performed.[24] A successful performance is one that is accepted as successful by the social group who sanctioned the practice (and this includes the performer sanctioning their own practices, although this may be experienced at a psychological rather than sociological level—but the effect is the same).

In this case, the relationship between social groups and sanctioning points us toward the issue of power dynamics, although a dynamic that is not, however, imposed outside the social group. There are no predetermined structures that compel actions that are sanctioned. And, further, there is no cultural overlay—the habitus—that somehow permits certain dispositions for agents in the field. Rather, the performances are sanctioned from within the social group.[25] Change is not brought about by a tectonic shift in structure but by the negotiation of alternative performances. The more authority one has, in relation to the other relevant members of the social group, the more one can push forward the acceptance of new practices for specific social groups (which is not to assume that all members of a particular collective have the same access to power).

This conception both allows us to see how the body is natural, or that a sexual practice is natural, and yet also allows us to explain change in these conceptions of naturalness. A checklist will shine light on this point: Birth control was considered unnatural, but now it is pragmatic for many; homosexuality was considered unnatural, but now it is permissible for many; sadomasochism was considered perverse, but now it is fun for many. Not all practices are considered natural, however. Pedophilia, for example, is now considered monstrous to many, whereas in other times in other societies (Hellenic antiquity, or when twelve was the legal age of consent in pre-1885 England, for instance), it was sanctioned under certain circumstances and treated as largely unproblematic. These fundamental developments were possible because of changing conceptions of what was to be sanctioned. Homosexuality, for example, did not become acceptable overnight. Rather, sanctioned performances of homosexual practice, inseparable from discourses of sexuality produced by sexologists, anthropologists, homosexual-rights activists, ethicists, journalists, lawyers, and so on, changed, and made their way into dominant social representations (through their being increasingly acceptably repeated).

The sexual body is only one aspect of an individual life—but one that is given an extensive role in the twentieth century because, despite its comparative secrecy,[26] sex is one of the most sanctioned elements of identity. Sexual acts, although deeply situated in culture, are reliant on the use of a natural body, in Barnes's sense.

THE CHANGING SEXUAL BODY

Following from this model of the body that is performed through a variety of practices, and is sanctioned depending on legitimations that fit the hierarchies of specific communities, is a way of negotiating the changes in representations of the sexual body in the twentieth century. These changes involved overcoming "limit experiences," as they have been termed by Michel Foucault.[27] Specific changes that I address in the following to illustrate developments in the sexual body include alterations in so-called normal sexual practices (including the emergence of safe-sex practices in response to sexually transmitted infections [STIs]), pubic hair, sexual body modifications, and physically training the body to respond in a sexually desired way. (There is a plethora of other legitimate source material—from learning to kiss to learning to enjoy anal sex—that could also have been considered for the same result.) In analyzing this material I privilege individual voices and popular representations rather than official texts such as medical and legal discourses. These knowledges, while operating as sites of power that legitimate or pathologize sexual performances, have dominated the historiography of the body to date.[28] Rather than concentrating on the construction of scientific knowledges, I analyze the exercising of power by directing attention toward individual performances. The sources for such a venture come from a variety of popular publications and Internet sites. It will be noticed that I privilege examples of the female body. This is because it is comparatively more malleable—and more prone to the effects of power and of masculine cultural dominance. I have attempted to balance this where other material was available.[29]

CHANGING SEXUAL PRACTICES

Sexual activity changed significantly throughout the twentieth century, the century that brought us the widespread adoption of contraception, which effected an almost-total divorce between sexual activity and reproduction in the West.[30] In the same century, sexual advice was spread to a wider audience than ever before.[31] Rather than being commonly considered perverse, this century saw some practices become common activities in heterosexual relationships. Examples include cunnilingus, fellatio, and anal sex.[32] A significant touchstone in the period being discussed is Theodoor van de Velde, a dominant figure of the 1930s and 1940s. His book, *Ideal Marriage: Its Physiology and Technique*, originally published in 1926, went through multiple editions and is still available on amazon.com in an edition published in 2000. Van de Velde initially opposed use of artificial contraception, although he changed his mind over time. Despite emphasizing heterosexual intercourse within marriage, Van de Velde was not overly concerned with reproduction. Rather, he was a strong believer in marital intimacy and advocated foreplay, including oral sex (particularly cunnilingus, or the "genital kiss" as he termed it), copulation during menstruation, and the use of a variety of positions for sexual intercourse. This advice, following on from a number of other sexologists in the early decades of the twentieth century, such as Havelock Ellis and Marie Stopes, was directed toward sexual pleasure through technique.[33] That Van de Velde was able to discuss these issues already shows a wider acceptance of various sexual acts in the first part of the twentieth century. Unlike some of his nineteenth-century counterparts, he does not couch his work in a language of overt Christian morality or in a medical rhetoric that treated all manifestations of the sexual instinct as either reproductive or pathological. Rather, in response to major changes in the way sexuality could be discussed openly and candidly, he could tackle issues that were probably widely practiced in a way that made them normal. By doing so, Van de Velde and others sanctioned sexual changes that emphasized the importance of sexual pleasure as necessary for a healthy relationship. This intimacy at the heart of a relationship became one of the defining aspects of Anthony Giddens's conception of modern relationships.[34]

A like change in the way a sexual practice is treated is exemplified by anal sex. Anal intercourse has gone from being considered an unnatural act,[35] illegal in many jurisdictions (and it remains illegal in a significant number today), to much more common sexual act, approaching mainstream—if its increased discussion in glossy magazines and mainstream pornography, and its ubiquitous representation in hard-core adult film, is anything to judge by. Although it has, of course, existed in heterosexual relations for many years, as evidenced by medical texts discussing possible infections caused by such an act, or in pornographic literature of previous periods,[36] the growing acceptance of anal sex as not simply a way of avoiding pregnancy shows that changes in conceptions of natural sexual acts, and their status in sexual relationships, involve a complete subversion of previously dominant discourses and their conceptions of sanctioned sexual acts. Now, anal sex is commonly discussed in women's magazines, on chat sites, and so on. These representations are a far cry from the signification of anal sex in films such as Bernardo Bertolucci's *Last Tango in Paris* (1972) or in Paolo Passolini's *Salo ou les 120 journées de Sodome* (Salo, or 120 days of Sodom; 1975) as formerly shocking depictions of sexual excess. This alteration in the sanctioning of an act shows that a widespread adoption of a practice changes its cultural value and hence provides for different performances. That change is occurring is also borne out by recent medical figures that show an increase in young women (between twelve and twenty-five years of age) self-reporting anal sex from 3 percent in 1994 to 5.5 percent in 2004 in a Baltimore sexually transmitted disease (STD) clinic.[37] A different publication, quoting a U.S. Government Department of Health and Human Services report, concludes:

> "For males, the proportion who have had anal sex with a female increases from 4.6 percent at age 15 to 34 percent at ages 22–24; for females, the proportion who have had anal sex with a male increases from 2.4 percent at age 15 to 32 percent at age 22–24." One in three women *admits* to having had anal sex by age 24. By ages 25 to 44, the percentages rise to 40 for men and 35 for women. And that's not counting the 3.7 percent of men aged 15 to 44 who've had anal sex with other men.[38]

These figures are also problematized by the fact that many religious groups will not allow medical practitioners to question young people about their sexual practices, except for heterosexual vaginal intercourse—a dangerous position in the time of AIDS and also one that fails to capture the dramatic increase in heterosexual oral sex among teenagers. Some groups consider it potentially corrupting to ask teenagers about homosexual experiences, as it might give them sexual ideas (although presumably they could also read about such activities in some of Paul's epistles?).

The most drastic change in sexual practices in the twentieth-century West must surely have been brought about by the contraceptive pill. Some have argued that the sexual revolution itself was brought about by the pill.[39] While we cannot say that the pill was solely responsible for female sexual freedom, it did allow for a more reliable divorce of sex from reproduction. Contemporaneous ideas suggest that such a divorce was already likely. The 1962 best seller, *Sex and the Single Girl*, by Helen Gurley Brown, American editor of *Cosmopolitan* magazine, stressed that the modern woman (the Cosmo Girl) should not consider having children until her thirties or even forties but rather should reach for fulfilling sexual partnerships that were not permanent. Her follow-up, *Sex and the Office* (1965), further argued that women were able to enjoy economic and sexual freedom—a position that the pill enabled to a large extent, especially in the 1970s when it was more widely available to unmarried women (although a flip side to this liberation from pregnancy was that even the most unlikely men came to expect sexual favors on learning that a girl was on the pill—think "Polythene Pam" from the Beatles' *Abbey Road* [1969]). This outwitting of biology followed in a long line of contraceptive advice—indeed, some of the arguments about relationships offered to the Cosmo Girl were to be found in the writings of Ellis and Stopes in the second decade of the century.

Other sides to this sexual utopia of the late 1960s included changes to family life,[40] an increase in STDs outside of the military and sex workers,[41] and higher rates of marriage breakdown.[42] Of these, the feature that has the major effect on the sexual body in recent decades has been infection. Although STIs have long had an impact on the way that sex has been practiced,[43] sexual life in the modern era has been marked by an awareness of sexual risks in an unprecedented way. The most important recent development that has changed the way that sex is practiced is the advent of safer sex in response to the HIV/AIDS epidemic that began in the early 1980s. This movement, which has involved public education and advocacy of changes in sexual practices (emphasizing the need for the prevention of bodily-fluid exchange by avoiding penetrative sex without using [male or female] condoms), also had the effect of encouraging open discussion of the risks of sexual activity. Such discussion, I argue, led to an increased circulation of sexual ideas, which in turn brought about changes to the sexual body.

Condom use has become ubiquitous in modern sexual relationships. The return to universities after the summer vacation sees prophylactic devices being distributed by the bucket load in some cities.[44] Sexually active people acting outside of monogamous relationships have added the use of condoms to their sexual repertoires. Whereas the population relying on the pill for contraception in the 1970s had little use for condoms and suffered the ravages of genital herpes and other problems as a result, the avoidance of STIs as serious as HIV and hepatitis C (among others) has meant a change in normal sexual practice, encouraged by wide-reaching public-health campaigns. Drawers next to the beds of the casually sexually active are now used for the close-by storage of condoms and lubricants. Toilet walls are covered in posters advocating the use of male (and female) condoms.

FIGURE 2.1: Eric Avery, *How to Use a Male and Female Condom.* Linocut on paper, 1995. Reproduced with permission of the artist. This sexual advice for saving lives is, unfortunately, not universally supported. On February 25, 1996, after a mother's complaint about her twelve-year-old daughter seeing the installation at the Brasil café (2626 Dunlavy, Houston), the Houston Police Department's Vice Squad ordered that Avery's print *How to Use a Male and Female Condom*, be destroyed (see *Houston Press*, March 7, 1996). It can only be hoped that the daughter picked up enough information from this brief glimpse to avoid catching a sexually transmitted infection in the future. See Avery's Web site, http://www.docart.com, for more, including other art/medicine actions for health.

MODIFYING THE SEXUAL BODY

In the preceding section, it has been argued that sexual acts have changed over the course of the twentieth century. Acts that were at times considered pathological are now held to be, if not totally ubiquitous, at least normal, and as a firm indicator of this trend, many of these acts are no longer legally proscribed in the West to the same extent (such as anal sex). Focusing on more than the development of sexual acts, it is important to address the concomitant modification of the sexual body in physical terms. I will use three examples to illustrate the changes in the sexual body through the twentieth century: pubic hair, sexual piercings, and female ejaculation. Other examples, such as tattooing, would also be appropriate for examining the transformation of the sexual body, but since a number of other chapters in this collection address such a practice, and since tattooing signifies more than the sexual, I do not address this particular modification. Permanent modifications, including breast enhancement, penis enhancement, and vaginal surgery, could also be addressed, although they are neither as common (except breast augmentations) nor as able to be done at home as the other examples, being serious surgical techniques.[45] Thinking about such surgical modifications would take on board Thomas Schlich's model of the surgical technical fix (see his chapter in this volume). I also do not consider non-Western sexual surgeries, such as clitoridectomy or circumcision, which are also considered in another chapter of this collection (that of Cole and Haebich), although these practices, too, are associated with a sense of self that is attached to modifications of the sexual body.[46]

PUBIC HAIR

To gauge the changes in the sexual body, and thus in sexual tastes, the case of pubic hair is worth further consideration. Pubic hair makes a particularly good example—it is private, is related to sexual activity via notions of attractiveness and sexual hygiene, and can be easily modified according to taste. Such modification practices are overwhelmingly, but not exclusively, used by women.[47] This situation is more than likely linked to the fact that many Western women already submit themselves to various regimes of hair control, such as shaving legs and underarms. This development in ideals of female beauty has its origins in North America, where attitudes toward hirsuteness have been uncoupled from ideals of hygiene and desirability, but in a global culture where many dominant images come from the United States, these ideals have found a ready reception in Europe and Australasia.[48]

To gauge these developments it is worth looking at the late nineteenth century, when attitudes toward pubic hair were far from those that appear dominant today. A touchstone for these tastes is the Victorian pornographer,

Walter. When in Constantinople, Walter was taken to a brothel in which an Armenian woman who had plucked her public hair worked. He described his experience thus:

> Cunt it was, but a slit in white flesh it really looked, for not a vestige of hair was visible. She had but a small clitoris (perhaps she'd had it cut off, I have since heard that such things are done in the East) and very small nymphae. The cunt lips puffed out and I thought, on carefully looking, that I saw signs of stubbly hair, but could feel none. The cunt looked in fact like a long cut in a lump of dough, with a little red line indicating the parting.

The next day, Walter met with an Italian prostitute, to whom he described his liaison with the pubeless Armenian. His Italian paramour agreed to procure another woman—a Greek—with no pubic hair. This allowed Walter to compare the two:

> I fucked her whilst the Italian, laying on the bed, showed me baudily her horsehaired sperm sucker, and I came to the conclusion that a hairy cunt in woman is much handsomer and more voluptuously enticing than a hairless one.[49]

Pubic hair had a sexual signification in Walter's time that has since been lost from public representations of sex, although the process was a relatively recent one, largely because pubic hair was out of the public eye for most of the century.[50] We see this if we look at nonpornographic representations of the female sexual body throughout the twentieth century (as overwhelmingly the artistic images of the nude have been female). For example, Amadeo Modigliani's nudes shocked the Parisian art world precisely because he openly depicted pubic hair on some of the models, a display that had already been achieved by Goya in *The Naked Maja* (1798–1805) and much less subtly in Gustave Courbet's *L'Origin du Monde* (1866). Pubic hair was considered the border between art and pornography for many. In 1966, Michaelangelo Antonioni's *Blow Up* was the first mainstream film to show pubic hair, in a scene in which blonde Jane Birkin and brunette Gillian Hills wrestle with (or are stripped by) protagonist David Hemmings on a large roll of backdrop paper, with quick glimpses of female pubic hair as a precursor to a ménage à trois. By Antonioni's *Zabriskie Point* (1970), hirsute sexual scenes abound—especially with the naked couples in a voluptuous, stoned orgy in the dust of Death Valley. Strangely, legal (softcore) pornography followed the lead of avant-garde directors. The American magazine *Playboy* had introduced pubic hair to their magazine in the August 1969 issue, with photos of dancer Paula Kelly. Although it was a photo of her bum, her pubic hairs reflected in a cleverly positioned mirror. The blonde Norwegian glamour model Liv Lindeland, also known as Miss January 1971, was the first to expose her pubes deliberately in the American top-shelf magazine, albeit it in a coy sitting position.[51] Meanwhile, the British publication of *Penthouse* had been showing pubic hair since 1966.

The American edition of *Penthouse*, *Playboy*'s main competitor, decided the only way that they could challenge *Playboy*'s market dominance was to be more explicit, which initially meant showing more pubic hair, with the posing becoming more explicitly sexual. Other pornographic magazines followed suit. Larry Flint's *Hustler* started the trend of extreme close-ups of genitalia, an action that had him prosecuted in 1977 for publishing a close-up photograph of a clitoris in his magazine. The thick bushes of pubic hair ubiquitous in the top-shelf magazines of the 1970s did not remain for long. By the late 1980s, pubes were becoming more trimmed to reveal more of the genitalia in close-up shots that bordered on the gynecological, in much the same way as Courbet had treated his anonymous model in 1866. This imperative exposure is now the norm in pornographic representations, as if the exposure of flesh free from hair is somehow closer to the real experience of sex. As one online article notes, "These days, the vast majority of women in porn have smooth-shaven vulvas, or close to it."[52] The same point is made more personally: " 'I was a Penthouse model in the early 1980s,' says retired porn star Kelly Nichols, 'and I posed with a full bush. No one in adult entertainment shaved back then. Now everyone does.' "[53]

This en masse hair removal coincided with skimpier women's swimwear, first in Brazil (hence the name of the thin strip of pubic hair left above the vagina) but soon across Western beaches, leading to a mainstream adoption of severe pubic topiary by the end of the century. The influence of pornography on pubic hair can be seen in ordinary newspapers, for example, the *Observer*'s piece from 2002 on "Pubic Relations," which discussed pornographic hairstyles in a candid manner that once would have been censored but is now accepted as commonplace.[54] This article implied that pornographic images were making it into the mainstream, implicating a lot of readers as users of the genre. Even if this were not the case, it would appear that pubic hair is now an unfashionable accessory. The Web site http://pubicshave.stores.yahoo.net captures some of the rationale for this war on pubes, announcing its purpose: "Do you want to learn to shave your pubic hair, project a cleaner image, or do you have medical reasons for shaving your body? This website is here to give you advice on how to remove hair from all over."[55] Interestingly, any effect that shaving had on sexual experience was omitted from the page, although in the manner of an open secret. The same Web page describes various categories: a "close trim," which is self-explanatory; and "Brazilian Style/taking it all off," which "can be a liberating and exciting sensation. Feeling your lover's touch down below can be exciting and interesting."[56] Various shapes, such as a triangle or a "landing strip," are lauded as "a fun way to express yourself; you might want to choose a pubic hairstyle that matches your personality[?!]."[57] Most interesting in this brief tour of pubic *coiffure*, however, is being "au natural [*sic*]": "Some people are into a giant bush. I guess it is a fetish for some. I suppose that everyone needs some maintenance once in a while, so even if you are going natural style, we can help you choose how to trim the hedges a little bit."[58] This portrayal of pubic hair as a sartorial, hygienic, or even psychiatric problem, something to be removed via an intensive regime of waxing or shaving, is a new phenomenon that has to do with changing ideals of fashion and sexuality, and conceptions of naturalness; in the North American context in particular, the purported issue of hygiene is repeatedly raised against being acceptably hirsute. "Superfluous hair," as it became known (which initially denoted leg hair and armpit hair), included pubic hair toward the end of the century.[59] Pubic hair is no longer a sexual signifier—indeed, its unmanaged existence implies precisely the opposite, a barren, unattractive sexual neglect.[60] In an age when it is increasingly on display, pubic hair has become something needing control. "Like the removal of hair from other body parts, the Brazilian wax is being promoted by consumer capitalism—particularly through beauty salons, popular magazines, and Web sites—as a way for women to increase their attractiveness and sexuality."[61] Perhaps the best example of this commoditization of genital hair was a 2003 ad campaign by Gucci, which had a female model leaning back against a wall with a man crouching in front of her, his proximity only a breath away from her body as if about to perform cunnilingus, pulling down her panties to reveal the trademark Gucci G shaved into her pubes—a play on the model's G-spot as well as an original product placement.

Despite their leading position at the initial uptake of depubing, it is not only women who have succumbed to the temptations of hairlessness. For comparison, here are a few contemporary examples from a women's Internet chat group, the "better sex" forum on http://www.femalefirst.co.uk/board. The discussion of the removal of (male and female) pubic hair, its pros and cons, is a perennial item on this list. Apart from articles on the actual details of *how* to exfoliate (and particularly how to avoid shaving rash—a liberal application of Visine is key), this site often features discussions about attitudes toward pubic hair, making it a good acid test of the issue. Both genders often detail their own exfoliation experiences on this list. On a thread entitled "Shoul [*sic*] men shave their pubic hairs," a variety of expressions were given, including by a woman who preferred defoliated sex with her partner: "I love a shaved scrotum hehehe. Seriously when me and my boyfriend are both shaved sex is just better, I can go down on him. . . . He can do me . . . and there is no mess."[62] A man (there are many on this list, and I doubt that they are all doing sociological research) offered an explanation of his depilatory practices in sexual terms: "Yes I shave and keep it clean. . . . I think its only fair, if the girl Im [*sic*] with keeping it clean and smooth for me so that I can go down or what not. . . . She deserves the same. Just like we guys don't want any hair when we go down, I think girls don't either."[63] And these hints about better sex have had an effect on at least one other male reader: "Hmm reading through this topic, I am seriously debating shaving around that area!"[64] Although many men have hitherto avoided shaving their pubic hair (along with other forms of hair removal—especially in the blue-remembered days when beards were fashionable), an increasing number of salons are offering male-specific exfoliation, including the self-explanatory "back-sack-crack" wax.

There are a number of personal issues at stake in the removal of pubic hair. Ex-stripper-turned-blogger Chelsea Girl, a veritable modern-day Walter, enjoys both the experience and the results of having her pubic hairs waxed: "Unlike just about every other woman I've ever spoken to about this Brazilian matter—and every man too, actually—I like getting waxed. I like the whole de-furried enchilada: the exhibition of my naughty bits, the warmth of the wax, the laborious process, the prepubescent effect, and the pain."[65] This algophilic aspect is even clearer in other sexual modifications, such as genital piercing, which are increasing in popularity. More significantly, Chelsea Girl also captures some of the issues surrounding this new trend for pubelessness. It is occasionally suggested (usually by the genitally hirsute, it seems) that removal of the pubic hair imitates prepubescence and that it somehow touches a primal pedophilic nerve in those who enjoy their partners without hair. Most people commentating on their own experiences in such online fora do not agree with this interpretation, citing instead heightened sexual sensation as the reason for depilation, as well as commending the "cleaner" benefits of hair-free oral sex. Exposure of the genitalia harks back to the reason for hair removal in erotic imagery, an increased visual exposure, a point that is clearly a particular aspect of some sexual performances.[66] The mystery of the genitalia has been removed along with the hairs. Hairlessness signifies an unhampered sexual availability—quite the opposite of the presexual, prepubescent body. It used to signify other things, from a fetish to an aspect of humiliation and control in a BDSM relationship.

To reiterate, performing the sexual body through pubic hairstyling has two main features: One emphasis is on the cleanliness of the area, the other on the heightened sensations that will be felt without hairs. Treating unwanted body hairs as unnatural, dirty, unhygienic, or a sexual nuisance is an example of the naturalization of a fashionable norm, as it uses ideas of cleanliness to frame the genitalia as odorfree and therefore uninfected and pure (a point that should not be underestimated in the time of HIV/AIDS and other STIs). While the performance of cunnilingus took considerable time to be accepted as mainstream rather than perverse, it has only recently been related to the removal of hair for its successful commission. The same is often said of performing fellatio. It would seem from those writing on the subject to various chat groups that presenting a shaved genital area is an invitation for sensitive and clean oral sex and that the existence of pubic hair now seems to discourage such commission, or so a number of responses on the "better sex" chat group would imply.

The repeated performances of hairless genitalia as hygienically preferable and sexually desirable have, in the last quarter of a century, changed attitudes toward the body and, more significantly, have impacted on the sanctioned performances of the sexual body. For a body to be desirable in this age, it has to be smooth. Hairlessness signifies sexual desirability as well as cleanliness to many. The other major reason for pubic-hair removal, increased sexual stimulation, likewise illustrates a modern trend in taking control of the body in order to live out a desired sexual performance. The sexual body is malleable; changes to it for heightened sexual sensation show that sexual embodiment can transcend the biological in pursuit of fulfillment, although this apparently unnatural modification is presented in such a way to support the idea that other biological processes—arousal, stimulation—can be better achieved. This same point is made clearly in the practice of sexual body modification through piercings.

SEXUAL PIERCINGS, SEXUAL CYBORGS: TECHNOLOGIES OF THE SEXUAL SELF

One of the major forms of sexual body modification that is more than hair deep is piercing (although branding, perhaps suspensions, and, to some extent, tattooing could also come under this head). Piercing the sexual organs has existed for eons, especially in non-Western cultures, although depending on the context the relatively recent relaxing of piercing laws has led to an explosion of possibilities, from the visible—eyebrows and, for sexual purposes, tongues—to the intimate (with a variety of possible genital piercings for both sexes). Piercing of other erogenous zones, particularly the nipples, is also common. Engaging in such sexual modifications is not gender specific or even dominated by one gender. The imaginative variety of piercings, as well as the mushrooming of establishments offering these services all over Europe, North America, and Australasia, suggests that adornment of the genitalia has become a widely practiced activity.

While sociologists focusing on facial piercings have largely argued that piercing is the result of a desire for group belonging, a form of urban tribalism,[67] piercing for purely sexual purposes is a common phenomenon—there are a number of people who have only their genitals pierced, for the principal reason of heightened sexual pleasure (in addition to signifying sexual experimentation, unordinariness, control of the body, and a desire to have their genitalia highlighted through adornment). In addition, some women pierce their genitalia to reclaim their bodies from the experience of past sexual abuse.[68] Piercing thus becomes a new technology of the self but at an individual level rather than as a part of a modernist discursive regime. Master piercer Elayne Angel of ringsofdesire.com notes that "I have many clients who are not otherwise pierced except for genital piercings. These folks are obviously NOT doing it for social reasons—It is all about the eroticism for them."[69] For this sexual market, Elayne Angel invented a new technique, the "Princess Diana" (see Figure 2.2), akin to the male urethral piercing, the Prince Albert (Figure 2.3).

The reason for such piercings is sexual modification of the body: increasing the body's biological abilities through minor surgical enhancement. Information supplied by Elayne Angel indicates that:

> The VCH is by far the most popular female genital piercing we do, for several reasons. As the name implies, the piercing is placed vertically, and women's genitals are built vertically. This means the jewelry [*sic*] situates comfortably between the legs and is not subject to much stress, unwanted pulling or irritation. . . . Further, most women are built with

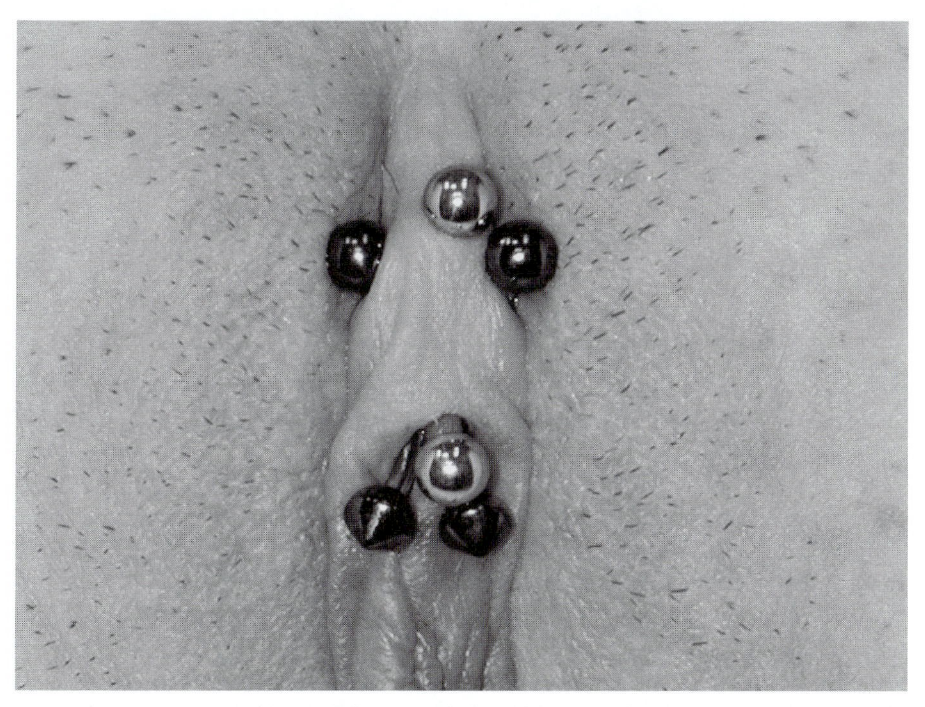

FIGURE 2.2: A Princess Diana and a vertical clitoral hood piercing, showing that piercing goes hand-in-hand with pubic shaving, thus preventing the jewelry being lost from view. Photo courtesy of Elayne Angel, http://www.ringsofdesire.com.

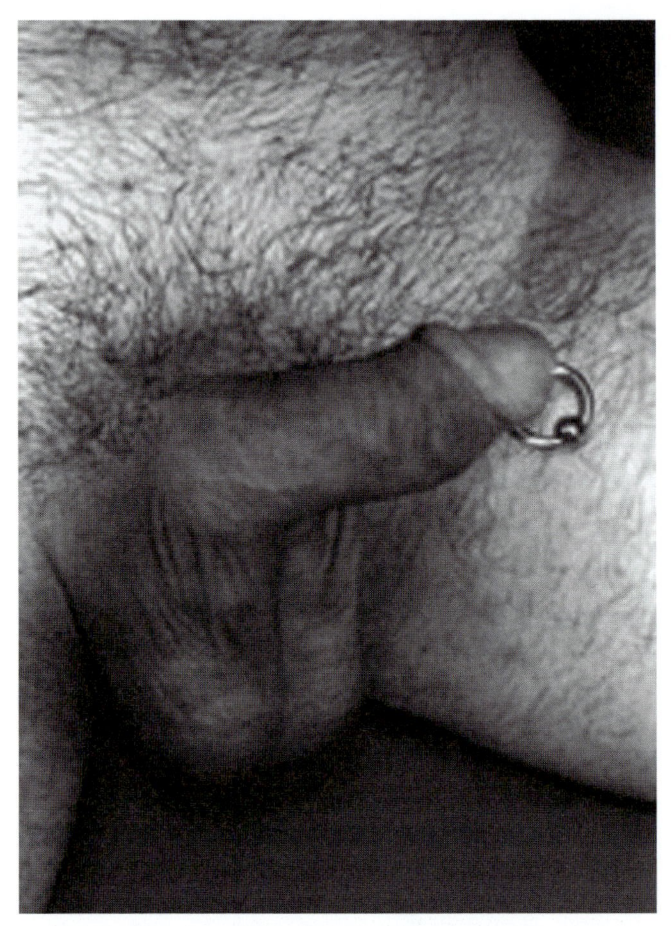

FIGURE 2.3: Prince Albert piercing. Photo courtesy of Totally Pierced, Sydney, Australia: http://totallypierced.com.au.

> some amount of hood tissue in which the jewelry [*sic*] can be placed for a safe, comfortable, attractive, and stimulating piercing. The piercing goes through only a thin bit of tissue above the clitoris (NOT through it) but the jewelry [*sic*] itself touches the clitoris. That means when there is any action in the area, the clitoris will receive more direct stimulation. Most women like this a lot. In fact, we have distributed a research study for the University of South Alabama regarding VCH piercings and female sexual satisfaction! The data has shown that this piercing DOES result in increases in sexual DESIRE and PLEASURE![70]

Put succinctly, Angel notes that "for the pleasure you derive from it, the momentary pinch is a small price to pay."[71] Angel's observations are backed up by medical investigations into sexual piercings. A study published in the *British Medical Journal* that examined sexual body modifications reported that some female participants experienced their first orgasm during vaginal intercourse after getting clitoris or clitoral hood piercings.[72] According to an article published in the *Lancet*, genital piercings have often been associated with homosexual men, particularly the "Prince Albert," or penile urethral piercing. This conclusion is less likely to be accurate today, with both women and men undergoing genital piercings for a variety of aesthetic and sexual reasons.[73] Indeed, it has been suggested that, like the trend for pubic-hair modifications, most body piercing is done for reasons of fashion.[74] Lawyers, too, have examined piercing in terms of the rights individuals have over their bodies, for, until recently, piercing salons were illegal.[75] Changes in the law have allowed for freer expression of the sexual self. In this bodily enhancement, the piercing signifies more than sexual experimentation; it suggests an individual who has a strong sense of his or her sexual self and an ability to take control of his or her sex life in an active way. Similar conclusions can be reached in terms of pubic-hair removal, and indeed if imagery on the Internet is anything of a guide, piercings are often found in those who denude their genitalia. This reflexive construction of the self, and its entwinement with the body, is a classic expression of postmodernity.[76] Performing the self through bodily modification can be seen as individuals taking control of their biology—becoming sexual cyborgs. When such modifications are sexual in nature, the performance becomes further intricated in the establishment of identity: The self becomes the pierced self, and when erogenous zones are pierced, it becomes the pierced sexual self. The technology and the body are inextricably linked in the sexual sensations and in the performativity of the sexual self.

Reasons for genital piercings are various and personal, but as noted sexual stimulation is often high on the agenda. Blogger Chelsea Girl describes the occasion when she had her clitoral hood pierced:

> I remember when Dan [the piercer at The Gauntlet salon on 5th Avenue, New York] was ready to plunge the needle through my hood, I locked eyes with [her friend] Alexis, exhaled sharply, and it was done. . . . There in the folds of my labia glittered a little gold hoop. I went home to the apartment I shared with Will like I was walking on vibrators. It was amazing, the feeling of this alien piece of metal rubbing incessantly against my clit. I was perma-wet. Will was home when I returned. He was excited to see my new accessory, and I peeled down my jeans to show him. He promptly bent me over his standard-issue black leather bachelor couch, placed a sheet of Saran Wrap, the poor man's dental dam, over my newly pierced clit, and licked my pussy until I came like the proverbial freight train. Astonishing doesn't even begin to cover that orgasm, that very first time, when, still raw and bleeding from Dan's sterilized needle I had my pussy licked.[77]

This extract from an engaging discussion of a contemporary sex life illustrates the modification of the body for sexual purpose in graphic detail. The "new accessory" gave a heightened feeling of sexuality and of sexual stimulation. Adopting this jewelry was more than an addition of a technology; Chelsea Girl was coperforming a sexual self with "a little gold hoop." Building on notions of coproduction of gender and technology, Chelsea Girl's adoption of the clitoral piercing extended the possibilities of sexual performance and thus extended her performance of self.

There are many opportunities for others to share their piercing experiences on http://www.bmezine.com. One of the more interesting discussions of a Prince Albert piercing comes from John, a fifty-one-year-old Texan who is married with children. For John, "life is very middle class and good, but I had always been tempted by things just a little bit on the edge." His choice of body modification was clear: "The Prince Albert was the one that really got me." One of the reasons for his choice was sexual. The sexual benefits conferred by this modification—apart from the pleasure in being a pierced man—was the reaction of John's wife, who profits from the piercing rubbing directly on her G-spot in their favored sexual position. To quote John's approval: "I can't say enough about how great it is."[78]

The sexually modified self is reflexively established through an extension of biology, for the purpose of increased sexual stimulation and sexual signification, and resulted in a transformed sense of sexual self. It is not possible to separate out biology and culture from such expressions of the self. Sexual body modifications rely on extending both physical and social possibilities and result in a performance of the self that is staunchly artificial.[79] Such performative extensions of the sexual self need not be technological, however, as the next section illustrates.

LEARNING TO PERFORM: FEMALE EJACULATION

Some body modifications are cosmetic, such as those considered in the preceding. Others—such as female ejaculation—are more biologically rooted, although they can still be shown to involve practices of the self. Conceptions of the female orgasm changed significantly over the twentieth century. For a long period, following its enshrinement in Freud's writings, it was thought that there were two types of orgasm: the "immature" clitoral orgasm and the "mature" vaginal one.[80] Although this notion was powerfully dispelled by William Masters and Virginia Johnson, who insisted all orgasms were clitoral orgasms in their 1966 best seller *Human Sexual Response*, the precise nature of the female orgasm has still been the subject of much debate. Hite, for example, in her 1976 *Hite Report on Female Sexuality*, argued that 70 percent of women who were unable to orgasm through sexual intercourse were able to masturbate themselves to orgasm and therefore were not "frigid," as Masters and Johnson had believed. This ability to take matters into one's own hands recurs in the following when the issue of sexual education and masturbation is addressed in relation to new forms of orgasmic performance.

Female ejaculation is a phenomenon with a very strange history, as doctors have sometimes considered it to exist and at other times not. Since the first anatomical descriptions of the clitoris in the sixteenth century, there have been representations of female sexual orgasm associated with an ejaculation of liquid. Realdo Columbo, the Paduan anatomist, noted in his *De re anatomica* (1559) that, on orgasm produced by a rubbing of the clitoris, "semen swifter than air flies this way and that on account of the pleasure."[81] This idea of female ejaculation survived, tied variously to notions of generation, for a long time—although it met a sudden demise right at the point when sexual science was paying increasing attention to female sexual response and to sexual physiology, being passed off as urination cause by spasmic release of the bladder rather than an ejaculate.[82] This view long dominated twentieth-century medical discourses, which considered it impossible for women to ejaculate during sexual orgasm. A number of anatomists and physiologists looked for a function of the urogenital sinus glands following their discovery by Alex Skene in 1880.[83] John Huffman, in a series of papers written in the 1940s, concurred with a number of other authors that these glands were homologous to the male prostate and detailed the potential pathologies that the glands may present. He also noted that these glands were mucous secreting.[84] This is not to say that these anatomists were convinced of the possibility of female ejaculation—indeed, they do not raise it directly in their works.

Not everyone was as puzzled by these glands. Ellis, for example, considered women to be capable of ejaculation: "The process of erection in woman is accompanied by the pouring out of fluid which copiously bathes all parts of the vulva around the entrance to the vagina. . . . There is, however, a real ejaculation of fluid which, as usually described, comes largely from glands, situated near the mouth of the vagina."[85] Since Ernest Gräfenberg's 1950 article on the role of the urethra in the female orgasm, female ejaculation has become an important topic in female sexual physiology. To quote Gräfenberg at length:

> The glands around the vaginal orifice, especially the large Bartholin glands, have a lubricating effect. Therefore they are located at the entrance of the vagina and produce their mucus at the beginning of the sexual relations and not synchronously with the orgasm. Sometimes the mucus is produced so abundantly and makes the vulva slippery, that the female partner is inclined to compare it with the ejaculation of the male. Occasionally the production of fluids is so profuse that a large

> towel has to be spread under the woman to prevent the bed sheets getting soiled. This convulsory expulsion of fluids occurs always at the acme of the orgasm and simultaneously with it. If there is the opportunity to observe the orgasm of such women, one can see that large quantities of a clear transparent fluid are expelled not from the vulva, but out of the urethra in gushes. At first I thought that the bladder sphincter had become defective by the intensity of the orgasm. Involuntary expulsion of urine is reported in sex literature. In the cases observed by us, the fluid was examined and it had no urinary character. I am inclined to believe that "urine" reported to be expelled during female orgasm is not urine, but only secretions of the intraurethral glands correlated with the erotogenic zone along the urethra in the anterior vaginal wall. Moreover the profuse secretions coming out with the orgasm have no lubricating significance, otherwise they would be produced at the beginning of intercourse and not at the peak of orgasm.[86]

The significance of this finding, and the attention to the G-spot named after Gräfenberg, changed the way that the female orgasm was considered. Gräfenberg located a spot on the inside of the anterior wall of the vagina that was responsible for intense voluptuous sensation. This view of a vaginal orgasm was criticized by Masters and Johnson, who argued that all orgasms were clitoral—and the dominance of Masters and Johnson's view (facilitated by some feminists' advocacy of the noncoital orgasm) came to silence much discussion of the possibility of nonclitoral sexual orgasms. This opinion has since been challenged and takes account of the experiences of numerous women who can orgasm in a number of ways depending on the stimulation.

This medical confusion surrounding the possibility of female ejaculation is mirrored in some women's own experiences and searches for information. A typical question to Columbia University's online sexual advice column, Ask Alice, reads as follows:

> I am a female and I have had these strange orgasms lately that I never had before with my partner of six years. Instead of your basic orgasm, I have very powerful ones that last forever and include a lot of liquid coming out of my vagina. I have never heard of a woman "ejaculating" but that's what it seems like. Is this normal or is something wrong with me?[87]

Despite much physiological research into the problem, there was a good deal of reticence in the medical world to accept the idea of female ejaculation. The research into female ejaculation by Beverly Whipple and her associates challenged this viewpoint in particular.[88] No longer are women who ejaculate having to live in fear that they are peeing the bed when they orgasm. Instead, ejaculation has become something to emulate for some, and perhaps even a new standard to aim for.

Much of the representation of female ejaculation has come from its increasing depiction in pornography, with a number of specific adult sites specifically devoted to it. Although films of squirting that abound on the Web must occasionally be treated skeptically, there is no sound reason to believe that all such images are fantastic and designed solely to show even more sexual ability in the actress. A number of performers have been filmed "gushing," in particular the self-styled "squirt-woman" Cytherea, winner of adult movie awards Best New Starlet (2005) and Best Specialty Tape—Other Genre (2005) for her film *Cytherea Iz Squirtwoman*.[89] This representation of female ejaculation has caused a problem for British film censors, who largely deny that it exists, seemingly not giving weight to a number of modern studies in sexology. Indeed, the British Board of Film Classification has on occasion banned the import and production of such images for their perceived urologic content. There have been strong reactions against this issue by the feminist group, Feminists Against Censorship, who argue that the British film censorship board are perpetuating views against female sexual expression (and even physiological potential) through their policy that emphasizes medical knowledge of women's experiences—despite the fact that this knowledge is also controversial.[90] This group is fully behind the circulation of images that show women to be equal to men in all ways—including ejaculatory potential.

Regardless of the occasional British Board of Film Classification treatment of female ejaculation as an elaborate form of urination, a number of how-to Web sites and books exist. The Web site sextutor.com notes that "you can learn how to unleash a tidal wave of orgasmic juices in your bed"[91] and proceeds to instruct the uninitiated who desires such an effect. Various Web sites also contain the same information written from a (often somewhat less literary) male perspective so that they may take their female partners to new orgasmic heights. Likewise, manuals such as the 2003 book *Female Ejaculation and the G-Spot* by Deborah Sundahl, who also holds classes for those who wish to acquire the skill in a more pedagogical setting, are common. A search at amazon.com throws up hundreds of titles under the term "female ejaculation," many of which are training manuals—so a market must surely be out there. It is interesting to note that these works often recommend the ideal setting for the initial exploration of ejaculatory potential is in solo, with either fingers or a curved G-spot vibrator—both suggesting that this act is primarily for the pleasure of the woman expanding her own sexual horizons and also neatly extending Tom Laqueur's history of masturbation, *Solitary Sex*.[92]

The case of female ejaculation is significant for this chapter in that it does not involve artificial body modification but does incorporate sexual technique and an enhancement of sexual capability. This is a new technology of the self.

To achieve such expansions of the sexual self, a climate must exist whereby information of intimate detail can be freely distributed, self-reflexive projects that place an emphasis on sexual experience and performance can take place, female masturbation and sexual pleasure are no longer be considered taboo, and a certain value is placed on these expansions of the sexual self. We in the liberal West live in a context in which these are possible. The body has become a project for many who hope to transcend their sense of self and their natural body via regimes of training and practice that allow for new possible performances that have also undergone changes in sanctioning—from something associated with shame (wetting the bed) to something equating to masculine sexual prowess through the expulsion of liquid during orgasm. Indeed, it is this visual display that makes sense of the increased presence of female ejaculation in pornography—the female equivalent to the "money shot," as the male ejaculation outside of the partner's body is known. Such sexual performances embody more than a hint of sexual equality. To ejaculate (and to acknowledge this as a possibility for women) is to enjoy a sexual possibility specific to our time.

CONCLUSION: PERFORMING THE SEXUAL BODY IN SPACE

This chapter has argued that increasingly over the twentieth century it has been possible to alter the body for sexual purposes through cosmetic, surgical, and physical modifications. These technologies of the self have expanded the possible performances of the sexual body. These changes were not undertaken in a social vacuum, however. This survey of the changes in sexual representations has relied on a variety of sources: magazines, chat groups, and other (adult) material on the Internet, as well as a number of medical and psychological sources. By way of conclusion, I touch briefly on the ways that the body has been represented, and specifically where it has been represented, over the twentieth century. These representations themselves become resources for sanctioned sexual performances—that the representations of the body in these various sites are enrolled in future uses of the sexual body, although acknowledging that past uses of the body do not determine future uses.

One of the key twentieth-century developments in the reconceptualization of the sexual body came from those interested in naturism. In Germany, adherents of the *Freikörperkultur* (FKK) argued that nudity was essential for health.[93] From the turn of the century, various people explored the body as a natural entity, as something that is healthier and more natural when it is completely exposed, either lying on the rocks in the sun along the Croatian coastline or bobbing up and down playing volleyball at a nudist camp.[94] This move was important not just for the natural health movement, which attempted to shed the bodily inhibitions of the nineteenth century along with their clothes; it was one of a number of ways that the body became exposed to others—although not in a *primarily* sexual way. Exposure is still an issue in many parts of the world; while contemporary men and women in Berlin might be happy to jump off their bicycles, strip, and lie around naked in the Tiergarten without a legal concern in the world, such liberation was not extended to the British walker Steve Gough, who hiked from Land's End to John O'Groats in the nude in 2003 and found himself in constant trouble with the police.[95] Naturists, however, are quick to point out that their embracing of a textile-free lifestyle is not sexual, despite nudity being so closely allied to sexuality in the wider "clothes-obsessed society."[96] To show that the choice not to wear clothes is a wholesome family affair rather than a perversion, many images show family and sporting activities in an asexual way.[97] These practices include the widespread northern European adoption of public mixed nude saunas, with additional, if select, opportunities available for those interested in naked yoga—with classes run in New York City by Isis Phoenix.[98] While not wishing to dispute the motives or reasoning of FKK devotees, it is not insignificant that many of the images from various summer camps show the removal of pubic hair in the same ways that I have noted in the preceding. The same pubic modifications can be seen in public saunas.[99] These naked spaces do not promote sexuality.[100] But they do show the variety of sexual modifications of the body to other patrons—from sculpted (or not) pubic hair, to sexual piercings (nipples, genital), to the enhanced attractiveness of physically fit individuals. They demonstrate the possibility of bodily sexual performances—what Gail Weiss has called "an ongoing exchange between bodies and body images"[101]—even without taking part in sexual acts. In turn, they regulate understanding of sexual possibilities, transmitting sexual possibility through the appearance of the passive, nonsexual body.

Public nudity—or public spaces in which nudity has been condoned, albeit in limited ways—has been a stimulus to other far more sexualized issues including skimpy fashion and strip clubs. From vaudeville in the 1920s, to the advent of nude shows that have removed the sense of erotic revelation (as conjured up by Roland Barthes in *Mythologies*), strip shows have made a liminal space for the performance of sexual titillation between the pornographic and the regular theater.[102] Strip shows bloomed in the Parisian demimonde, frequented by artists and the avant-garde, and the boundaries between stripping and prostitution have not always been clear-cut. Nudity in public, and the relationship between nudity performed for sexual purposes and the wider acceptance of both sex and nudity in public, have reduced public reactions of a moral nature (in the liberal West). The real effect on the sexual body has been one of emulation. The more widely distributed these sexual spaces become, the more they change sexual practice through exhibiting possibilities. This situation has become even more expanded with the establishment of spaces for sex. A particular example is the sexual space provided by gay sex clubs, as well as other homosexual meeting points such as cruising beaches and parks.[103] Such spaces for sex have a large role in moving sex from a private to a public sphere. The effects on individuals have been a wide exposure to other bodies and a massive expansion in sexual possibilities. Other less salubrious effects have been sexual risks associated with STIs, which have produced changes in sexual practices—most notably safe sex—as discussed in the preceding. The greater the possibility for interactions between sexual bodies, the greater the possible repertoire of sexual performances—from piercings and other modifications to sexual acts that push the limits of experience. Or to put it another way, the more experience one has, the more one can do with one's body.

This argument is not limited to gay subcultures. A variety of semipublic sexual performances for heterosexual individuals have proliferated, from swingers' clubs to "dogging" (public sex, often in British parking lots, with a circle of spectators—many of whom are out "walking the dog" at night). If one considers the growing acceptability—and availability via the Internet—of pornography, it can be seen that an increase in the repertoires of sexual practice relates to depicting a greater variety of sexual acts.[104] This can be seen in the case of sexual body modifications, just as it can be seen in the exhibition of physiological processes, such as female ejaculation. The great availability of images of sexual bodies and sexual acts performing specific sexualities (and sexual acts) leads to wider sanctioning possibilities for sexual performativities.

Not all performances of the sexual body in public or virtual spaces are pornographic. Performances of everyday genitalia, both male and female, can be found on Betty Dobson's educational Web site: http://www.bettydodson.com/subgenit.htm. Here individuals send in photographs of their genitals with covering stories that typically give their sexual histories and explain their ideas about their genitalia (from how to masturbate, to feelings of shame at the appearance, to boasting about their sexual capacities). These stories show how sexual knowledge (gained from both sexual images as well as other sources of sex education) have a direct impact on attitudes toward the sexual body and also show that people decide to modify their bodies—physically as well as through undertaking new action or varying the way they masturbate. The exhibition of the self in public in such a manner has altered the way people experience their selves at a fundamental level. This has been taken further by the German site http://www.wie-ist-meine-muschi.de, where many women (or their partners?) upload explicit photos of their genitals for rating by the wider Internet community (there are sister sites devoted to rating submitted *arschen* [asses] and *titten* [tits] for those so inclined). Other significant new developments in amateur pornography (from Australia) in which the subjects represent themselves, often from feminist-allied perspectives, with an emphasis of photographic quality rather than sexual explicitness, include ishotmyself.com (showing naked photos sent in by models themselves); beautifulagony.com (showing self-submitted film of the faces only of people reaching orgasm); and ifeelmyself.com (with more explicit sexual material, largely including women masturbating but directed by the women themselves).[105] These public self-exposures are clearly enabled by the technological society in which we live. The same argument runs alongside the increase in acceptance of pornography—when it is easily and freely available, more people can satisfy their initial curiosity, despite the fact that it is possible to see exactly what people have been viewing—a problem that does not exist to the same extent in the less-technologically provided forms of pornography.

One of the ways to understand these developments of the self relates back to Foucault's notion of limit experiences.[106] Bodily and sexual changes take place when individuals adapt currently sanctioned performances to their own lives, performing them in ways that extend the previously sanctioned possibilities, first through proliferation (making something normal through its wide acceptance and use), and second through the will to overcome limits imposed via social sanctioning processes (enjoying new sexual experiences, pushing sexual boundaries). This transcendent model for sexual change also exhibits one of Foucault's main concerns: power. In this sense, Foucault's micropolitical analysis of power shares much with the performative model of bodies when used to discuss the body as a social institution, as discussed in the preceding.

One of the features of this chapter has been the extended attention given to the female body. The distribution of power in society has made the female body more plastic. In some of the cases discussed here, this is because only female bodies have certain capabilities—such as biological potential for multiple squirting orgasms. But in the bulk of these cases, from the removal of pubic hair to taking responsibility for contraception, women's bodies are the ones that have had to adapt more readily to the newly sanctioned performances of the sexual body. This is not to say that male bodies are not also trapped in cycles of repetitious representation of acceptable masculine norms—the construction of the male athletic body is a good example.[107] But male bodies are more fully situated in the prime position—meaning that the heterosexual, adult, virile, nonaging male body is held up as a natural icon that does not need to be modified to the same extent to be sexual. Of course, there are transgressive takes on such a body, for example, when gay ideals are embodied in the same athletic body—in a sharp move away from feminized ideals of homosexual bodies that had been promulgated in various medical and other normative discourses. The performativity of the sexual body, then, is not able to be taken outside of power networks any more than it can be removed from either culture or biology. All three are embodied. All three are present in the sexual body as it changes over time.

This survey of cultural manifestations of the sexual body has deliberately steered away from official discourses such as those produced by medical or legal fields. That is not to say that such fields have not impacted strongly on the sexual body in terms of regulating possible practices and sanctioning those that cross socially acceptable boundaries. To appreciate the sexual body through the twentieth century, one has to focus more on how people have managed their performances and how these performances have met with either approval or negative sanctioning (whether this be at the individual, the intimate, or the social level). This procedure has been made infinitely easier by the current practice of discussing the sexual body in detail, especially on the Internet in blogs and on discussion lists, as well as in more commercialized representations. The sexual body has become increasingly displayed in the twentieth century as social mores relaxed and sex became increasingly central to conceptions of the self and likewise increasingly distanced from religious and moral doctrines that dictated sexual behavior. Expanding secularization has led to amplified sexualization—indeed, one might consider that the slackening of sanctions against sexual expression was one of the key aspects of the continued move away from religiosity in the liberal West. The more we explore our bodies, the less we need "higher" powers to dictate what we do with and to them. Constantly repeated images and descriptions of the uses of the sexual body, a heightened value put on experience and pleasure, and the spread of information to facilitate sexual expression has made the sexual body of the twentieth century possible.

CHAPTER THREE

The Technological Fix and the Modern Body

Surgery as a Paradigmatic Case

THOMAS SCHLICH

In the twentieth century individual bodies became the locus for dealing with issues that had previously belonged to other realms of social life and culture, such as politics, morals, or religion. By defining problems as belonging to the bodily sphere, they became amenable to medical treatments, for example, surgery, drug therapy, or genetic engineering. All these technologies consist in circumscribed and controlled interventions into the bodies of individuals. Continuing and increasing a trend that started during the previous century, they all can be understood as a technological fix to originally nontechnological problems. They are carried out by highly specialized experts, and they come with a particular kind of science-based knowledge about the body.

This had consequences for how the body was viewed. Historians and anthropologists have found that medical knowledge and practices have modified the ways in which individuals in Western societies tend to think about health, illness, health care, and their bodies.[1] Popular views of the body were shifted by new ideas from scientific and medical discoveries; as, vice versa, scientific body concepts received impulses from changes outside of medicine. In what follows I focus on the strategy of the technological fix and analyze its significance for concepts of the body. More specifically, I use the example of surgery and demonstrate its implications for body images in terms of reification, alienation, fragmentation, commodification, and disembodiment. Surgery is the field in which the principle of solving problems by bodily interventions has been put into practice in its most advanced form. Not only is the surgical approach ubiquitous in all domains of modern medicine, surgery has even become *the* model for how medical problems can be solved in a technological way.

THE CENTURY OF SURGERY

Originally, surgeons had not been in a position to offer attractive solutions to medical or societal problems. The art of bodily manipulation was seen as a manual craft, separate from the learned medical profession and less prestigious. For the most part, its practitioners were dealing with emergencies, like broken bones, and minor manipulations at the body's surface, such as removing skin tumors or lancing boils. Traditionally, surgical intervention was associated with the horrors of pain and infection. Even more important, surgery seemed to make no sense in the treatment of most diseases anyway. Until the early nineteenth century the body was seen as a functional whole, interacting with its external environment. People believed that diseases originated from disruptions in the balance of body humors (fluids) caused by the sick person's way of life or some other environmental factor. Diseases could be treated by changing the environment or one's way of living or, more technically, by restoring the humoral balance through emetics, purgatives, and bloodletting. Opening up the belly and cutting out some little part of the intestine, as today's surgeons routinely do in cases of appendicitis, would have looked absurd.[2] As medical historian Christopher Lawrence stresses, the simplest surgical practices employ a theory of the body and disease, either explicit or implicit. Even pulling a tooth, he explains, "implies a theory of the local origin of pain and the relative harmlessness of removing a body part."[3] The rationale that made modern surgery possible developed during the nineteenth century, and it soon came to dominate modern medicine in general. It is based on the idea that the human body is a composite of organs and tissues with particular functions. Diseases are lesions of those organs and tissues that can affect them at the structural or functional level. Surgery can rectify those disorders by removing the diseased structures or restoring their function.[4]

The spread of the new body concept, together with the new professional status of surgeons as university-trained doctors, provided the basis for the expansion of the field. Once anesthesia and asepsis had liberated surgeons from the problems of pain and infection, they could start to explore the potential of their newly acquired abilities. In fact, surgeons set up a whole network of control technologies, including anesthesia and asepsis, which gave them the power to manipulate living organisms to an unprecedented degree.[5] "Within less than half a century," Theodor Kocher, the first surgeon to receive the Nobel

Prize, stated in 1909, it had "become possible to expose all the organs of the body, brain and heart not excepted, without danger, and carry out the necessary surgical measures on them."[6] By the early twentieth century, surgeons were not only fully acknowledged as doctors and scientists, they were even considered modern heroes. They were able to repair bones, remove tumors, and restore complicated internal body structures. Their spectacular successes outshone all other branches of medicine. The operating room was the glimmering center of the modern hospital, a place where miracles were expected to happen.

To the extent that surgery had become safer and more reliable, operative treatment was no longer restricted to emergencies. People started to see it as a routine treatment option. No intervention embodies this new attitude better than appendectomy.[7] In 1890 most American doctors and patients would not have thought of surgery when seeing a case of appendicitis. In 1920, in contrast, appendicitis was clearly a surgical issue for the American medical profession and the general public. And, as hospital statistics in the first decade of the twentieth century suggest, patients were increasingly willing to undergo an appendectomy for this reason. With increasing surgical skill and sophistication, the balance between the risk of surgery and the risk of the disease itself seemed to warrant a more aggressive treatment strategy. The public gradually became used to the idea of surgery. Whereas historians have found "still a considerable fear of surgery on the part of the American patient" at the beginning of the century,[8] the Swiss-born medical historian Henry E. Sigerist described the popular fascination with surgical progress in the United States in 1933: "The American patient wants to be cured as quickly as possible. Faced with the choice between surgery and prolonged medical treatment, he will unhesitatingly choose surgery," he reported.[9] Patients had picked up on the technological imperative. The twentieth century was to become the century of surgery.[10]

In the following sections I look at how modern surgery emerged together with a new view of the body. The cases I have chosen might look marginal at a first glance. They are, however, particularly useful for understanding some general issues. I will start with trauma surgery as an example of the political dimensions of the technological-fix strategy. It is an instructive case for demonstrating how pursuing a technological approach can mask the political character of the original problem. The next example, psychosurgery, illustrates how, in certain historical contexts, the strategy of the technological fix became particularly appealing to practitioners. However, it is also an example of how a technology can fall into disgrace—though often to be replaced by another technological fix, in this case psychotropic drugs. Cosmetic surgery then introduces another dimension into the story. It shows how pursuing the technological fix in the form of surgery can lead to a view of the body as being a "thing" that a person "has" instead of partly constituting the person as such. This development can be called *reification* and *alienation*. The split between person and body becomes even more striking in the three subsequent examples, which concern the surgical construction of sexual identity. These examples also demonstrate how the availability of a technological fix restructures the problem it is used to solve. In the case of sex reassignment, technology even creates the specific—and technically solvable—form of the problem, transsexuality. All three examples demonstrate that the technological fix of a problem and its definition are linked through a feedback loop.

The subsequent section deals with transplant surgery. Transplantation is paradigmatic for how modern high-tech surgery leads to a conceptual fragmentation of the body. Fragmentation in turn can lead to commodification, a development for which transplantation is also exemplary. The sections on transplants will be followed by two nonsurgical examples of medical technology and knowledge and their influence on body concepts. Both of them, immunology and genetic engineering, have, however, close historical or conceptual links to the rise of surgery. They are also both concerned with the relationship between the body and personal identity and how the essence of the person is being projected into certain body constituents or body systems. Another body system that became seen as the site of personal identity is the nerve system, in particular the brain. The brain-death concept illustrates how this localization of personhood helped to render moral questions into a technological format. In another step, personhood is being projected into selected brain structures instead of the whole organ. Finally, in another progression, one can even consider the information contained in those structures as the real essence of the person. In the ultimate consequence, personhood would then completely lose its association with the body.

THE TECHNOLOGICAL FIX: SOLVING PROBLEMS WITHIN THE INDIVIDUAL BODY

Once it was accepted as safe and efficient, surgery became implicated in a variety of societal contexts. Sometimes the societal implications are no longer obvious. Trauma surgery, for example, looks like a straightforward, utterly nonpolitical field of surgical activity. It has, however, an implicit political dimension. Trauma surgery emerged only in association with modern society's compromise of tolerating a certain rate of accidents in industry, traffic, and sports, on the one hand, and providing medical care for the victims, on the other.[11] This relationship is strikingly illustrated in the records of the German discussions about traffic safety in the 1930s. The trauma surgeons who participated in these discussions were opposed to imposing speed limits on motor traffic. Speed limits would impede economic development and technological progress, they claimed. And if accidents should happen, modern surgery would be able to take care of the victims, provided society was prepared to bear the costs of a sophisticated trauma-surgery service. This means that traffic accidents were redefined as a problem of the injured individual. Individualizing the issue in this way deflected causal, moral, and political responsibility away from traffic policy as such. It contains the implicit decision that the individual traffic victims have to bear the burden of the risk that is linked to the spread of this new technology, and they bear it literally with their bodies. The individual body thus became the site where the negative side effects of motorization would manifest themselves and where they could be dealt with.[12] A political problem was successfully turned into a surgical one.[13] The occurrence of accidents now appears to be a natural fact—historians have called this the *naturalization* of accidents[14]—and the development of modern trauma surgery looks like a necessary and inevitable response to that fact. Its political origins have become invisible. It is an example of how the technological approach makes it possible to evade and circumscribe politics and to create the false impression that technology can be completely separated from other human and social practices.[15]

Clearly, surgery, like other technologies, is involved in decisions about the way we live and how we organize society. But this is not obvious to most people. In tandem with surgery's success, its cultural and social implications have become invisible: "In surgery," Lawrence rightly claims, "the fiction that medicine has nothing to do with politics reached its purest expression. Surgical intervention could be represented as the inevitable, scientific solution, in comparison to which alternative solutions seemed inferior."[16] The impression of superiority and necessity is characteristic of technological solutions in general. Once it exists, a technological fix quickly appears "not only preferable but singular, necessary, and inevitable," as Alice Dreger has noted in the context of sex-assignment surgery.[17]

REAL MEDICINE: PSYCHOSURGERY

In the course of the century, ever more social and ethical problems were transformed into "a biomedical format for which the physician"—or, more specifically, the surgeon—"is the resident expert," as Karl Figlio has put it.[18] A case in point is psychosurgery. Even though psychosurgery might seem a very idiosyncratic field, it demonstrates some characteristic features of the technological-fix strategy with particular clearness. Psychosurgery is the attempt to solve mental disorders by surgical intervention. As such, it was part of an extended process of projecting social issues into a particular organ of the individual's body, the brain. In premodern societies deviant behavior was interpreted in a number of different ways. It could be a moral or religious issue. In some contexts disturbed people were integrated into society. In others they were segregated and neglected. A modern strategy of dealing with this problem is to declare strange behavior an illness and attribute it to the medical domain. This happened in the course of the eighteenth, nineteenth, and twentieth centuries, and it opened the way to different forms of treatment, either pedagogic attempts to improve the "lunatics" by discipline, regularity, and work or physical means, such as baths, drugs, and bloodletting.

Once the cause of mental illness was located in the brain, it seemed to be a logical strategy to try and solve the problem by fixing that organ. First attempts at psychosurgery in the late nineteenth century remained isolated experiments.[19] Only from the 1930s to the 1950s did psychosurgery become a widely used treatment option, especially in North America. The paradigmatic psychosurgical operation is lobotomy. The procedure became popular after the Portuguese Egas Moniz undertook a series of twenty operations on psychiatric patients suffering from different kinds of problems in 1935. He reported good results in cases of agitated depression and schizophrenia.[20] Moniz was not satisfied with justifying his interventions by clinical observations alone. Going back to the localistic rationale of all surgery, he also offered an anatomy-based explanation of his technique: He theorized that mental illness was a result of a "fixed loop" occurring in the nerve pathways of the neocortex, specifically the frontal lobes, which were the seat of "psychic activity." Severing the white matter that connects the frontal lobes to the sensory areas would interrupt the feedback loops that were overstimulating the patients. The visible short-term results in patients deeply impressed the first neurosurgeons. Some of them started to use the technique on chronic schizophrenics in the setting of mental hospitals. In the United States, in particular, this class of patients was seen as the most pressing problem for the mental hospitals because no adequate treatments were available for their condition. In the years following World War II, the use of psychosurgery reached its heyday, peaking at over five thousand such operations performed in 1949 alone, the year Moniz was awarded the Nobel Prize.[21]

The context of lobotomy's widespread use was a somatic approach to the treatment of mental disease that was already established at the time. Lobotomy was seen as a "natural extension" of existing somatic therapies and, in some ways, as their fulfillment. The most important of those treatments were the various forms of shock therapy. The principle was to trigger convulsions and states of unconsciousness in the patient, through either injection of metrazol (a form of camphor), insulin-induced hypoglycemia, or direct application of high voltages to the head.[22] On a number of accounts, surgery proved to be a particularly attractive variety of somatic treatment. The immediately verifiable effects of surgical intervention, for instance, seemed to demonstrate that brain physiology and human behavior were connected "in a way that was understandable to laboratory science and open to practical intervention by trained physicians."[23] This meant that psychiatry could claim to have a material basis. It also meant that through psychosurgery the field would be able to catch up with other areas

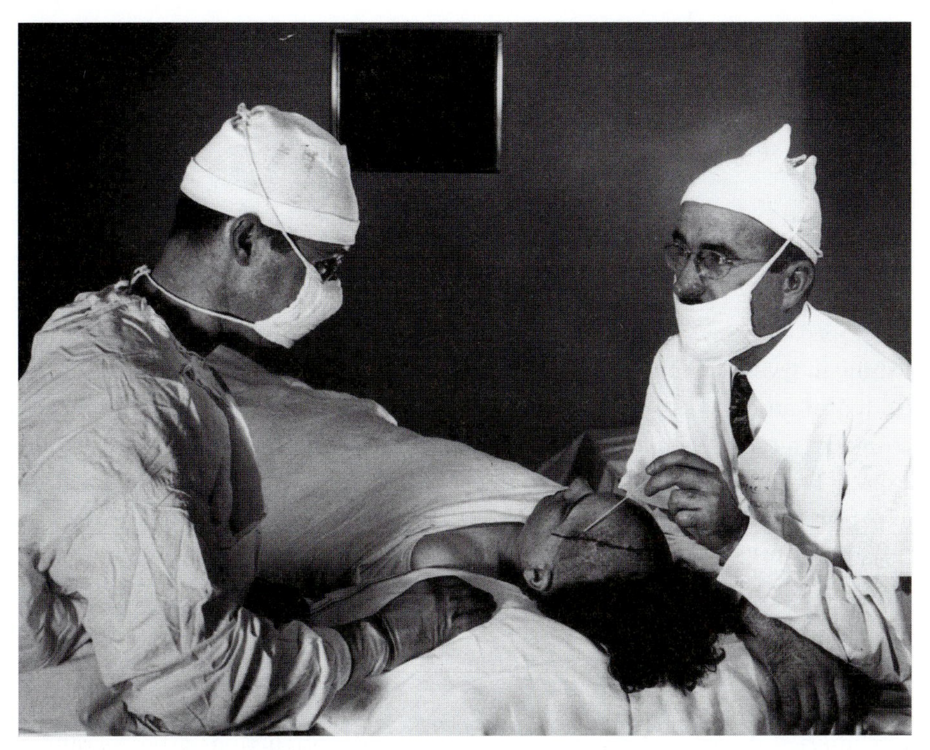

FIGURE 3.1: Physician Walter Freedman and neurosurgeon James Watt examine a lobotomy patient in 1942. Courtesy to Special Collections and University Archives, The Gelman Library, The George Washington University.

of modern medicine. At that time surgery was seen as the epitome of modern medicine. Thus, the best strategy for becoming accepted as a part of "real" medicine was to become like surgery. As Jack Pressman states in his analysis of the lobotomy era, modern psychiatry seemed to have advanced at last to the point where the treatment of mental illness through repairing the brain was "as mundane as when surgeons repair other malfunctioning body parts."[24]

However, as the use of lobotomy accelerated, the unease generated by such a drastic intervention grew too. It was still unclear what lobotomy actually did to the individual. The issue of scientific validation could no longer be avoided. Follow-up studies produced discouraging results.[25] With the advent of tranquilizers in the 1950s, lobotomy fell out of favor. Clinical trials showed that the drug chlorpromazine seemed to have the same effects as a successful lobotomy without the irreversible brain damage and disorientation associated with the surgery.[26] But the demise of psychosurgery did not indicate a rejection of the technological fix. Drug treatment of mental disorders was just a more sophisticated version of the same strategy—a type of reversible lobotomy, as contemporaries saw it. One technological fix was replaced by another. And like any technological fix, the use of psychotropic drugs can be seen as avoiding the real, underlying issues. According to critics, the exclusive focus on controlling the mind via the body is a dead-end street. Instead, doctors should work with more holistic models of mental disease that do justice to the complexity of human experience.[27]

IMPROVING THE BODY: COSMETIC SURGERY

At around the time that psychosurgery became obsolete, cosmetic surgery developed into one of the fastest-growing medical practices in the world. As I show in the following, this branch of surgery represents another technological fix for an originally nontechnological problem. Considering its popularity, cosmetic surgery is much less marginal to modern surgery than one would expect: During the second half of the century, the field underwent a startling expansion in terms of numbers and range of interventions on the face and the body.[28] And the growth has not stopped: In 2004 alone, 9.2 million cosmetic plastic-surgery procedures were performed in the United States, 24 percent more than in 2000.[29]

Originally, the techniques employed in cosmetic surgery were mostly developed for problems that were clearly attributable to the medical sphere, such as war injuries, congenital deformities, or breast reconstructions after therapeutic mastectomy.[30] Those are areas of application for what is called "plastic" or "reconstructive" surgery. In contrast, "cosmetic" or "aesthetic" surgery is seen to follow a basically nonmedical rationale. However, to the extent to which cosmetic surgery gained recognition as a legitimate medical specialty, the boundary between plastic and cosmetic surgery became increasingly blurry. The medical and social reasons for changing one's appearance can be seen to lie on a continuum: For example, the saddle-nose deformity that occurs as part of a chronic syphilis infection caused serious social problems for the individual suffering from it. Correcting that deformity also eliminated the social stigmatization it entailed.[31] In World War I, facial wounds from war injuries were immediately recognized as both a surgical *and* a social problem. At that time, surgeons were very open about the social function of their activity. The explicit goal of facial surgery was to recreate a face that would enable its wearer to function in society, for example, as a family breadwinner.[32] Focusing the discussion in this way on the benefit for the patient, cosmetic surgeons early in the century were able to medicalize their activity by transforming the problem of being unhappy with one's appearance into a medical diagnosis. They posited that "individual well-being depends on mental as well as physical health, and a mentally healthy person is one who feels comfortable functioning in the society in which she or he lives."[33] And they were able to offer a fast and clean way of achieving this state of well-being, a "magic bullet," as it was called in parallel to modern medicine's quest for this kind of highly selective remedy. A face-lift, for example, can be portrayed as such a magic bullet—the quickest and "most effective solution for an individual patient who is convinced that her sagging chin line is making her life miserable."[34]

But whether a face-lift was seen as a solution to anything at all depended on the cultural context. Medicalized or not, the problems for which its practitioners provided a technological fix were clearly cultural. They had to do with individual, social, and cultural identity. The ways in which these aspects of identity are linked to one's external appearance (or not) depends on the cultural context. Cosmetic surgery could thrive only in a culture that understands physical beauty "as an external, independent—and thus alterable—quality." Good looks had to be considered as a quality that is not inherent to the person but as something that can, if necessary, be achieved by artificial, technological means or even something that can be bought and sold—in other words, a consumer commodity.[35] Another condition is that external appearance had to be of importance for one's identity and social standing. In premodern Western societies, other qualities such as one's ancestry or religious denomination were far more important than one's visual appearance. Only in modern, urban environments are identity and status derived to such a large extent from visual self-presentation.[36] Such an attitude already assumes that the body is to a certain extent external to the person who uses it for self-representation. It is something the person *has*. And it can be optimized to better represent that person.

Cosmetic surgery thus evolved in close relationship with cultural changes that made it possible. At the same time, it also *contributed* to those changes. As Elizabeth Haiken has analyzed, the public's assumptions about beauty shifted in parallel with the emergence of the correlated surgical technique. Surgery was part of that shift: By giving and receiving a lift, the surgeon and the patient affirm the culture that made the operation desirable. The availability of a quick fix reinforces the cultural preconditions that had created the problem in the first place.[37] This mechanism is particularly striking in the case of surgery for ethnic markers. As part of the range of cultural assumptions, *racist* biases have manifested themselves in cosmetic surgery from the start. Placing markers of ethnicity in the categorical grid of normal and abnormal had the effect of coding them as deformities.[38] "Jewish," "black," or "Irish" noses and Oriental eyelids became abnormalities in need of "correction."[39] By associating different aesthetic "problems" with different ethnicities in this way, cosmetic surgeons reinforced the existing racially based standards of beauty.[40] The effects of this feedback mechanism actually go beyond mere reinforcement. The spread of cosmetic surgery has changed the expectations of external physical normality for everyone in the Western world and has contributed to the beauty cult as a striking cultural phenomenon of the twentieth century.[41]

The whole process has an important material dimension: In their interaction surgeons and patients are not only transforming the way people *think* about external appearance, but actually changing the way many people look.[42] Also, the feedback mechanism between expectation and surgical fulfillment depends crucially on the availability of effective surgical technology: The expectation makes sense only if there is a chance of fulfilling it. Only in a context where the (re)construction of physical normality exists as an actual technical possibility, Bernice Hausman rightly states, can the desire to be *physically* normal be seen as a normal attitude: "Where that possibility does not exist, such a desire would be counterproductive and possibly antisocial—certainly not indicative of a 'rational' or sensible approach to disfigurement."[43] The availability of a functional technology is clearly an integral part of the feedback loop that links the perception and definition of problems with their technological solution.

TECHNOLOGY AND GENDER POLITICS: THE SURGICAL CONSTRUCTION OF SEXUAL IDENTITIES

The existence of highly sophisticated surgical technology was also central for the emergence of another focus of surgical therapy for social ailments: sexual identity. *Sex-assignment surgery* has been in use since about the midcentury for babies whose genitalia are judged to be ambiguous.[44] Since attribution to one of two sexes is one of the fundamental categories in many societies, an ambiguous sexual identity can cause all kinds of social and cultural problems. Interestingly, intersex experts agree that intersexuality is not primarily a medical problem but rather a social one. Yet they aim to resolve this social problem "by technologically bringing intersexuals into the categories of 'male' and 'female'."[45] However, surgery itself plays a part in defining the problem in the first place. As Dreger has shown, the availability of sex assignment has narrowed the scope of the anatomical variations in genital organs that are accepted as normal. At the same time, the technological approach has pushed alternative ways of dealing with intersexuality to the background. Because of that restriction in the scope of dealing with intersex conditions, many of those who underwent sex-assignment therapy are critical of it. They feel colonized by medicine and criticize the very principle of redefining a social problem as a surgical one. Instead of surgically modifying the body and subjecting them to the adverse side effects of surgery, they demand that doctors and society help them and their social environment to live with their intersex condition, for example, by allowing for more ambiguity and multiplicity. Sex-assignment surgery is just one example of the whole range of techniques used to create physical normality. Using the case example of conjoined twins, in another study Dreger analyzes the politics of anatomical normalization on a more general level. She demonstrates once more that surgery is only one of several possible strategies to deal with physical difference and that it is not necessarily the one that is favored by those who are affected.[46]

However, as the example of cosmetic surgery has already made clear, the initiative for surgical normalization does not always come from the medical establishment. Things are more complex, as the history of *sex-reassignment surgery* shows. The establishment of this procedure actually depended on the precedent of sex-assignment surgery in various ways. On the material level, sex-assignment surgery had produced sophisticated surgical techniques for genital reconstruction that could be used for reassignment as well. Conceptually, the existence of sex assignment stabilized the expectation that sexual anatomy has to be unambiguous. More significantly, its precedence helped to establish the assumption that the *sex* of the body is not necessarily identical with the *gender* of the person.[47] Sex-assignment and sex-reassignment surgery thus both assume the possibility of a separation between person and body. Accordingly, sex reassignment is called for when an individual feels himself or herself to have a body that does not correspond to his or her real gender. The very fact that the body plays such a huge role in performing gender identity served as the starting point for medicalizing this role conflict. "The situation was reinterpreted in the mould of a body-person split as a conflict between a female or male soul being captured in the wrong body," social scientist Stefan Hirschauer states. "This helped gender deviants to avoid moral condemnation. They were not morally wrong they just had the wrong body. A body now could be experienced as 'wrong' because it can be corrected."[48] The technological fix here converts a moral problem into a technical one.

Even though the origins of the concept of transsexuality seem to go back to early twentieth-century European scientists, transsexuality as a diagnostic category apart from transvestism, intersexuality, and homosexuality gradually stabilized only in the 1950s.[49] At that time, new technologies in endocrinology and plastic surgery made sex reassignment a practical option.[50] The existence of those techniques produced transsexuality as an "identity category for a diverse group of sexual deviants and victims of gender role distress."[51] To be sure, males who dressed or lived as women, and females who dressed or lived as men, existed in earlier times. But transsexuality as the demand to be *made into* the other sex originated in the twentieth century.[52] Technology became part of the medical definition of the problem: When the diagnosis of transsexualism was introduced in the 1950s, it was restricted to those persons who demanded sex change.[53] Thus, the crucial element in creating the new category was "the possibility of a material response to the conviction of being the other sex."[54]

We have here an example of how technological developments made new discursive situations possible and opened up new subject positions. Hausman has analyzed this connection: As significant as the ideology of gender might be for constraining the semiotic economy of the body, she argues, it is also necessary "to recognize the body as a system that asserts a certain resistance (of constraint) upon the ideological system regulating it." This resistance of the body is directly addressed by medical technologies. Medical technologies can change the body's capacity to signify sex and thus affect its relation to gender ideologies.[55] Once more, the technological approach strengthens dominant cultural assumptions. In providing a rite of passage between sexual identities, sex-change surgery implicitly reaffirms traditional male and female roles: "Individual patients unable or unwilling to conform to the sex roles ascribed to them at birth are carved up on the operating table to gain acceptance to the opposite sex role," critics Dwight D. Billings and Thomas Urban wrote in 1982.[56]

Here, too, the technological imperative relegated alternatives to the background. According to Billings and Urban, transsexual therapy "pushes patients toward an alluring world of artificial vaginas and penises rather than toward self-understanding and sexual politics. Sexual fulfillment and gender-role comfort are portrayed as commodities, available through medicine."[57] It is questionable whether this promise can be fulfilled, though: Male-to-female patients especially are described as caught up in an escalating series of cosmetic operations to approximate more closely an idealized female form. The range of surgeries includes breast implants and operations to reduce the size of the Adam's apple, nose reconstruction, injection of Teflon to modulate vocal pitch, reduction of the thickness of ankles and calves, and shortening limbs. Silicon is used to alter the contours of the face, lips, hips, and thighs.[58] Some patients even hope that doctors could enable them to bear children.[59]

THE POLITICS OF THE TECHNOLOGICAL FIX AND THE BODY

These examples show that surgery reflects a modern tendency to prefer a technological fix to other possible strategies of problem solving: liposuction instead of diet, liver transplants instead of drinker's cures, brain surgery instead of psychotherapy, trauma surgery instead of safe driving, and—to put it polemically—war surgery instead of diplomacy. Some of these technological solutions have fallen into disgrace, lobotomy, for instance. Others have seen an enormous expansion in the second half of the twentieth century, such as cosmetic surgery. And as all examples demonstrate, the use of a technological solution leads to a redefinition of the original problem: Trauma surgery helped in the reinterpretation of traffic accidents as a problem of the individual accident victim instead of the society that tolerates those accidents happening; on a different level, the focus on changing the body in cosmetic surgery, as well as sex-assignment and -reassignment surgery, changes the cultural standards of appearance, and so on.

We have seen how the technological approach projects moral, political, and cultural issues onto the bodies of individuals. Doctors can then treat them as purely technical and individual problems that can be solved by intervening into the individual's body. We have also seen that in the context of this kind of bodily intervention, the body is being treated like a thing, separate from its owner. As Figlio has phrased it, the traditional notion that the self extended into a unique and inviolable corporeal volume became less self-evident. Instead, the self was increasingly seen as only loosely possessing a body. "This body could then be brought to the physician for servicing by a technique which need have no regard for the self."[60] The person becomes dissociable into a self plus a body, a body that can be treated like any other kind of object or machinery. As the extreme cases of cosmetic surgery and sex reassignment illustrate, in some cases the body may even be altered to accommodate the self. The individual's body is thus no longer a given and relatively unchangeable feature of the person but, rather, an object that can and should be repaired, modified, and optimized.[61]

Along these lines, modern medicine went even further. Problems were not just projected onto the individual's body but onto a particular *part* of the individual's body. There, on the level of circumscribed body structures, these problems become amenable to a technological solution. Thus, in addition to the dissociation of self and body, the body becomes dissociable into its elements, as is shown in the following.[62]

FRAGMENTATION: TRANSPLANT SURGERY

The field emblematic of this new attitude toward the body is transplant surgery. Transplantation not only represents the idea of a technological fix for a large variety of medical problems but also implies the divisibility of the body and the exchangeability of its body compounds more than any other surgical technique. Originating in the late nineteenth century, modern transplant surgery represented a new approach to treating diseases: Previously, within the framework of a more holistic environmental and humoral understanding of disease, replacing an organ would not have made much sense. Treatment consisted mainly in reestablishing the balance of humors in the body or changing one's way of life and one's environment. In the course of the nineteenth century, scientists and doctors started to look for the cause of diseases in the failure of particular organs. For this purpose, they literally separated organs from their bodily context: Surgical interventions in humans and planned experiments in animals showed that after the removal of particular organs specific disorders would emerge. Vice versa, reinsertion of those organs could reverse these disorders. It was in this context that Kocher performed the first organ transplantation, a thyroid transplant, in 1883.[63]

FIGURE 3.2: Portrait of Dr. Carrel, *Chanteclair*, no. 138, Paris, 1906. Carrel, who was famous at the time for his transplant experiments, is depicted as a magician creating all kinds of chimeras. Wellcome Library, London.

However, after World War I, surgeons gradually realized that transplants between different individuals did not succeed. As a consequence, organ transplantation was temporarily abandoned. Not until 1945 did scientists and doctors restart transplant surgery. In the early 1960s, the introduction of chemical immune suppression initiated a new phase in the history of transplant surgery. The availability of more selective immunosuppressive agents such as cyclosporine since 1982 accelerated the pace of growth of the field. Since that time, organ transplantation experienced an enormous boom in the range, number, and combinations of tissues and solid organs.[64] Transplantation became the technological fix of choice for the growing number of medical problems that were redefined as being caused by organ malfunction. Instead of complicated and unreliable measures for rebalancing the body and its environment, a circumscribed intervention by a highly specialized expert sufficed to solve the problem.[65]

FIGURE 3.3: *Undreamt of Perspectives, or: Surgery in the Year 2000.* "Hearts and livers are available in the department for innards, on the fourth floor, madam—for self-assembly in hygienic packages, on shelf six, straight ahead, madam—" This cartoon was published as a reaction to the report of the first successful heart transplant. *Der öffentliche Dienst*, February 9, 1968. Courtesy of vpod.

SPARE PARTS: COMMODIFICATION

The growth of transplant surgery since the 1980s went along with changes in the view of the body. Sociologists Renée Fox and Judith Swazey, who have followed modern transplant medicine from the 1950s, observed the emergence of what they call "spare parts pragmatism," with the "replaceable body" as the ultimate vision.[66] The idea that dysfunctional and worn-out body parts can be replaced carries with it the promise of overcoming "the biological and human condition limits imposed by the aging process . . . and our ultimate mortality,"[67] they claim. According to this view, transplant surgery, as the theologian and ethicist Paul Ramsey so strikingly put it, makes people vulnerable to "the triumphalist temptation to slash and suture our way to eternal life."[68]

As a side effect, the body's boundaries and its relation to the self are being reconstructed. The spare parts pragmatism associated with transplant medicine redefines body organs as separable from the donor's person. Bodies are thus not only replaceable, they are also divisible: A functioning body can in principle be assembled from tissues and organs of different origins. This is just one instance of how modern biotechnology in general, according to Paul Rabinow's analysis, tends to fragment the body into "a potentially discrete, knowable, and exploitable reservoir of molecular and biochemical products and events."[69] The resulting body "fragments" assume a new status. Fox and Swazey observe that organs, for example, stop being living parts of a person. Instead, they are increasingly thought of as "just organs."[70] As such, they can take on an economic value and be bought and sold. This commodification of body parts is emerging within the discourse of scarcity that is attached to transplants. The idea that the demand for organs far outnumbers the supply is based on the numbers surgeons calculate for theoretically desirable transplants. The resulting discrepancy with the rate of donations has led to a situation that critical observers characterize as "battle for body parts" in which "human organs are openly described as 'scarce' and 'precious' goods that frequently 'go to waste' . . . when, in fact, they should be 'recycled' for social reuse."[71]

To alleviate the assumed scarcity, proposals have been brought forward to expand the market supply of transplantable human body parts. Some of them, such as bioartificial organs and xenotransplantation, concern technological changes. Others aim at redefining the threshold of death and thus personhood, so that anencephalic and non-heart-beating cadavers can be donors. Still others try to increase the organ supply through presumed consent laws for organ donation. The most controversial suggestions advocate a direct market approach or in a mitigated version "forms of 'rewarded gifting' to the surviving kin in the form of estate and income tax incentives and assistance with burial fees."[72]

Transplant surgery thus illustrates how the new body concepts that are guided by the technological imperative of modern medical knowledge and technology have an inherent potential for the fragmentation and ultimately commodification of the body.[73] As the example of blood transfusion demonstrates, other technologies also had a commodifying effect.[74] The spread of the technological approach has more generally resulted in an "expanding desire for cadavers and skeletons, blood, organs and other transplantable tissues, microscopic ova and sperm, and, most recently, genetic material," a desire that has been answered by "a proliferation of the marketability of human body parts."[75]

At the same time, transplant surgery also demonstrates that new body constructions emerging from biomedical technology and knowledge are not automatically accepted. Despite massive publicity campaigns, the rate of organ donations in Western countries seems to stop at a level that is far from its theoretical potential. Apparently, the notion that body constituents can be removed and exchanged as needed is not shared by everybody. There seems to be a "cultural rejection" of those aspects of modern medicine, as anthropologist Donald Joralemon calls it in parallel to an immune rejection.[76] Among other things, this cultural rejection affects the new ways of separating the body from the self.[77] The fashion in which the physical body and personal identity are related to each other can vary with different cultural contexts. Transplant surgery is one example of how medical knowledge can interfere with this relationship. In the following sections I discuss the additional examples of immunology, genetics, and the neurosciences. While being related to surgery in specific ways, these cases also illustrate the interaction between body concepts and nonsurgical knowledge and technology.

IMMUNOLOGY: THE SCIENCE OF SELF AND NON-SELF

Transplantation is based on the assumption that body parts are interchangeable in principle. A heart is a pump, and it can function as a pump in one body as well as in another. But, as I have pointed out earlier, surgeons and scientists had to acknowledge early in the twentieth century that individual identity cannot be ignored. The failure of transplants between different individuals seemed to be due to each individual's specific biological identity on the molecular level. Maintenance of that biological identity was ascribed to the same mechanism that was responsible for the organism's defense against infectious agents, immunity.[78] These discoveries resulted in the construction of a body concept based on immunology—the science of self and non-self, as one of its founders, Frank Mcfarlane Burnet, has characterized it.[79]

Looking at popular body images, anthropologist Emily Martin has examined how immunology has influenced the way people feel about their bodies in relation to health and disease.[80] She has observed a shift of "emphasis from the outside of the body, with its envelope of protective skin, to what happened inside. By the early 1990s, discourses on the immune system had become central to the body image in relation to health and illness. People were now exhorted to take care of their immune system as a means of protection against ill health and disease."[81] Martin found that new scientific facts were incorporated into public discourse and personal views, as, vice versa, societal assumptions and values found their way into scientific research. In this case, scientific claims about immunity reflect and simultaneously enhance the notion of individual flexibility, she claims.[82]

GENETIC ENGINEERING

Biological individuality is also the main theme of genetics. According to the genetic worldview, the essence of personal identity is located within the DNA. This is why, on a popular level, people often think that somebody's clone would be identical to that person.[83] The genetic material is also implicated in the cause of diseases. Therefore, DNA offers another target for a type of medical intervention that follows the surgical pattern. In surgery, disease equals deranged body structure. Healing means repairing that structure, for example, by cutting and rearranging it. This idea could be transferred to the genetic material in the nucleus of the cell, which was also seen as a structure and which could also be cut and rearranged. Since people are used to the surgical approach, the idea "that it will be possible to 'fix' disease by 'fixing' genes sounds plausible, and it is easily understood by the public and the media."[84] In this way, the rise of modern surgery can even be interpreted as a condition of possibility for genetic engineering. Here, it seems, the surgical strategy of circumscribed structural manipulation has merged with the idea of the magic bullet, the pure, hyperselective drug.[85] This is not an isolated case. Surgery often provides the imagery for future technological fixes: "Dissecting the genetic component of our common killers should lead to a better understanding of their pathophysiology, and it may even be possible to tailor-make drug regimens to suit individual patients' genotypes,"[86] a leading British doctor explained in 2000. Instead of organs and tissues, now cells and molecules are the material to be worked on. It is no longer simple surgery but "the fields of molecular and cell biology" that are expected to "spawn developments that will be of enormous value in the clinic."[87]

Genetic engineering offers a refined version of the same technological-fix strategy that surgery has so well established in modern culture: In her prospect about a world past cosmetic surgery, Haiken envisages a "future, in which babies are made to order." Features that are now objects of surgery, such as the size of noses and the longevity of hairlines, will then be determined in advance on a molecular level. In the future, Haiken continues, "postbirth alteration" by cosmetic surgeons might be "remembered nostalgically as the primitive practice of well-meaning but technologically handicapped medical practitioners."[88]

Genetic engineering works within the same conceptual framework as surgery. As in surgery, "framing ill health as genetic and promoting individual genetic or pharmacological solutions pushes the problem back to the individual." The idea of genetic manipulation is therefore "attractive to those who do not want to deal with complex social and economic determinants of health. Genetic explanations allow policymakers, employers, and others to avoid awkward questions about employment or tax policy, educational or child-care policies, and inappropriate workplace practices."[89] This function characterizes genetic engineering as a typical technological fix.

DISEMBODIMENT: THE BRAIN AS THE SITE OF THE PERSON

As important as the substance DNA is in genetic discourse, the essence of the genetic person does not lie in the material substrate but in the way it is arranged, in other words, in the information it contains. The idea of information has been a key concept in the scientific understanding of the organism in the twentieth century. Twentieth-century biology conceptualizes the basic vital functions as the storage, transfer, modification, and translation of genetic information. In 1948 Norbert Wiener, the founder of cybernetics, provided a panorama of modern scientific body concepts. He depicted the biological organism of the eighteenth century as a mechanical apparatus. The body of the nineteenth century he saw as a steam engine. The twentieth-century organism, however, he characterized as a medium of control, understood in terms of message, code, and information.[90] Medical scientists and doctors increasingly think of the biological components of the body in terms of the information they contain.[91]

Besides the genetic material, another focus for understanding the workings of the body in terms of information is the nerve system, in particular the brain. The localization of mental problems and personal traits in the brain made it possible to locate personhood in that body structure too. Thus, it was not a big step to claim that irreversible and complete damage to the brain amounted to the death of the person, even when the rest of the body was still alive. In most Western countries, the condition of *brain death* was declared a defining symptom of the individual's death in the late 1960s. Linking the death of the person to the death of one of its body structures in this way was the result of two developments: First came the progress of intensive-care technology in the 1950s. The improvement of artificial respiration techniques allowed bodies to be kept alive even in cases where the brain had already completely ceased to function. As a consequence, doctors wondered whether they were allowed to put an end to such a condition by turning off the respirator. According to the brain-death concept, this is permissible. However, the decisive influence came arguably from another development, the growth of organ transplantation with its need for tissues and organs. Defining brain death as the death of the person made it ethically acceptable to remove living organs from a brain-dead patient. In this situation, brain death became institutionalized as the death of the person across North America and in much of Europe.[92] The death of a person was thus equated with a specific biological change in a circumscribed part of the body. Whereas traditionally the definition and determination of personal death had been dealt with on a variety of cultural and societal levels, it had now become a matter technological diagnostics applied by highly specialized experts. A morally charged problem had been turned into a technically manageable one.

As indicated in the context of transplants, this kind of redefinition is not universally accepted. In a globalizing world local particularities keep limiting the reach of science-based universalistic claims about human nature. Thus, in Japan, the idea of the person is usually not located in the brain but diffused throughout the body. And people do not just abandon this idea because they are told otherwise by scientists. Even in Europe brain death is still problematic in a number of national contexts, for example, in Germany.[93]

At the same time, interpreting brain death as personal death has not been the last word on this issue. Some scientists and doctors go beyond the whole-brain death concept and project personhood into ever smaller structures. Proponents of the concept of *partial* brain death claim that a person is already dead when those parts of the brain that are necessary for consciousness have been irreversibly damaged. Such a being, they postulate, will never again be able to think, feel, or act.[94] Others go even further and equate personhood with the storage and processing of information in the brain. This information is no longer necessarily bound up with the material substrate of brain cells. Wiener assumed that all brain functions could be reproduced by electric systems.[95] Accordingly, it would be totally immaterial for the issue of identity whether a functioning system was made up of nerve cells or microchips: Personal identity would be maintained even without the brain. The information from the person's brain just has to be transferred to a computer, and it will continue to unfold its activity there.[96] A person's feelings and thoughts could be "copied" onto a computer, in the same way data are copied onto a disk, and mailed over great distances, as futuristic visionaries explain. Computer scientists even dream of linking the downloaded consciousness of individuals with each other on immortal computers. Individual identity as we know it would dissolve into data space.[97] The fact that these visions are possible at all is a sign of changing concepts of the body and its relation to personhood and the self. Here, in another step, the fragmentation of the body inherent in modern medical concepts has led to the vision of total disembodiment.

CONCLUSION

The example of surgery has allowed me to spell out some of the most significant effects of modern medicine on body concepts. It shows on a concrete level how modern medicine's technological-fix approach has made the body a primary site for dealing with a whole range of issues that had not previously been considered bodily problems. One effect of this development has been the unprecedented growth of the medical enterprise in the course of the century. Another effect has been the redefinition of the original problems as technological ones. The resulting positive feedback loop has shaped modern culture in many areas—traffic, beauty ideals, gender identity, and so on. As a result, the body has become more significant as a place of problem solving and manipulation. At the same time, the body is increasingly seen as an object, separate from the person as such (who owns a body rather than is a body), divisible, and, in a paradoxical turn, even dispensable.

But surgery is only one example of the technological-fix strategy in twentieth-century medical knowledge and technology. As already indicated, a specific and well-directed intervention into the body's structures and functions can also be achieved by drugs or other biotechnologies. The inherent potential for the reification, alienation, fragmentation, and commodification of the body through medical knowledge and technology can also be observed outside surgery. Anthropologist Lesley Sharp has identified the post–World War II era as the watershed period for technology-induced changes in attitudes toward the body. Around that time a whole range of (nonsurgical) technologies were introduced, such as dialysis, the respirator, immunosuppressants, cybernetic systems, sonography, fiber optics, magnetic resonance imaging, and genetic and immunological diagnostics. According to Sharp, these developments in knowledge and technology together might lead to a paradigmatic shift in how people envision their bodies. They may learn to regard themselves as "both technological subject and object, transformable and literally creatable through biological engineering. In essence," she concludes, "certain biotechnologies now encourage self-objectification."[98] The commodifying effect of transplant surgery can also be seen as part of a more general phenomenon: the "incredible expansion of possibilities through recent advances in biomedicine, transplant surgery, experimental genetic medicine, biotechnology and the science of genomics *in tandem with* the spread of global capitalism and the consequent speed at which patients, technologies, capital, bodies and organs can now move across the globe," as Nancy Scheper-Hughes has phrased it.[99] The combined effect of diverse biotechnologies, surgical and nonsurgical, might be sufficient to cause a revolutionary reformulation of our concepts of body and person: Joralemon envisions in his paper on "organ wars" how "perhaps transplanting organs together with genetic engineering, artificial reproduction, therapeutic cell lines, and mechanical implants will be enough to overwhelm the intuitively unambiguous connection we feel to our bodies."[100]

However, the influence of scientific knowledge and medical technology represents only one side of the equation. Medical technologies also provoke criticism. As Joralemon points out for brain death, it is "important not to underestimate the cultural force behind the idea that self and cell are not entirely separable, that it is not only the brain in which the 'I' resides."[101] The uneven acceptance of the brain-death concept on a global scale also points to the significance of local cultures and their specific potentials of resistance to universal claims about concepts of body and self. Maybe people resist technological body definitions because principles such as fragmentation and commodification contradict the way they normally experience bodies in their particular culture. For whatever reasons, it is clear that changes in body concepts inspired by new medical knowledge and technologies are not necessarily accepted. Some of them get modified and adapted, for example, the immunological view of the body. Others are being rejected in some contexts, such as brain death and organ transplants or gender-assignment surgery. It may well be that the reformulation of people's concepts of body and person won't be revolutionary but partial and spotty, unpredictable and idiosyncratic. Nevertheless, whether rejected or accepted, appropriated or imposed, medical knowledge and technology are never politically and culturally neutral. As part of the general transformation modern societies are undergoing, they have an enormous influence on how people feel and think about their bodies.

CHAPTER FOUR

Diseased Bodies in the Modern World

ANNA CROZIER

Diseases, and responses to them, changed drastically in the twentieth century. Disease was (increasingly) not the inevitable cause of early death it had been in the pre-pharmacological, pre-preventative medical worlds; different diseases rose to prominence (the tuberculosis and syphilis of the premodern world gave way to AIDS, cancers, and genetic diseases), as did the variety of means used to deal with diseases. Knowledge about disease—detection, treatment, nosology, avoidance, eradication—rose to become one of the hallmarks of the modern age. And such knowledge of disease was not simply the product of medical breakthroughs on a submissive population; everyday people fought disease through prophylactic measures, encouraged by health campaigns for hygienic living, healthier eating, fitter lifestyles, better housing, prophylactic medical interventions, genetic counseling, and safer sexual relationships. At the center of these adoptions of medical regimes—and in some cases resistance to them—is the body. It is the way that diseases have impacted on the body, not just in terms of how they affect patients physically, but how diseases encourage new corporeal norms, that is the focus of this chapter.[1] The body may be the site of disease, but it is also the location of resistance, prevention, and regeneration, as well as the site of death. As this chapter argues, to control disease, the body itself is controlled through a variety of practices.

To examine the diseased body in the modern age, a number of parameters need to be outlined. This chapter is not concerned with issues such as the advent of sophisticated nosologies or medical breakthroughs; nor does it specifically tackle public health measures, except insofar as they have brought about new corporeal possibilities. Likewise, it is not concerned with other specificities of medical practices. Such topics have been the focus of much medical history that has focused on diseases.[2] Rather, this chapter will select a number of cases from throughout the twentieth century that illustrate how the disease is enacted in the body; how modern bodies are the product of temporally specific practices surrounding both actual diseases and the risks posed by diseases. To focus on such diseased actions in such a way is to historicize the body. "The body is moulded by a great many distinct regimes; it is broken down by the rhythms of work, rest and holidays; it is poisoned by food or values, through eating habits or moral laws; it constructs resistances."[3] The interactions between disease, health, and illness constitute other such regimes that can be used to situate the body in time. In this way we can address the body as "a site on and around which political and ideological meanings cluster, brought to bear by a complex constellation of social forces and networks of power."[4]

One of the key factors in the historicity of corporeal reactions to disease has been the twentieth-century trend toward medical reductionism in the West. To quote medical historian David Cantor, "diseases came to be seen as entities, quite distinct from their manifestation in the individual patient, and therapeutics came to be directed towards the disease rather than the person."[5] This reconceptualization of disease within a biomedical weltanschauung has had significant consequences for the body that need to be outlined before turning to specific examples. The main development of this standpoint has been the manner in which medicine has specialized and moved remarkably away from holistic conceptions of health toward specific practices and treatments—even varying ontologies and epistemologies. The rise of clinical sciences has, therefore, done much to reconceptualize corporeality through a variety of means, including temporality, genetics, and reproduction.[6] This is not to suggest that medical holism dropped out of the picture completely in the last century, of course,[7] but rather that the intense reductionist study of diseases in the twentieth century and earlier produced wide-ranging effects on the conceptualizations of the body and health. Such intellectual structures are played out in the power relations of everyday life, with corporeal factors joining political and ideological struggles in the shaping of the modern world. The body has become an entity for clinical ontology, a process that began in the nineteenth century when the groundwork for the analyzed body was set. The way the body is conceptualized also changed in parallel to the way the mind is conceptualized. For example, developments in psychiatry at the beginning of the twentieth century increasingly reduced the mind to a single constituent part that could be "treated." Therapies subsequently not only sought to locate tangible external causes for mental problems but also, increasingly, in the latter half of the twentieth century offered drug therapies (for example, selective serotonin reuptake inhibitors) particularly targeted at specific physiological locations of the brain.[8] The same trend can be seen in the increasing conceptualization of mental disease as genetic—something that grew out of earlier models of degeneration that had abounded in nineteenth-century psychiatry.

Another important factor in considering disease in this chapter derives from Annemarie Mol's ethnographic work on medical ontology, and specifically from her insistence that the body is multiple.[9] This suggests that no single disease acts on the holistic body but rather that both the body and the disease have multiple ontologies—that is, different practices "enact" disease and the body in different ways.[10] So when Mol talks about atherosclerosis, a major cause of health problems in the Netherlands, where she conducted her ethnographic research, she does not imply a single entity but rather a disease that is different in the consulting room, in the patient's life, in the diagnostic laboratory, in the pathology department, in epidemiological assumptions, in medical statistics, and in medical policy. What atherosclerosis is differs depending on where it exists. This spatiality can be found in other diseases, too. In addition, such ontological differences also have a historical dimension—so the disease and the body are different diachronically as well. This commitment to multiplicity is central to this survey of the way the diseased body is dispersed throughout the twentieth century. This standpoint suggests that when one considers disease, one must be specific about what is meant—we must be aware that diseases are not universal categories but rather are enacted through a variety of mutually influencing practices.[11] In this chapter, bodily practices formed through interaction with disease entities and knowledges of disease are the primary focus, and as such this is a partial view of diseased twentieth-century bodies.[12]

DISEASE AND THE CONTROL OF BODIES: CONTAGION, IMMUNIZATION, AND EPIDEMICS

Control of people through public reactions to disease is by no means a new concept (see, for example, Leviticus 13:1–8). Prominent examples from the medieval and early-modern period have shown how leprosy and plague were treated as reasons to prevent certain social interactions, by locking away lepers and preventing movement between towns and regions during times of plague.[13] Such issues were, of course, exacerbated with the increase in trade and emigration through the nineteenth century and became internationally discussed concerns of politics and statecraft, as can be seen in, for example, the diplomatic efforts to deal with Asiatic cholera by the international sanitary conferences that met throughout the nineteenth century.[14] The control of the movement of bodies affected by disease was also particularly evident in quarantine, which again was produced largely through the necessities of trade and travel.[15]

A number of acts (since 1711 in England) were passed that established the state's power to isolate individuals considered dangerous through their suspected carrying of disease. A notable example of this principle being carried forward is the treatment of women suspected of being prostitutes in the later nineteenth century under the contagious diseases acts.[16] Such attempts to defend society from disease were situated within the body.

Although miasmatic theories of contagion initially underpinned these legal and social means of controlling the movements and actions of diseased bodies, it is significant that movements toward this sociomedical control intensified at precisely the same time as new ideas about germs and contagion were becoming accepted within scientific circles.[17] These modern theories of contagion informed the ways diseases were treated and the ways in which bodies were conceived of at a number of levels, from the personal to the political. Within these theories of contagion were contained a huge array of ideological commitments about the limits and predispositions of disease manifested through (changing ideas of) race, sexuality, class, cleanliness, and conceptions of personal hygiene. Such associations were especially pronounced during the twentieth century and can be seen in the huge variety of images and discourses used to control the movement of disease.

In popular discourses and images, the general public was encouraged to take part in the control of disease by adapting to new bodily norms that would prevent the spread of germs—hand washing, covering up of coughs and sneezes, hygienic cooking practices, clean clothes, proper toilet habits, condom use, mosquito nets for sleeping, and so on. The threat of disease, and its impact on the body through the construction of hygienic norms, is significant in the shaping of twentieth-century corporeality.[18] The body thus became a synecdoche for disease.

Notions of public responsibility to prevent disease are still prevalent; for example when sudden acute respiratory syndrome (SARS) first became a potential international public health crisis, a vast amount of information was rapidly spread that addressed issues of travel to foreign places, especially Southeast Asia, and discussed the conditions in which people lived in these places in terms of their risks for disease. Although no pandemic has yet emerged, bodies were controlled through the media and through medical advice as to where to travel and how to behave to avoid risks. We see these practices through the official advice given to the masses regarding how to respond to pandemics.[19] Such issues are beyond the personal. Large state and international organizations, such as the U.S. Center for Disease Control and the European Centre for Disease Prevention and Control (ECDC), both of which are mandated to research potential health threats and to advise on specific actions that should be taken in times of crisis (actual or immanent). The ECDC in particular has the aim to "strengthen Europe's defences against infectious diseases." Such

FIGURE 4.1: A handkerchief being blown away by a sneeze. Color lithograph, 1946. Issued by the Ministry of Health and the Central Council for Health Education: Printed for H.M. Stationery Office, January 1946. Wellcome Library, London.

a task is achieved through surveillance of existing and emerging diseases in order "to develop authoritative scientific opinions about the risks posed by current and emerging infectious diseases."[20] As a state body of the European Union, the ECDC uses the media as well as official publications to inform the public of actions that are recommended. Even at this political level, we see the reductionist stream of medical thought at play, with diseases largely identified as infectious pathogens to be dealt with via specific methods. These discourses are only a part of the reality of the diseased body, but they cannot be ignored, as they position the body and disease in a very specific sociopolitical context.

Other examples of such reductionism include the eradication of disease through specific targeting. The exemplary case in this respect is smallpox, which was agreed to have been eradicated after intensive surveillance in December 1979.[21] Smallpox had been one of the deadliest human diseases: Between three and five hundred million people are believed to have died from it in the twentieth century. It was brought under control by intensive vaccination. Immunization was the central technique adopted in many cases of disease control. A number of childhood diseases were vaccinated against in the twentieth century, including the introduction of a number of specific vaccinations: diphtheria (1923), pertussis (1926), tuberculosis (1927), tetanus (1927), and yellow fever (1935).[22] These interventions on the body saved lives not through cures but through the successes of vaccination as a collective preventive action, where the reality of the disease changes through coordinated action around a particular medical technology.

The preceding is not to suggest, however, that vaccination was a universally adopted issue. Already in the nineteenth century, grave fears about such medical interventions were seen. Indeed, as Nadja Durbach has shown, a strong strand of anti-vaccinationism existed in England.[23] More recently, we have seen such skepticism directed toward medical policy in the responses to the measles, mumps, and rubella (MMR) "triple-jab" vaccination that is offered by the British National Health Service. In this case, a media storm brewed out of a paper published in the *Lancet* in 1998 that suggested a link between the MMR jab and the onset of autism.[24] Such fears led to a number of parents fail-

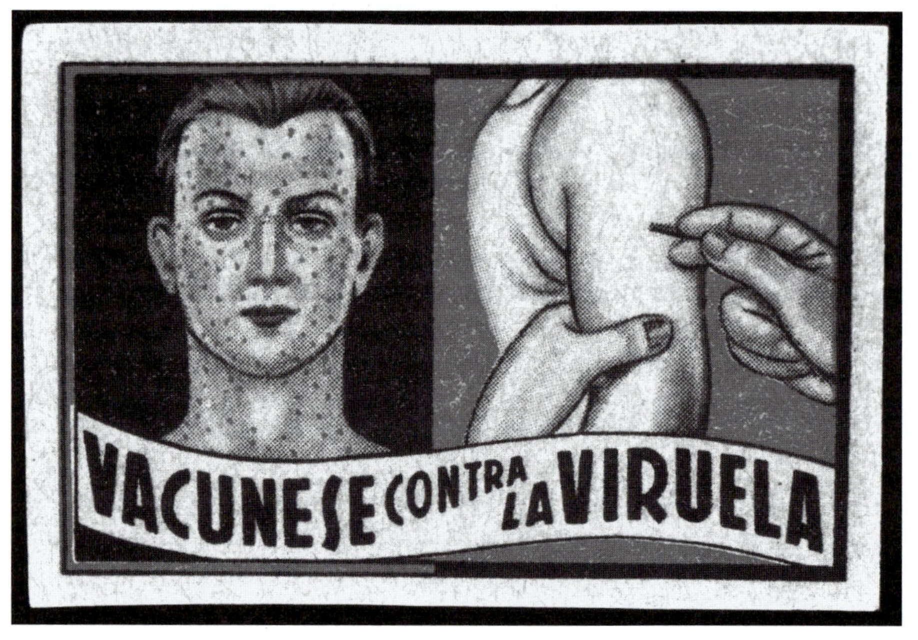

FIGURE 4.2: *Left*, Face of a man suffering from smallpox; *right*, vaccination against smallpox, 1940. Wellcome Library, London.

ing to properly immunize their children, even though a number of the authors publicly withdrew their support from this conclusion.[25] To control disease, docile bodies are required; when they resist, medical fears are circulated concerning the potential health risks to society.

Today, we see similar forms of coercion of the population through other health issues such as smoking prevention, restraining alcohol consumption, avoiding unprotected casual sex, and following dietary recommendations. All of these examples are related to the potential risk of disease. The so-called obesity epidemic is a prime example of such corporeal control, with increasing media exposure raising concerns over the expanding size of the nation. An interesting feature of this debate has been the way in which obesity has changed from simply being a case of corpulence caused by one's inability to control one's desire to eat to a disease entity in its own right, with genetic predisposition and glandular problems, as well as a number of lifestyle factors, being regularly presented as part of its causation. This turn has dramatically affected the ways that fatness has been conceptualized within modern society, frequently being no longer so much self-inflicted as a specifically medical problem, in some important degrees beyond even personal control.[26] This development does suggest that in some cases diseases are not reduced to single-factor pathogens and that some diseases are treated more holistically. The common feature, however, is the ways in which the body is medicalized, and these medical norms provide the justification for bodily control.[27]

COLONIAL AND POSTCOLONIAL DISEASED BODIES

As mentioned earlier, increased opportunities for travel and trade created urgent imperatives to manage disease—via the control of moving populations—particularly through early measures to implement quarantines. Imperial endeavors from the nineteenth century and earlier had the same effect. Contact with new and exotic environments necessitated a variety of corporeal responses, as the colonizers sought to manage and control the new worlds they entered and the different health challenges that faced them there. The inequitable power structures inherent within colonialism meant that differentiations were made between the white colonial bodies and the black or brown colonized bodies, with the former being upheld as encapsulating normative, so-called civilized behavior and the latter deviant or digressive behaviors.[28] Diseases that affected these bodies were managed differently accordingly by the colonizing powers, even though many of these interventions were resisted or subverted, making the diseased body, as David Arnold has shown in colonial India, the physical location of contestation and dispute.[29] By the twentieth century, the specialism of tropical medicine was established and was being increasingly transmitted to the developing world for local adaptation on the ground by health-care practitioners and their populations. These medical encounters were often inequitable in their distribution, hegemonic in their assumptions, and insensitive to local cultures and have been the subject of much important postcolonial scholarship.[30]

The British experience of colonialism in Africa provides examples of differential manipulations of diseased bodies. In the early period of modern colonialism, from the second half of the nineteenth century until World War I, attention focused on colonial health so far as it touched colonial authorities or the productivity of the indigenous and immigrant workforces. In these early years, tropical medicine was preoccupied with vector-borne diseases rather than sanitary measures or public health issues, which really became a focus of concern only after World War I. In this context little heed was paid to indigenous understandings of disease, its causation, its effects on the body, and its treatment. Instead, colonial authorities universally used Western medicine as a rationale to intimately control people's lives—not only through moving or altering where they lived, or farmed, but also through personal and physical violations, such as enforced medication or treatments.[31] The tenets of bacteriology particularly provided new rationalization for segregation and surveillance of local peoples: Leper asylums were founded, the mentally ill were segregated, controls were placed on the movement of plague victims and their families, large tracts of land were cleared (and residents displaced) to combat sleeping sickness, vaccinations (such as against smallpox) were enforced; and medically "dangerous" traditional practices, such as circumcision, were publicly decried.[32]

Such interventions were rarely passively or gratefully received; issues of colonial state power and oppression created a situation whereby the bodies of the colonized were constructed as manageable and comprehensible through the better judgment and superior medicine of the ruling power. Western medical institutions, whether hospitals, asylums, or rural clinics, became the chief locations to control, educate, and discipline unruly diseased bodies. The production of disciplined bodies—justified through the language of therapy and cure—meant not only the elimination of disease but also the regulation of lifestyles.[33]

In these colonial constructions, indigenous bodies were represented as the primary sites of diseases, even though diseases were seldom race specific. But not only were the indigenes presented as harboring the vectors and pathogens of diseases, they were often presented as childlike and unable to properly take care of their bodies, needing to learn appropriate, "civilized" routines of health care. Warwick Anderson (among others) has argued for the "colonial poetics of pollution," pointing to the rhetorical associations made between ill health and allegedly poor native health habits. The central focus of the colonial "civilizing mission" centered on reforming and educating indigenes to comply with certain regimes of personal and domestic hygiene.[34]

This manipulation of the diseased colonial body as a ramification of the power structures of colonialism, entwined with the assumed ascendancy of biomedicine, affected the whole demographic spectrum of those caught up in the web of the colonial encounter. While indigenes were seen as the sites of pathogens, contagion, and filth, the colonizers were regarded, in stark counterpoint, as innocent victims, exposed through the circumstances of migration to unfamiliar diseases and climates. On the one hand, centuries of exposure meant that indigenous peoples were considered to have developed some racial immunity to many of the medical dangers that threatened European health and safety. On the other hand, their persistent unhygienic behavior meant that they embodied the threat of contaminating Europeans who came in contact with them.

This view of the dangerous, pathological indigenous population, coupled with a long-held belief in the problematic effects of hot climates, molded the way Europeans' high mortality statistics in the tropics were understood and articulated. The idea that white bodies in the tropics were especially susceptible to tropical diseases was therefore represented as resulting not only from them being unused to the pathogens that caused these diseases but also from the way the tropical climate degenerated European constitutions, as persistent theories of acclimatization emphasized.[35] This translated into concrete medical advice from respected tropical medical authorities, which justified many aspects of white lifestyle, such as situating their housing well away from the settlements of local people.[36] As late as 1938, the Royal Geographical Society still advised European émigrés to the tropics to "wherever possible camp at some distance from a village" because "the chief reservoirs of infection are the natives, and more especially the children."[37] In fact, considerable aspects of European life were advised to be thus regulated, with health guidebooks offering advice on how to wash food (or supervise its washing), what to wear, how Europeans should treat servants, and how they should act in the tropics to reduce the possibility of becoming diseased.[38] These theories consistently condoned racial segregation and emphasized local peoples as the embodiment of disease.[39]

One colonial disease was notably specific to Caucasians, however: tropical neurasthenia. This condition was understood as a somatic problem directly akin to the fashionable American and British diagnosis of neurasthenia, except that the tropical version was explicitly triggered by the stresses of the new tropical environment and remoteness from civilization. This was a remarkably common cause of invaliding from the tropics, which manifested itself through a wide variety of symptoms, including lethargy, ennui, and stomachaches and headaches.[40] Interestingly, this medical diagnosis was not only an indication of greater colonial anxieties (particularly as a means of distancing colonizer from colonized, as it was conceived of as a "civilized" disease that affected only the white middle and upper classes) but also a practical means of controlling the behavior of the colonizers, by neatly categorizing certain behaviors under this diagnosis but also invaliding people from the colonies because of it.[41] Thus, the somatic disciplining of white bodies occurred just as the disciplining of diseased indigenous bodies did, although with less overt force.

After World War II, the increasing internationalization of medicine, demonstrated through the growth of large international agencies such as the World Health Organization (WHO, established 1948) or United Nations Children's Fund (UNICEF, established 1946) meant a greater focus on developing transnational health and welfare policies. Although these, and other, organizations were established specifically to address inequalities and to provide fairer, more equitable health care, in practice their singular focus on reductionist bacteriological understandings of disease assured the cultural dominance of the Western biomedical model. As the West defined the health agenda of the developing world (and arguably still does today, though to a lesser degree), indigenous medical systems were further sidelined and the diseased bodies of the vast populations within the developing world continued to be manipulated, constrained, and constructed through the dominance of Western biomedical models of health and disease.[42]

One way that this was obviously played out was by the vertical health campaigns that characterized the later colonial and early postcolonial periods. New hopes were created via developments such as DDT for use against mosquito habitats and streptomycin and isoniaziol for use against tuberculosis. Most famous was the worldwide eradication of malaria promised by the WHO by 1957, which involved extensive DDT spraying, a policy that effectively touched millions of bodies that would not have been touched by Western medical interventions before. Precisely because of these violations, the body became powerfully symbolic as the site of the power struggle between colonial medicine and indigenous peoples' somatic and psychological freedoms. The most famous anticolonial doctor, Frantz Fanon, summarized the violation of the nation in terms of the cultural and physical contraventions Western biomedicine had made on his body: "My body was given back to me sprawled out, distorted, recoloured."[43] The medicine itself threatened to "whiten," and therefore pollute and warp, the cultural understandings of disease and its personal and community ramifications.

It was not really until the time of decolonization and the 1960s and 1970s that comprehensive reform of international agency perspectives really began, although arguably a Western cultural imperialism still exists today, with many Western products relating to health and cleanliness dominating the marketplace of modern African and Indian societies.[44] In a similar vein, the Western mass media still very much control the way diseased bodies in the former colonial worlds are portrayed internationally, with images of famine and AIDS often dominating the headlines rather than images of health and success. Persuasive

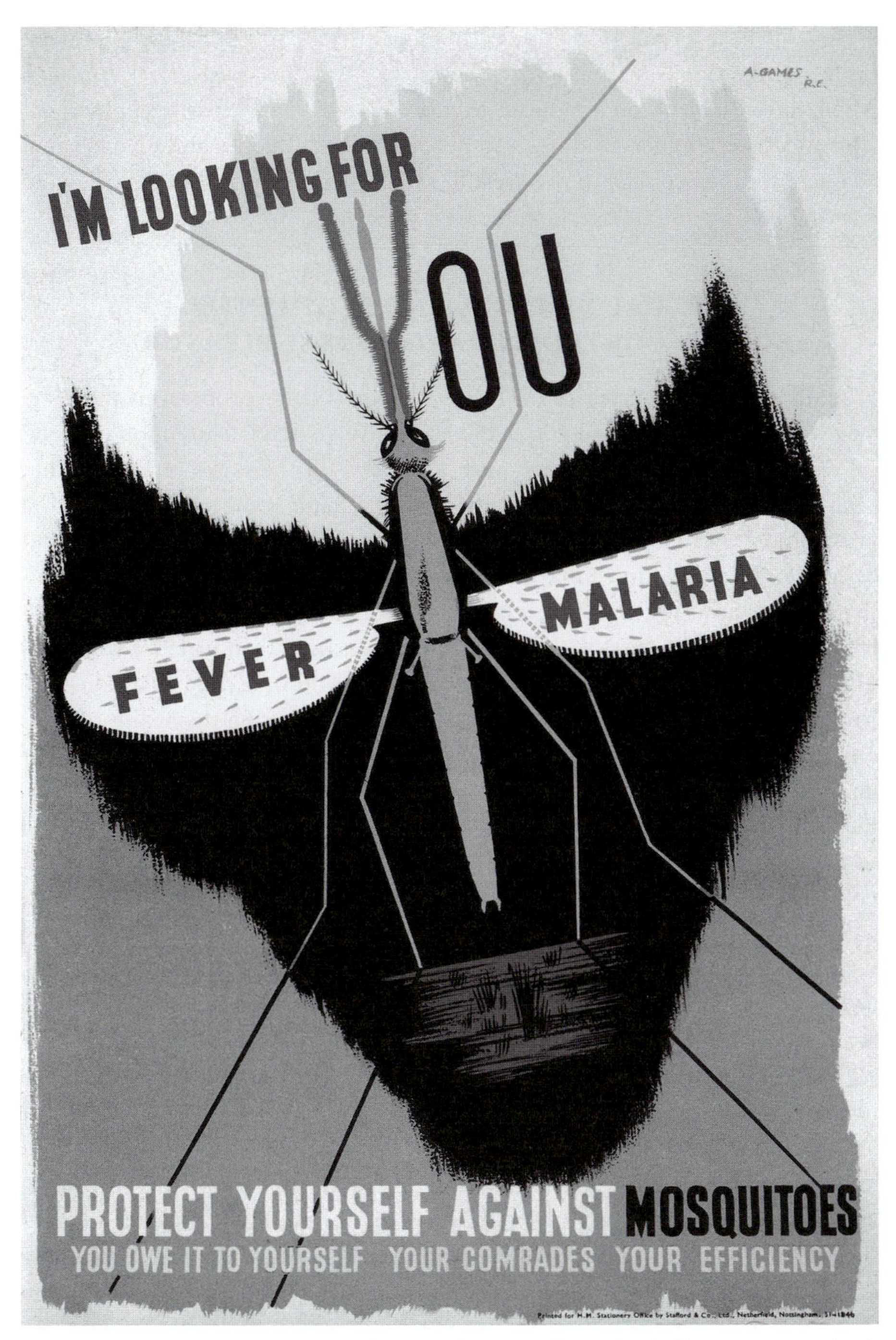

FIGURE 4.3: The malaria mosquito forming the eye sockets of a skull, representing death from malaria. Color lithograph after A. Games, 1941, H.M. Stationery Office, London. Wellcome Library, London.

arguments have been posited about homogeneity of images of poor populations and the diseases from which they suffer. Africans, particularly through AIDS, have often become reduced to essentialized images, easily comprehensible to Western understandings and devoid of the subtleties and nuances that characterize the problem in the West.[45] Africa becomes a land of diseased and starving bodies, not an emerging modern society.

SEXUAL DISEASES AND REGIMENTED BODIES: CONTROLLED CORPOREAL PLEASURES

Sexually transmitted diseases (STDs) were a particular problem in the nineteenth century. Syphilis and gonorrhea were given extended attention in medical circles.[46] Indeed, syphilis was such a scourge on the states of Europe that a number of laws were passed controlling the main source of transmission: prostitution.[47] These laws were developed largely because of the impact that casual sex leading to infection was having on the armed forces. It is also worthy of note, however, that much medical research into chemotherapy was directed toward STDs, first with mercury therapies, and then in 1909 with Paul Ehrlich's Slavarsan (arsphenamine, "compound 606"), a drug that could effectively begin the control of this disease that was affecting the health of various nations.[48] Sexual diseases, then, became in some ways paradigm cases for understanding disease and the way it affects the control of bodies. Partially, this is because sexuality in the nineteenth century was increasingly considered something of a "secret of the self," that is, a fundamental part of ones identity, able to be situated in broader medico-moral debates about public health.[49] As social circumstances changed (caused in part by the uncertainties of two world wars as well as other cultural changes surrounding religion, public morality, homosexuality, the sexual rights of woman, etc.), control of the spread of STDs became an increasing part of the ways in which bodies and sexuality were regulated in the twentieth century. We see this, for example, in the wartime posters warning against "loose" women.

Relative control of STDs also led to other possibilities—including that of less-stigmatized casual sex, in which the liberation from the risks of disease and from pregnancy played no small part in the wide-ranging social changes of the 1960s. Freedom from the fear of contracting a potentially life-threatening disease such as syphilis cannot be underestimated in the increasing relaxation of sexual mores in the West. It was not until the 1970s, with the increase in STDs such as herpes, that such issues of sexual health were put back on the agenda. This situation was, of course, drastically changed in the time surrounding the emergence of AIDS, first among gay men, intravenous drug users, and hemophiliacs, and then more generally in the casually sexually active population.[50]

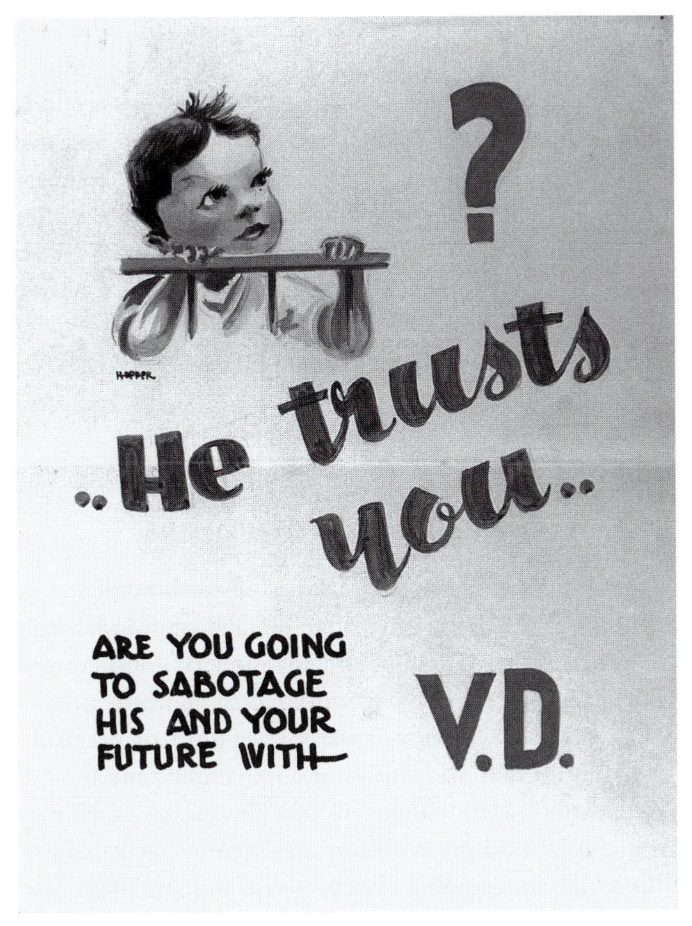

FIGURE 4.4: *He trusts you.* Original artwork (signed Hooper) for anti–venereal disease posters, 1943–1944. From the papers of Major General Sir Ernest Cowell CMAC/466/48. Wellcome Library, London.

Responses to the AIDS crisis varied. In the United States, money was channeled away from it under the Reagan administration. In contrast, in Australia and northern European countries, more effort was put into wider sexual education that dealt directly with the new risks of AIDS. This standpoint did not advocate prohibition or sexual abstinence but rather encouraged practices based on knowledge of safety—such as condom use, not sharing needles, and so on. Such public health campaigns, while acknowledging sexual activity was widespread, nevertheless impacted on such practices in profound ways, thus changing modern sexual habits. These issues are not confined to the West, with Africa in particular raising many issues related to HIV/AIDS public health education. And it is in African discourses around sexual health and disease that we see precisely the assumptions of race, contagion, and otherness that were raised above when discussing quarantine.[51]

The control of sexuality in the twentieth century can be seen to have moved away from the realm of moral control and toward health and the body as a method of surveillance and a mechanism for establishing regimes of power. The sexual body, as it developed through the century, was policed through one of the side effects of the pleasures it sought. STDs were thus central to the control of more than just a medical problem. Sexual mores have continually developed in response to attitudes toward disease as well as broader notions of sexual hygiene. In this way the body, sexual practices, and pleasure have been shaped through disease control.

GENETIC DISEASES AND BIOCITIZENSHIP: DEEPER SURVEILLANCE OF BODIES

Genetic explanations for diseases have always existed in some form or another. Indeed, discussions of the heritability of all manner of diseases from cancer to mental illness were rehearsed well before any modern notion of genomics had emerged.[52] But only comparatively recently have diagnostic practices become so available that they have changed our conceptions of pregnancy, motherhood, and the family. Of course, issues of cosanguinity and the prevention of marriage have existed for all manner of societies, as shown by various prohibitions on sexual relationships in cultures, as seen by the vast field of kinship studies within social anthropology. Such restrictions are often also enshrined in religious texts, notably the Bible (e.g., Leviticus 18:6–7). But the diagnostic possibilities of screening before conception, and monitoring the fetus during pregnancy via genomic techniques, have become a part of the modern way of having children in the West.[53] Risk groups, when identified either through screening or other technologies, are supplied with counseling. Debates take place in political and religious institutions concerning the ways in which appropriate responses to genetic diseases can be best maintained—from the time abortion becomes possible through to ways in which laws prohibit some genetic manipulation of the embryo. Such debates around reproductive technologies, and the practices of having children that are associated with them, are indicative of the ways in which diseases can change behaviors as fundamental as childbirth.[54]

Genetic counseling makes an especially suitable case study for considering how notions of disease affect practices at a number of levels. A panoply of factors come to play when risks surrounding so-called genetic diseases are dealt with in modern medical facilities.[55] These include a variety of models that determine how best to assess the risks to a patient, or a fetus, as well as a number of ethical and religious concerns that are taken into account. There are two main forms of genetic counseling: prenatal, based on a variety of screening methods, and adult, where the genetic risk of contracting a disease is calculated based on familial and other evidence and which can result in prophylactic surgery in some cases, such as radical mastectomies for women determined to be at risk of breast cancer despite showing no symptoms.[56] In both of these cases, the specter of disease can precipitate major medical interventions, including abortion in the case of prenatal screening. The role of counseling here is to help the patient/parent assess the risks of the disease and make an informed choice about the course of action that can best be followed. Complex issues of human rights, disability rights, and medical management all come into play in these issues. The impact of biotechnological knowledge in these cases, and the ways they have shaped medical practice, has in effect created genetic patients from disease potential and through normative ideals of the body and health.

A flip side to the ways in which current genomic and biotechnological knowledge has altered bodily practices such as childbirth is also seen in the ways in which specialized responses to diseases are now possible—or at least are hotly debated at present. A case in point here is research into stem cells. Stem cells are unspecialized cells that have the ability to develop into other cell types. It is claimed that research into stem cells and related therapies will produce better medical responses to Parkinson's and other degenerative diseases, as stem cells can be cultivated to replace otherwise degenerated neurons.[57] Stem-cell research is hotly debated. It relies on stem cells derived from human embryo bone marrow that had been gathered in the practice of other reproductive technologies, especially in vitro fertilization (IVF).[58] This political attention results not least from the criticism of IVF by Catholics and others who oppose abortion, as IVF and abortion involve similar ethical issues.[59] And there are reports of highly dubious practices surrounding the illegal trade in bone marrow harvested from aborted fetuses in the Ukraine, which have only served to keep stem-cell research a political topic.[60] As such, disease affects people not only in terms of sickness but also in profoundly political ways.

Discussion of modern genomics and related biotechnological interventions is only just beginning. Conceptions of disease and the negotiations of risk reduction and of modern medical care are highly politicized. In effect, conceptions of humanity are changing with the emergence of a genomic identity in recent years. This emergence has a variety of impacts on how people conceive of diseases and their responses to them.

CONCLUSIONS: APPROACHING DISEASED BODIES

In this chapter a variety of examples have been used to consider the ways in which disease becomes a point of control for the body. Such measures of control operate at a variety of levels, from the state implementing restrictions on movement, preventing the spread of diseases through official channels; through the use of the media and other sources of public information to shape bodies—as perhaps best seen in terms of the changes with regard to safe-sex practices in the post-AIDS world; through to personal responses to disease, including points of resistance (from not employing the triple MMR vaccination, to smoking cigarettes or drinking too many glasses of wine, to not wearing condoms in the passion of casual sexual encounters, all in the face of the best current medical advice). In all of these cases, bodies are controlled because of their vulnerability to disease and the risks that are associated with such health concerns. Control in this sense is a collective action and is maintained through sanctioning those who do not comply. Bodies show the mark of disease without actually having contracted the disease—but instead through their docile reception of regimes of medical power that pathologize and control.

This chapter also raises a number of issues that require further exploration. The first of these revolves around conceptions of corporeal multiplicity. Despite the ways that historians and sociologists largely refer to "the body" as if it is a stable entity, it is best conceived as fragmented and multiple. Often, only a specific body part or process is the focus.[61] And the meanings of the diseased body also depend on the specific practices that consider it—from the varieties of methods for dealing with disease within medical practices, to the differences of experience of disease in separate cultures and even between individuals. Diseased bodies are also temporally and spatially bound. Disease is linked through time to other practices that become present. It is also situated spatially via synchronic practices.[62] It is therefore important here to emphasize a lack of reductionism when locating the diseased body in history—and instead to celebrate the multiplicity of diseased bodies.

Diseased bodies are a site of power and a site of resistance. A struggle emerges between the individual and the individual's construction by regimes of power—from political structures impacting on issues such as citizenship and travel (disease control disallowing people from certain movements or migrations) to personal struggles to cope with impending illness and suffering. The diseased body is not just a biological struggle between the sick and their pathogens but a struggle in terms of politics, freedom, and the choice one has in one's own death.

CHAPTER FIVE

Popular Beliefs

DAN O'CONNOR

In Amsterdam, in 1928, when some of the athletes in the first-ever women's Olympic 800-meters final finished the race sobbing and exhausted, few people were terribly surprised.[1] In fact, there was something rather reassuring about the whole sorry spectacle of exhausted, sobbing women staggering around the cinder track. It affirmed what most of the spectators had been privately (and not so privately) thinking: that women's bodies just weren't cut out for that sort of exertion. The 800 meters would not be an Olympic event for women again until 1960. Yet by 1983, at the inaugural World Athletics Championships in Helsinki, track-and-field fans were thinking something a little different. For not only had the Czechoslovakian athlete Jarmilla Kratochvilova just won the women's 400 meters final (breaking the world record in the process), but she had, quite staggeringly, won the 800 meters final just the previous day. The belief now was not that women couldn't run that far but that they couldn't *possibly* do it that fast. As the *Times* of London put it, there were two theories "equally repugnant, but growing in popularity" that sought to explain Kratochvilova's success: the first, that she "has taken drugs to improve her performance. The second, quite bluntly, is that she is not a woman."[2] By the latter, they meant, of course, that she was a man. Fast-forward again to Seoul 1988, when no one thought Florence Griffith-Joyner, America's new double Olympic champion at 200 and 400 meters, was a man. No way, not in that outfit, not with those fingernails. But they did believe she had taken drugs—and lots of them. After all, you don't dismantle the old 200-meters world record by about half a second without *some* help, especially when that record had been held by two East Germans. Neither Kratochvilova nor Griffith-Joyner were ever found to have taken any sort of performance-enhancing drugs, but the allegations were to dog them for the rest of their lives—which in Griffith-Joyner's case was a mere ten more years. She died of a heart attack in 1998, an event that many commentators believed to have been brought about by her past drug use. The British heptathlete Jane Flemming recalled Griffith-Joyner's record-breaking runs. "They were awesome," she told the BBC. "Very few of us found them credible. I think this is a wake-up call. There'll be a lot of people running to their doctors today."[3] And there, in her tacit accusation of widespread drug abuse among athletes, Flemming enunciated one of the most popular beliefs about the body in the age of change: that it had become capable of extraordinary things, transcending its natural limits, but that it had required medical help to do so.

This chapter examines popular beliefs about the body in the United States and the United Kingdom in the twentieth century and asks how they were affected during this "age of change." The chapter consists of three parts. Each of these parts is, in some way, connected to the idea of limitation and the use of medical technology to transcend it. The first of these continues with my opening stories and looks to the world of competitive athletics to elucidate changing beliefs about what, in terms of the Olympian ideals of *citius, altius, fortius* ("faster, higher, stronger"), the human body was capable of doing. The questions of gender and race that this elucidation necessarily begs lead on to a consideration of the boundaries of the human body. This section looks at how phenomena such as transsexuality, aesthetic surgery, organ transplantation, and other "enhancement technologies"[4] affected beliefs about the phenotypic boundaries of the body. The focus on boundaries then leads me to consider beliefs about chronological limitations, about how long the human body is supposed to last, and whether death really is the end of the corporeal form. The final section explores how mass-media interactions with the human body, specifically television and cinema, came to alter beliefs about how the body could be seen, envisioned, and apprehended.

A QUESTION OF SPORT

Feminism did more than any other social or intellectual trend to encourage new beliefs about the human body in the twentieth century. Initially, these changes focused on the female body, particularly the way its "sex" limited its material potential for work. From wide-held beliefs in the early 1900s that women ought not to take "too much" physical exercise for fear it would exhaust them mentally and emotionally, British and American societies slowly moved toward a conception of the female body as perfectly capable of undertaking the same physical tasks as men. At every Olympiad since 1948, the number of events open to female competitors increased, to the point that in Sydney in 2000 women were not believed eligible to compete only in the boxing arena. In this time women had swum the British Channel (breaking the male-held speed record in the process), run marathons, competed in Ironman triathlons, landed triple-axel ice jumps, formed their own football leagues, demanded (and for the most part achieved) equal prize money in tennis, and raced cars at LeMans. Such dramatic changes were effectively enshrined in the 1972 American law Title IX, which legislated the belief that women deserved equal access to sporting facilities because women were equally capable as men of using them. By the end of the twentieth century, these beliefs were not simply a matter of faith but of recorded fact—the world free diving record was held by a woman, as were several world endurance road-running marks.

Yet these changes in belief did not occur unprotested. When, at the very height of the Cold War, a monstrous regiment of Russian women arrived on the international athletics scene, their remarkable achievements were greeted first with astonishment and then with suspicion. Take, for example, the case of Tamara Press, the Soviet shot-putter and discus thrower, who between 1958 and 1966 won three Olympic gold medals and four European titles and broke sixteen world records, most of them her own. It was quite an extraordinary career, the previously unconquered pinnacles of which demanded some sort of explanation.[5] It was an explanation that popular beliefs about the human body were happy to supply: Tamara, like so many of her teammates, was notoriously

FIGURE 5.1: Runner in action (D.G.A. Lowe) showing correct style. Wellcome Library, London.

masculine in appearance. "She could play tackle for the Giants," mused *New York Times* sports writer Bob Daly, referring both to his home football team and to Tamara's six-foot-two-inch, two-hundred-plus-pound frame.[6] The gridiron became a recurring trope in American media coverage of the Soviet women. Tamara was "big enough to play tackle for the Chicago Bears," wrote another journalist, Richard Daly, who, in tune with the somewhat-wry tone that accompanied most reporting on the Soviet female athletes, opined that "and at the rate the Bears are going this season, they could probably use her, too." And not just Tamara but also her champion pentathlete sister, Irina, "who is about the size of a running back."[7] Again and again the Soviet women were "burly," "muscular," "powerful," "husky," "hefty,"—in short, masculine. Popular belief at the time ran that for a female body to perform to that level, it probably wasn't female at all.

And so, when the organizers of the 1967 European Athletics Championships decreed that there would be chromosomal sex tests for all competitors, few media commentators were terribly surprised that certain Soviet "women" failed to turn up. Though it was for "firmly stated reasons of family illness," no one believed that Irina and Tamara (and their teammate, long jumper Tatyana Lyschenko) were absent for any other reason than that they would have failed the sex test. And the popular press was vindicated in its conviction when, a year later, the world's "outstanding woman sprinter," Polish star Ewa Klobukowska, was revealed by the same test not to be a woman at all.[8] That she was not actually revealed to be a man (merely intersexual, with a few too many chromosomes) didn't matter. "Such persons," wrote medical journalist Jane Brody, "appear to be sexually immature females, often tall with underdeveloped breasts,"[9] which was merely one more way of saying that she, like the Press sisters before her, did not have a female body. If it had been beyond belief that the sisters had had female bodies, it was now proven fact that Ewa did not.

And yet records continued to be broken in women's track-and-field without any other competitors ever being dismissed for sexual deception. In fact, the medical technologies behind the sex tests became so sophisticated as eventually to prevent anyone from believing that a male body could get away with competing with female bodies. Thus, people started believing something else: not that these remarkably successful female athletes were really men but that they were taking so many androgenic performance-enhancing drugs that they may as well have been. Yet this belief was rarely, if ever, confirmed during the Cold War period—very few Soviet-bloc women athletes were ever actually convicted of doping offenses. While this may have had as much to do with the cupidity of the hyperdiplomatic International Olympic Committee as it did with the pace of medical testing development, it reflects a rather ancient trend in beliefs about women's bodies: that the men who held those beliefs often did so without any particular reason or proof. Of course, "now we know" (as the saying goes of Western knowledges of former Communist nations since the fall of the Berlin Wall) that there were massive state-enforced sports doping programs enforced all across countries behind the Iron Curtain—with the East German model being perhaps the most infamous.[10] Lost in much of the Western smugness about these revelations was the fact that the athletes themselves were, for the most part, unwitting victims who had, as very young women, been given "vitamins" by trainers and coaches who ran almost every aspect of their lives. East German women athletes, many of whom have since successfully sued for compensation in the light of these revelations, were forced to seek approval from coaches if they wished to have a baby, go on vacation with a boyfriend, even marry. The East German doping programs show, if anything, that Communism was no bar to another ancient belief about women's bodies—that men could do what they liked with them for political purposes. There is little difference between the mind-set of the fifteenth-century Hapsburg duke who marries a thirteen-year-old girl from a different country in the hope of cementing alliances and bloodlines and that of Dr. Manfred Ewald, chief of the East German sports program, who referred to his nation's medal haul at the Montreal Olympics of 1976, achieved by the mass poisoning of his female athletes with huge amounts of male hormones, as "a victory for socialism."

Yet beliefs about the athletes from East Germany and the Soviet Union were informed not merely by sexism but also by a mostly unspoken racism. Popular beliefs about the human body compelled many inwardly to ask how these East Europeans could be not only competing with but even beating African Americans? As John Hoberman has shown, sports commentary and sports science have always entertained a pervasive if implicit belief that, quite crudely, black people are better at sports than white people.[11] As a popular belief about the human body, it is merely the athletic manifestation of the ancient prejudice about the bestial black man and the rapacious black woman—both figures far stronger physically than their white counters. This sporting belief has been grounded in yet another ancient belief about the human body—that the male is taken as the normative and that the female should be compared to it. Thus, because only one white man since World War II had won the Olympic 100-meter sprint title and because by the 1970s world-level 100-meter finals had become almost exclusively black affairs, and because only black men had held the 100-meter world record from 1983 on, it came to be believed that black men, specifically those of Afro-Caribbean descent, were faster than white men. Thus, believing this about the male body permitted the same belief about the female body, even when the evidence screamed otherwise. White, East European men weren't faster than their black counterparts, so their womenfolk could hardly be, either. Nonetheless, the East German and Soviet women's teams regularly defeated their American rivals. In the face of the comparative lack of proof of

Communist doping programs available to commentators and the general public at the time, it seems that only the racist assumption that black people are naturally better at sports than white people permitted Western belief in such programs. It was the only explanation possible in the discourse. Even after the fall of the Berlin Wall, this view prevailed when female athletes from another Communist nation, this time China, began dismantling records at the long-distance events—records previously held by Soviet women. It was impossible, ran the popular belief, for these new records to have been achieved without the aid of drugs. Yet for fifty years previously, the men's equivalent events had been dominated to the point of hegemony by athletes from East Africa—particularly Kenya and Ethiopia—and barely anyone ever accused them of being on drugs because black people were believed to be naturally faster than whites and Asians. And so, when the most remarkable of the Chinese records, Jiang Bo's 5000-meter mark, was broken by a Turkish athlete in 2004, there was much satisfaction in the world of professional athletics, because things had been returned to a more believable state. The woman responsible was Elvan Abeylegesse, born and raised in Ethiopia.

Of course, perhaps the most famous doping case in history was that of a man: Ben Johnson, the Canadian sprinter. Johnson, the son of Jamaican immigrants, had won the 100-meter title at the 1988 Seoul Olympics in a world-record time of 9.79 seconds, knocking an eyebrow-raising 0.14 of a second off the previous mark. A test of his urine sample revealed that he had been taking stanozolol, an anabolic steroid and banned substance. No one was particularly surprised. Johnson had all the classic signs of steroid abuse: obscenely bulging muscles, jaundiced eyeballs, and a receding hairline. His archrival, Carl Lewis, had repeatedly made coded comments to this effect. But it was the massive difference between the new and old records that went furthest against popular belief—that the human body improved its capabilities only gradually—in the case of sprinting records, by a hundredth of a second not by a tenth. The current 100-meter world record, held by the Jamaican Asafa Powell, is actually 0.02 seconds quicker than Johnson's illicit mark, but it came at the end of a series of incremental improvements of 0.01, 0.02, seconds over a period of fifteen years and was thus rendered believable. This belief in gradual improvement gained increasing currency, particularly in the second half of the twentieth century, with the growth of what Dorothy Porter has dubbed "the reification of the fit body," as reflected in the massive boom in the billion-dollar gym and dieting industries.[12] Where the men and women of early-modern Europe may have believed in sudden corporeal transformations by witches and saints alike, their twentieth-century descendants adhered to a more measured conception of bodily improvement. Be it following the Weider method of bodybuilding (and being one of the six million people who subscribed to his *Muscle and Fitness* magazine) or going on the Weight Watchers diet (along with the million other people who visited the clubs every week), the prevailing belief was that bodily change took time, effort, and commitment. Vital to this belief was the notion that given time, effort, and commitment, bodily change *would* come, that the human body could get progressively better if one put the work in.

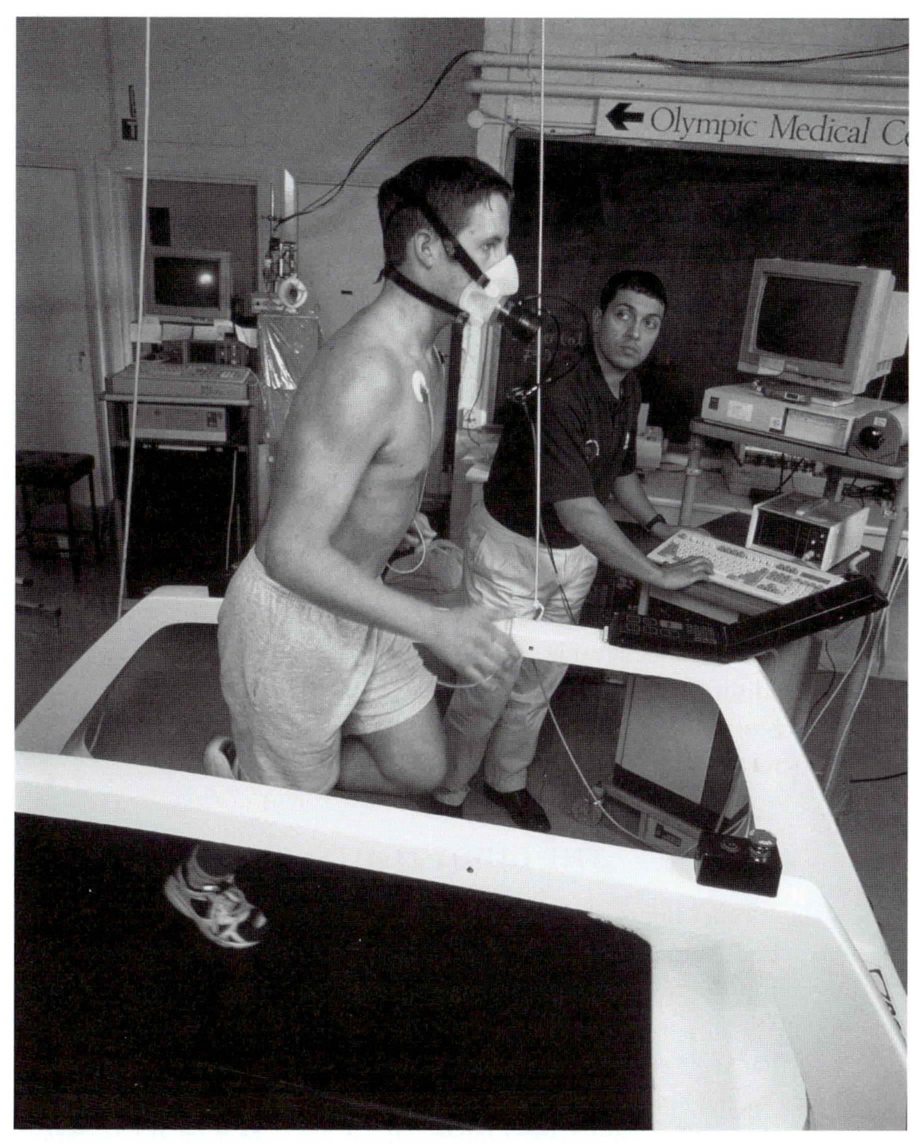

FIGURE 5.2: Male athlete having cardiovascular fitness tested on a specially equipped running treadmill at the British Olympic Medical Centre (BOMC). Heart rate, ECG, oxygen uptake, and work output are all constantly monitored. Lactic acid level, another measure of aerobic fitness, is analyzed at regular intervals using skin-prick blood tests. Wellcome Library, London.

There were, of course, limits to what it was believed that the human body could achieve. The fictional realms created in the comic books of both the Marvel and DC publishing houses since the 1950s provide us with an insight into these limits. In short, whatever Spider-Man, Wonder Woman, and the X-Men could do, it was popularly believed that the human body could not. Be it flying, controlling magnetism, being indestructible, or being invisible, the amazing abilities of the superhero's body acknowledged, by their very grounding in "fantasy and science fiction," the natural proscriptions placed on the human body. Thus perhaps the most interesting superhero of all was Batman, who had no superpowers whatsoever and was merely a billionaire in a rubber suit who managed his crime-fighting duties through a combination of extreme hard work, physical training, and cold, hard cash. Tired and exhausted after a tough night battling the forces of criminality in Gotham city, the bruised and bandaged figure of Bruce Wayne was a far more believable body than the impervious, immaculately coifed Clark Kent, whose world-shaking adventures as Superman rarely left him any the worse for wear as he strolled casually into work at *The Daily Planet.* This hardly stopped millions of fans from enjoying the comics, movies, and television shows based on these characters, but it did plant their corporeal achievements solidly in the realm of the preposterous. This was particularly true for female superheroes who, unlike their entirely nonfictional equivalents on the East German track team, achieved these remarkable physical feats without surrendering their believably female bodies, as the overtly sexualized figures of Supergirl, Rogue, Mystique, and Catwoman attested. The "real" physical achievements of Communist athletes rendered their female status unbelievable, while the "real" femininity of the superheroines rendered their achievements fantastical. Belief in the limits of the human body was regulated, it seems, not merely by the laws of physics and biology but by the strictures of social construction as well.

THE GIDDY LIMITS

Beliefs about other limits of the human body concerned just where and when the human body started and stopped. Throughout the twentieth century, medical and cultural developments inspired a range of alterations in popular beliefs about the physiognomic limits of the human body. Barbara Duden has suggested that the "enclosing" of the human body from the external world (the ending of the human-body-as-microcosm model) may be counted as one of the hallmarks of the modern, Enlightenment era.[13] Indeed, the twentieth century for the most part sustained a popular belief, in the West at least, in *homo oeconomicus*, the body as a discrete economic unit. Yet phenomena in the second half of the period did go some way to undermine the hegemony of this conception. A variety of "new" pathologies and conditions were identified that seemed to destabilize the notion that one owns one's own body—what Julia Kristeva has called *corps propre*.[14] The most obvious example was transsexuality, publicly manifest in the tabloid sensations surrounding first Christine Jorgensen in the United States, then April Ashley in the United Kingdom. Here were a group of people, small in number but great in newspaper column inches, who were denying ownership of very specific parts of their body—penises, breasts, testicles, wombs. Helped in no small part by Jan Morris's elegant autobiography *Conundrum* (1972) and Renée Richard's campaign to be permitted to play in the women's U.S. Open tennis tournament (1976–1977), popular discourse came to entertain the notion that, perhaps, for some people, their *real* body ended (or began) before (or after) it actually seemed to. Yet the popularity of this particular belief went only so far, as was seen toward the very end of the twentieth century in the case of body identity integrity disorder (BIID). BIID sufferers (known among themselves as "wannabes") had the unalterable urge to have one or more perfectly (physically) healthy limbs removed on the grounds that it simply, unshakably, did not feel as though it were a part of their body. This was far more difficult for popular discourse to come to terms with than transsexuality (which itself had hardly had an easy time of it), if only because of another long-standing belief about the human body: that any changes to it ought to result in socially acceptable amelioration. It was easy to believe, in social terms, that dieting made you better (thinner) or that weight training made you better (stronger) and politically vexed to say that there was something socially negative about changing into a woman or a man. BIID surgery, however, would have resulted in the human body being deliberately crippled. In a discourse in which elevators, hearing loops, guide dogs, and wheelchair ramps seek to make disability "disappear," this was believed to be socially unacceptable.[15] Thus, the British National Health Service, after considerable psychiatric review, provides transsexual surgery, but it will strike off and prosecute any surgeon who amputates a healthy leg. Different bodies may have ended in different places, but those places had to make social sense.

In the twentieth century, the social sense of amputation linked it frequently to warfare, particularly the two World Wars and the global spread of perhaps the nastiest of all military brain waves, the land mine. Ever since the brand-new techno-carnage of World War I, which showed that legs no longer had to be cut off by surgeons after infection but could be blown straight off by artillery, the loss of a limb has popularly signified a war injury, be it in combat (thereby rendering the victim brave and patriotic) or in peacetime's uncleared minefields (thereby rendering the victim tragically innocent). As Joanna Bourke has shown, those who were disabled in World War I were not deemed to have lost their masculinity along with their legs. Far from the dismembered male body being believed to have lost its potency, it was, in fact, the focus of much admiration and sympathy, with the newly "limited" body signifying a properly manly willingness to suffer in the defense of one's country.[16] In many ways, the laudatory treatment of these war veterans marked a change in popular beliefs about the disabled body, from a mixture of pity and contempt to a comprehension that society ought to do as much to enable the disabled body as possible—with precisely those "disappearing" wheelchairs, guide dogs, and so forth that made BIID seem so unreasonable. And so public buildings were built with ramps and lower light switches, and television shows were broadcast with subtitles and sign language. Other devices designed to enable bodies in the twentieth century (contact lenses, sophisticated prosthetic limbs, and cochlear implants, for example) begged further questions about just precisely where the body ended: Was a body purely constituted by biological matter, or was a hearing aid or an artificial heart part of the human body? Xenotransplantation—the use of animal organs in humans—went further yet in challenging popular beliefs about what counted as a human body and where that body might end. "Ashes to ashes, dust to dust" began to make less sense when shiny platinum cardiac valves could be found in the middle of decomposing corpses.

Popular culture both responded to and preempted these developments, particularly in the form of the cyborg body—a fixture of science fiction throughout the period. The cyborg, as theorized by Donna Haraway, is an admixture of the biological and the technical, an admixture that undermines the belief that the one is necessarily opposed to the other.[17] One of the early examples of the popular fictional cyborg was Robert Heinlein's short story "Waldo" (1950), in which a disabled man is able to manipulate his own prosthetic legs by remote control—a scenario that medical technology has since made unremarkable. More commonly, though, in popular culture the cyborg was not merely an enabled human but an enhanced one. Martin Caidin's novel *Cyborg* (1972) was based on actual research done by the U.S. government into human-machine combinations. The top-rated television show that it inspired, *The Six Million Dollar Man*, opened with the telling voiceover: "Steve Austin, astronaut. A man barely alive. We can rebuild him. We have the technology. . . . Better than he was before. Better, stronger, faster."[18] The popular fictional cyborg, then, was a manifestation of a further belief about the human body—that medical technology could enhance its capabilities beyond those of the unaltered body. Whether this was manifest in something as innocuous as Audrey Eyton's best-selling diet book *The F-Plan* (which, when published in 1982, made explicit its grounding in the science of nutrition—F stood for Fiber) or in something as wildly sinister as the U.S. government's MKULTRA program (which, between 1953 and 1972, aimed to create "supersoldiers" through the deployment of psychotropic drugs), it was a standard of popular culture throughout the century.[19] For example, as both Elizabeth Haiken and Sander Gilman have pointed out, the development of cosmetic surgery was based almost entirely on the popular belief that medicine could make the body more beautiful, more desirable, more efficient, more acceptable.[20] From the Cybermen on the BBC show *Doctor Who* (1963–1988), to Pamela Anderson's breasts on *Baywatch* (1992–1997); from Paul Verhoeven's *Robocop* (1987) to Jean-Claude van Damme in *Universal Soldier* (1992); from Popeye (1929–present) eating his spinach to the mathematically perfect Lisa in the movie *Weird Science* (1985), the technologically augmented body was probably second in popularity only to the dead body of detective thrillers as a corporeal trope in Western television and film.

Yet if transsexuality, BIID, cyborgs, and pneumatic blondes went some way to undermining belief in the physiognomically discrete model of *homo oeconomicus*, Duden's schema nevertheless retained much of its credibility and, indeed, found new ways to express itself. The term *personal space* was unique to the second half of the twentieth century and can surely be properly understood only in the context of highly individuated human bodies. Similarly, the legalization of abortion was founded, in many Western nations, on the belief that a woman's body was entirely her own and that what she did to it was no one's business but hers and her physician's. The U.S. Supreme Court ruling *Roe v. Wade* (1973), which made termination of pregnancy legal under federal law, was explicit in grounding its decision in the right to corporeal privacy.[21] This belief could also been seen in such seemingly disparate phenomena as the

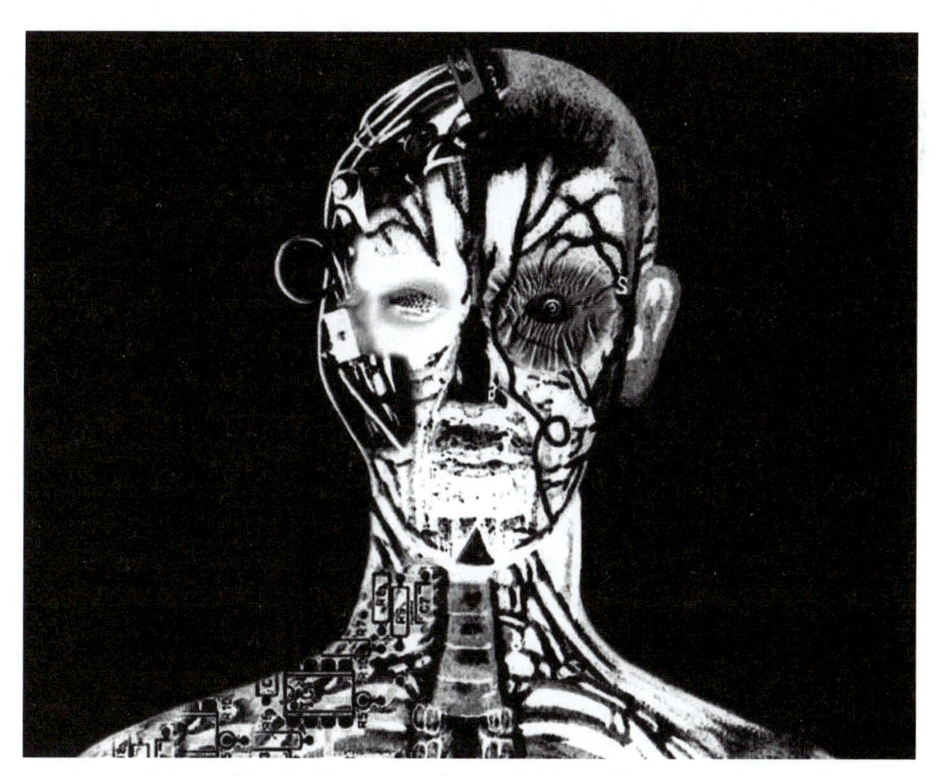

FIGURE 5.3: Dianne Harris, *Man or machine?* (2002). Wellcome Images, London.

campaign against secondhand smoke and the arguments in favor of remunerated organ donation: in short, that no one should experience an uninvited effect on his or her body. *Roe v. Wade* was particularly intriguing, though, because the very arguments about corporeal privacy which that that case appear to have been used to prosecute mothers whose children were born with fetal alcohol syndrome.[22] This, of course, illustrates some absolutely key questions about the *chronological* limits of the human body, about when it started and when it stopped. Even after the legalization of abortion in most Western nations, a profound ambivalence remained about the fetus's life status. Religious conservatives held that life began at conception, while social liberals insisted on viability as the hallmark of life. Throughout this key battle in the so-called culture wars, it is interesting to note that neither side has ever been able to prove its point, relying instead on unverifiable belief. "Do you believe in abortion?" became much the same sort of question in the twentieth century as "Do you believe in Father Christmas?"—an unprovable issue of faith, the answer to which clearly marked the responder as part of a particular social group. Indeed, so critically divisive had the belief become that, by 1998, a character on the American television show *The West Wing* was able to announce of American national politics that "What we've got is two corporate parties: one pro-life, and one pro-choice."[23] Belief (or disbelief) in the privacy of the female body had become a far greater social signifier than any sort of economic or social political standpoint.

At the other end of the human life span, even so seemingly definitive a statement as the "definition of irreversible coma" arrived at by the Ad hoc Committee of the Harvard Medical School to Examine the Definition of Brain Death in 1968 failed to homogenize Western beliefs about the chronological end of the living human body.[24] This was reflected that same year by the International Declaration of Sydney, which suggested that death was not a specific moment but that "the irreversibility of the processes leading to dying" served as the chief signifiers of the temporal limits of the body.[25] In a way, both the Sydney declaration and the Harvard committee were speaking to the increasing belief, enunciated a few years previously in a medical textbook, that the moment of death had become "arbitrary rather than actual"—something to be negotiated by physicians and relatives rather than simply measured by a machine.[26] The improvements made in "life-support" technologies meant, in effect, that a brain-dead body could be kept "alive" for years after the diagnosis. Once again, these mechanical-corporeal interfaces challenged popular beliefs about the body. Ought it to be kept alive artificially? Was it even "artificial" (or "not natural") to have a machine breathe for the body when machines were already helping bodies see, walk, eat, hear, process urine, and pump blood around the body? Of course, just as medical technology could prolong life, it could also expedite the process of death. Euthanasia (or assisted suicide) remained illegal in both Britain and North America in the twentieth century, but the laws were challenged again and again by what came to be called the "right to die" lobby.[27] The iconic figure of this movement was Jack Kevorkian, known to the tabloid press as Dr. Death. Kevorkian advertised himself in the late 1980s as a "death counselor" and assisted over 100 terminally ill patients in ending their own lives. Kevorkian's procedures were curiously cyborg in nature, employing machines he had invented for the specific purpose of assisted suicide. The Thanatron ("death machine") administered a lethal injection, while the Mercitron ("mercy machine") fed carbon monoxide through a gas mask attached to the patient. The machines, with their echo of common execution methods in the United States, struck directly at what Kevorkian considered to be hypocritical American beliefs about the end of the human body. Primarily, he questioned the power of the state to take a life when the individual was not empowered to take his or her own life. What Kevorkian was railing against, of course, was a further belief about the human body that had survived the passing of the centuries—that the state has a monopoly on violence.

Another major feature of popular twentieth-century beliefs about the chronology of the body was the increased compartmentalization of that chronology. The notion that the timeline of the body could be demarcated well beyond the traditional "three ages of man" became increasingly popular after World War II. Perhaps the most striking new temporal category for the human body was the "teenager," hitherto unknown but by the early 1960s the dominant figure on the pop-cultural landscape. The body of the teenager, popularly understood to suddenly be riven by hormonal change, was believed to be both dangerous and in danger. These themes were arguably first captured in the James Dean film *Rebel Without a Cause* (1955), which saw Dean's character, Jim Stark, arriving in a new town, disrespecting his elders, seducing girls, and fighting boys. Brooding, sexy, and packed into a tight white T-shirt, the film portrayed Stark's body as seemingly the driving force behind his socially unacceptable behavior. The teenaged body was believed to be the locus of rebellion against tradition and authority, whether it was young men growing their hair to identify as hippies or young women piercing their noses to identify as punks. Again and again throughout the second half of the twentieth century, the dangerous and endangered body of the teenager served as a theme in popular culture. In S. E. Hinton's notorious novel *The Outsiders* (1967), long, greasy hair marked characters out as members of a dangerous youth gang, while in the anonymous "diary" *Go Ask Alice* (1971), it was the central character's preoccupation with boys that delivered her to the party where she first takes the drugs that will ruin her life. In the movie *Footloose* (1984), the lithe body of Kevin Bacon's character encouraged preacher's daughter Lori Singer to indulge in the illicit passions of rock 'n' roll dancing, while in *Risky Business* (1983), Tom Cruise's teenage hormones lead him down the farcical path to opening a brothel in his parents house while they are on vacation. By the end of the twentieth century, the movie *American Pie* (1999) had established that teenaged boys could not be trusted not to masturbate into apple-based desserts, and the television show *Buffy the Vampire Slayer* (1997–2003) was making it clear just how dramatic the changes in a teenager's body could be. Above all, the teenaged body was believed to be at the mercy of their hormones and, crucially, they would "grow out of it" just as they had grown out of the childish habits of thumb sucking and picky eating.

Other ages of the human body took their place on the cultural stage in the twentieth century—the toddler, the senior citizen, the twenty-something and, most recently, the "tweenager." Discourses about the interplay of time and the human body sustained such popular beliefs as the ticking biological clock in women and the male menopause in men. Both these particular beliefs seem like hangovers from the notion of the hormone-driven teenaged body: a human body entirely at the mercy of its chemical development. From the beginning of life through its adolescence and right to its very end, the twentieth century sustained a never-ending conversation about the body, about what it was and what it could do. This conversation was hardly peculiar to the twentieth century (the preceding five volumes ought to have made that clear), but the nature of the conversation was. For in the twentieth century, popular belief was liberated from the spoken word and the written text, which had rendered the exchange of belief so slow. In the twentieth century, the exchange of popular belief became immediate, global, and unceasing because the body had been televised.

LIFE AS A METAPHOR FOR TELEVISION

By the end of the twentieth century there were more television sets in the United States and the United Kingdom than there were households. For the six decades or so of mass-broadcast television, the human body had been regularly transformed from a locally restricted, flesh-and-blood entity into a series of pixilated dots visible to anyone with a television set. The effect this had on popular beliefs about the human body was double-edged. On the one hand, it helped to encourage belief in some of the more remarkable feats of corporeal achievement. It is difficult, for example, to imagine much credence being given to Neil Armstrong's first lunar walk without the accompanying moving pictures being beamed around the world. On the other hand, an increasingly sophisticated viewing audience meant a growing skepticism about the veracity of such globally broadcast bodies. This was seen not merely in the fact that throughout the last three decades of the century, a large percentage of the American population retained the belief that the moon landings had been faked by the Central Intelligence Agency and Disney. The very same special-effects technology that made *The Six Million Dollar Man* possible as television fiction also made it possible for people to believe that lighting, makeup, camera angles, and (later in the century) computer generated imagery (CGI) were responsible for the way a "real" body looked on television. The first televised presidential debates in America, between John F. Kennedy and Richard Nixon in 1960, testified to the power of this belief. Kennedy's media advisors ensured that he was attended by a proper makeup artist before the broadcast—something Nixon's people neglected to do. Kennedy, then, appeared calm and healthy, in contrast with the sweating, pale-faced Nixon. Americans listening to the debates on the radio believed, by a rather small margin, that Nixon had come out on top. Television viewers, however, overwhelmingly felt that Kennedy had won. Belief in the corporeally transformative powers of television had allowed the Democratic strategists to encourage people to believe that Kennedy alone was presidential material. By the end of the century, this lesson had long been learned, and the 2000 debates were micromanaged to the very last detail, with both Republican and Democratic strategists fervently believing in the importance of the way their candidate's body came across on television. And so, although it was a phenomenon unique to the twentieth century, the televised body reflected the traditional belief that the way a body appeared would dictate the way it acted.

Both television and its elder sibling, cinema, continued to encourage belief in the importance of the appearance of human bodies. This was not merely a concern with what a body looked like but what that body appeared to be doing on the screen. One of the earliest, and perhaps most influential, indicators of this belief was the Hays Code, which was designed to regulate the content of American cinema. First adopted in 1930 but seriously enforced only after 1934, the Hays Code strictly defined what the human body could not appear to do on the silver screen.[28] Among the code's many rules were those that "excessive and lustful kissing [and] lustful embraces . . . are not to be shown" and that "actual childbirth, in fact or in silhouette, is never to be presented." Similarly, "complete nudity is not to be shown" and "undressing scenes are to be avoided," while "the treatment of bedrooms must be governed by good taste and decency." It is clear that, while various violent crimes were also prohibited by the Hays Code, it was above all a reflection of the belief that the human body should not appear naked in public and that it certainly should not appear in any sort of sexualized fashion. These beliefs were rigorously enforced by the Production Code Administration, and even multimillionaire movie moguls such as Howard Hughes were not immune to their censoriousness. His 1943 Western, *The Outlaw*, was banned because it was little more than a lengthy panning shot of Jane Russell's breasts. Yet by the late 1950s, attitudes were changing. The appearance on cinema screens of a sexualized human body, or a human body engaged in violent conduct, was no longer believed to be beyond the pale. Movies that violated the code were no longer condemned to box-office failure—indeed, they were often likely to be more successful. Beliefs about the human body had changed, and so the codification of outmoded beliefs no longer served. In this way, the abandonment of the Hays Code, and its replacement with a series of age-based film classifications in 1968, was a similar development to the gradual entrance of women into the full complement of Olympic sports: Shifting beliefs had caused institutional adaptation. Much the same thing had happened to bodies in British cinemas by the early 1980s, when the British Board of Film Censors changed its name to the British Board of Film Classification, reflecting the popular belief that though bodies may appear sexualized or violently on screen, access to those appearances ought to be regulated by age. Such regulations were typically isomorphic with the new chronological categories for the human body, with teenagers in particular being explicitly targeted by the "15" and "18" ratings in the United Kingdom and the "NC-17" mark in the United States.

Despite the apparently liberalizing trends in beliefs about the way bodies appeared, certain bodies still had the power to shock Western society out of its comprehensive comfort zone in the twentieth century. Typically, such images dealt with the suffering human body. The appearance of the emaciated, fly-covered children with bloated, empty stomachs who filled Michael Burke's BBC reports on the Ethiopian famine in 1984 seemed to ask the West just how far it believed it was responsible for the well-being of human bodies it was able to apprehend or believe in only because of television. Throughout the twentieth century, such visible "body shocks" had served to render believable those phenomena that, if otherwise reported, might not instantly have gained full credibility. The assassination of President Kennedy, for example, was made almost immediately, globally comprehensible by the release of Abraham Zapruder's video footage showing his death. For many, it was Nick Ut's still image, culled from newsreel footage, of howling children fleeing a napalm strike that made the previously abstract Vietnam war come to corporeal life. Even so, the suffering of such bodies was not always believed to be a necessary component of television news. Just as the Hays Code was (for the most part) self-imposed by the film industry, the Western news media regulated how the human body could appear on their broadcasts. The dead or suffering body was believed to be a proper subject for broadcast only when the newscaster warned of "images that some viewers might find disturbing," while the actual moment of death—particularly in warfare—was, with a few exceptions, never shown. Even in 2002, not a single Western news network would show the video footage of the American journalist Daniel Pearle being beheaded by the Taliban in Pakistan. There remained a pervasive belief that suffering, dying bodies retained the power to traumatize those who saw them. There was also, in the general unwillingness to broadcast the death of a human body, a retention of the belief in respect for the corpse, a belief as old as human civilization. This is not to say, however, that the human body could not be cut up live on television and have its insides displayed to the viewing public. By the late 1950s, live medical operations had become a staple of American television and were believed to be both educational and demonstrative of the march of American medicine.[29] They were also, undeniably, entertaining—often featuring a heart-warming story or a dramatic recovery. Indeed, the drama of live surgery meant that real operations were soon replaced by more manageable fictional ones in shows such as *M*A*S*H** in the 1970s and *St. Elsewhere* (United States) and *Casualty* (United Kingdom) in the 1980s. The surgeries in these shows were not particularly graphic but by the time that *ER* had become America's top-rated drama in the 1990s, audiences were accustomed to the bloody, twitching patient's body, believing it not merely to be a legitimate site of drama but also to be a plot device for the drama of the doctors' and nurses' lives. *ER* was a consciously "adult" show, not because it broadcast blood and guts but because its characters appeared to do the sort of things that would have violated the Hays Code had it applied to television.

Each of the preceding sections has highlighted some of the ways in which twentieth-century beliefs about the human body were often old wine in new bottles. Technological innovation often gave age-old corporeal conceptions novel spaces in which to work themselves out. Simultaneously, however, those innovations also drove some great shifts in what was popularly believed about the human body, especially what it was capable of achieving, where its boundaries were, and how it was able to appear. These beliefs were not, of course, uniform. They were also not nationally bound but rather the products of global interactions via television, air travel, and international education. Technology alone, however, cannot explain why people believed what they did about the human body. National political ideologies, religious creeds, personal philosophies: all these ways of thinking had an effect on the belief systems of individuals and the communities in which they lived. Writing critically about belief is fraught with contradictions because the application of rational analysis to what, in many cases, boils down to an article of faith tends to be an exercise in anisomorphism: The two simply do not map onto one another. The American political debate on abortion is a case in point. We cannot always be sure *why* people believed certain things, but we can—as I have tried to do in this essay—explore the things they did believe in order to understand their world more clearly. Popular beliefs about the body in the twentieth century often reflected far wider, far less corporeal beliefs. That is why the American women's swimming team at the Montréal Olympics in 1976 believed their East German conquerors to be drugs cheats who looked like men: not because there was any evidence of doping, nor because the East Germans were especially more masculine than the Americans, but because the Cold War climate of American culture insisted that Communism was so pernicious an evil that it would sacrifice even so sacred a thing as femininity in its fight to destroy freedom. Popular beliefs about the body were often just popular beliefs about the world writ small.

CHAPTER SIX

Beauty and Concepts of the Ideal

CHRISTOPHER E. FORTH

What a culture finds pleasing about the body often reflects the aesthetic views and self-images of those who dominate that culture and control its representations. Thus there is nothing innocent, neutral, or self-evident about the concept of "beauty," which comes to the beholder laden with reigning ideas about race, ethnicity, age, class, and gender. Yet just as beauty ideals function as boundary markers that divide people from one another, techniques of beautification hold out the promise of transforming oneself to achieve, or at least approximate, the aesthetic ideals that might facilitate membership in this or that group, and thus to enjoy the status and material benefits that such membership may confer. Beauty and concepts of the ideal thus constrain the individual while providing, within limits, avenues for self-creation. This seems to be an inescapable aspect of the modern condition, where, as Anthony Giddens explains, body care is "a core part of the reflexive project of self-identity. A continuing concern with bodily development in relation to a risk culture is thus an intrinsic part of modern social behaviour."[1] As this chapter shows, however, the twentieth century's ideal body is not merely a matter of *appearance* but is equally bound up with notions of *performance* that have increasingly demanded that individuals extend their physical capacities beyond what would have been considered "normal" or "healthy" in earlier centuries, notably in the domains of sexuality, labor, and athletics. The "natural" and the "cultural" are thus blurred as bodily appearance and performance become increasingly subject to technological interventions and enhancements. This chapter addresses the most dominant transformations in Western bodily ideals since the 1920s, paying special attention to the face and head, slenderness and the ideal shape, muscularity and fitness, and the problem of bodily performance in sexuality, labor, and sport, where the "natural" limits of the body are pressed even further into undecidability.

LOOKING AT THE FACE

The Great War, which had ripped bodies and minds to pieces for four years, played a significant role in encouraging a new emphasis on beautiful and whole bodies, the exact opposite of those often associated with the war itself. Handsome and well-built youths who rushed off to the "adventure" of armed combat might have reasonably counted on dying young, but the romantic hope of also leaving behind a beautiful corpse was frustrated by the shocking damage that artillery shells, machine guns, and poison gas could inflict on the body. This was especially true of heads and necks, which were particularly vulnerable in trench warfare. Despite the publication of books like Ernst Friedrich's pacifist *War against War* (1926), which through several editions featured graphic photographs of the mutilated faces of German soldiers, the 1920s witnessed a considerable denial of the effects of war and an aesthetic sensibility that was offended by reminders of human ugliness. This was manifested most vividly in the pleasure-seeking ethos that characterized mainstream urban culture of the time, which emphasized youthful good looks and slender figures. Still, the unaesthetic effects of war could not be completely denied. An estimated 11 to 14 percent of French war wounds affected the face and, if the patient survived, required sometimes massive reconstruction to help remedy the "double violence" this wrought both to his body and sense of self. In neither France, Germany, nor America was reconstructive surgery conceived as anything less than essential for reintegrating wounded men into society, even if ultimately they could at best "pass" only as members of the fraternity of what the French called *les mutilés de guerre*.[2] This emphasis on facial reconstruction as a social and psychological necessity provides a way of approaching the importance of a pleasing face in the twentieth century, for in the decades to come increasing numbers of civilians would seek to transform their appearances in a variety of ways.

The Western world has long been fascinated with the beauty of the face, which has often been considered a way of accessing the moral and personal qualities of the inner person. The ethnocentric and frequently racist underpinnings of Western approaches to the face are apparent in the anthropological depictions of African, Asian, and indigenous peoples since the eighteenth century, where external beauty was measured against Greco-Roman standards and taken as reliable evidence of mental and moral qualities.[3] Of course, many

Europeans failed to measure up to their own ideals, and, from the late nineteenth century on, changes in urban modernity placed an even greater emphasis on personal appearance. An unavoidable fact of city life, face-to-face encounters began to play an increasing role in business encounters. During the same period, attention to appearance, manners, and fashion became critical tools for members of the working class and petty bourgeoisie seeking to make their way up the social ladder. If in earlier years "character" was thought to be an inner quality that manifested itself in the face and body, from the 1920s on references to this inwardness became less common as "personality" was cited as a key to sexual, social, and professional success. Possessing and maintaining a youthful and pleasing face was thus a key component of the impression management needed for professional and romantic success.[4]

Hollywood and other mainstream films certainly reinforced this notion, not least through the innovation of close-up shots, and kindled among women in many countries the belief that being pretty, slender, and charming was the key to happiness.[5] A reliance on artifice was further promoted through the commercial cosmetic products that became more widely available and acceptable during this time. Once frowned upon as immoral, these products posed a decisive challenge to traditional concepts of natural beauty. Historically, the beautiful body was one that presented itself as being inherently appealing while concealing the effort expended to create that beauty. Bound up with the imperative to purchase and apply makeup in a way that seemed "natural" was a broader cultural development whereby men and women were trained to see makeup as a normal aspect of the female face.

This subtle naturalizing of the artificial undercut the very notion of a purely natural beauty; yet cosmetic use could have other unintentional consequences. Ironically, cosmetics designed to enhance beauty and conceal flaws also probably promoted adolescent acne, a skin disorder that had long been associated with dirt, microbes, and even illicit sexuality. In 1920s and 1930s America, teenage girls seeking smoother skin could take advantage of facials at most beauty shops, where their skin would be stimulated and any bumps or blackheads removed or peeled away. Dermatologists who condemned such practices responded in the 1930s by recommending estrogen injections and even X-ray treatments to eliminate unsightly pimples. From the 1940s on, acne was the "plague of youth" and a menace to anyone whose professional and romantic success hung on an attractive appearance. Although cultural shifts in the 1960s and 1970s often promoted more "natural" beauty and condemned the artifice of cosmetics, the beauty industry responded by offering cosmetic products promising to create that natural look.[6]

While the male use of cosmetics had mostly gone out of fashion in the late eighteenth century as a sign of "effeminate" noble manners, the growing emphasis on achieving a winning personality made it essential for a man to take

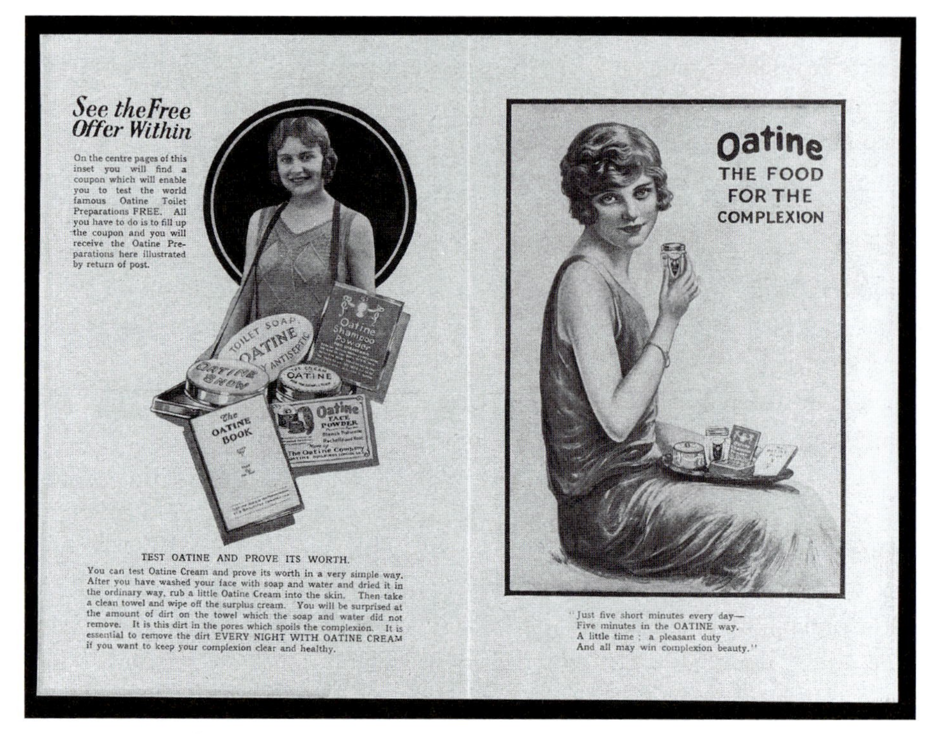

FIGURE 6.1: Leaflet and coupon for Oatine Cream and Oatine Snow facial cream for night and day use. Wellcome Library, London.

particular care of his face, an imperative that made it possible for some American men to experiment discreetly with cosmetics. As a writer for *Fortune* magazine noted in 1937, in big cities certain barbershops afforded men a very similar treatment to what women might expect from a beauty salon, "without seeming faery (except perhaps to the barber)." Another business writer reported on a successful and otherwise "HE-MAN" salesman who carried a small cosmetics kit with him and even used eyeliner to enhance his appearance. Such assurances did not alleviate the embarrassment that most men experienced about such practices, though there was arguably a conspiracy of silence between men who wished to look more attractive and the shops that provided the appropriate services and products. Although explicit cosmetic use would remain rather covert among most men, from the 1930s on a preoccupation with appearances was promoted explicitly in men's magazines that devoted considerable advertising space to an array of products promising to enhance the face and hair, including shaving cream, razors, and hair products. The American magazine *Esquire* is often described as a trendsetter in this regard, but from the 1930s on one can discern similar tendencies in publications like *Adam: Revue de l'homme*, *Man: The Australian Magazine for Men*, and, in Britain, *Men Only: A Man's*

Magazine. However steeped these ads were in "masculine" imagery, it is clear that care for the face and hair was considered a matter of professional and romantic survival for men as it had been for women. Brief challenges in the 1960s and 1970s in the name of natural beauty for men and women did not do away with this idea, even in African American culture where the "black is beautiful" campaign encouraged a celebration rather than rejection of ethnic distinction.[7] If anything, the urge for personal transformation has grown stronger since the 1980s as cosmetics companies successfully marketed moisturizers and other skin-care products to growing numbers of men.[8] Although a tension still persists between "rough" ideals of manhood and their more urbane incarnations, the type of men who have been recently dubbed *metrosexual* merely represent a further development of trends established decades earlier.

The surgical alteration of the face remains the most extreme solution for those wishing to more closely approximate socially accepted ideals, and one of the most controversial. While some moralizing still circulates around this practice, to dismiss cosmetic surgery wholesale as a matter of vanity is to ignore the element of social survival inherent in the twentieth-century search for beauty. During the 1920s and 1930s, many early innovators of the procedure, like the German-Jewish surgeon Jacques Joseph, were ethnically marginal individuals who recognized the dangers of having one's "otherness" readily apparent in distinctive facial features. Joseph is notable for the many rhinoplasties he performed on Jewish patients seeking to "pass" in an increasingly anti-Semitic environment and thus to compete with non-Jews on more equitable terms.

The desire for socioeconomic and romantic success also motivated individuals to alter their appearance. France's leading cosmetic surgeon of the 1920s, Dr. Suzanne Noël, was an early advocate of eyelid surgery and face-lifting who recounted numerous cases where clients came to her because they felt their looks prevented them from finding either suitable work or marriage partners. The growing professionalization of cosmetic surgery during the next few decades contributed to its legitimacy, just as the expanding global reach of American visual culture fueled the demand for surgical facial alterations among certain groups, including East Asians seeking eyelid surgery, African Americans wishing to change their noses or lips, and Caucasians fearing they too closely resembled these and other minorities. By the 1990s Dutch women who sought but were unable to afford cosmetic surgery even found themselves covered under the national health insurance system, as long as their appearance could be "scientifically" determined to fall "outside the realm of the normal." Regardless of these developments, mixed reactions to cosmetic surgery persist, and the growing (but still quite small) percentage of men seeking to surgically alter their faces still risk being stigmatized as effeminate for their preoccupation with appearance.[9]

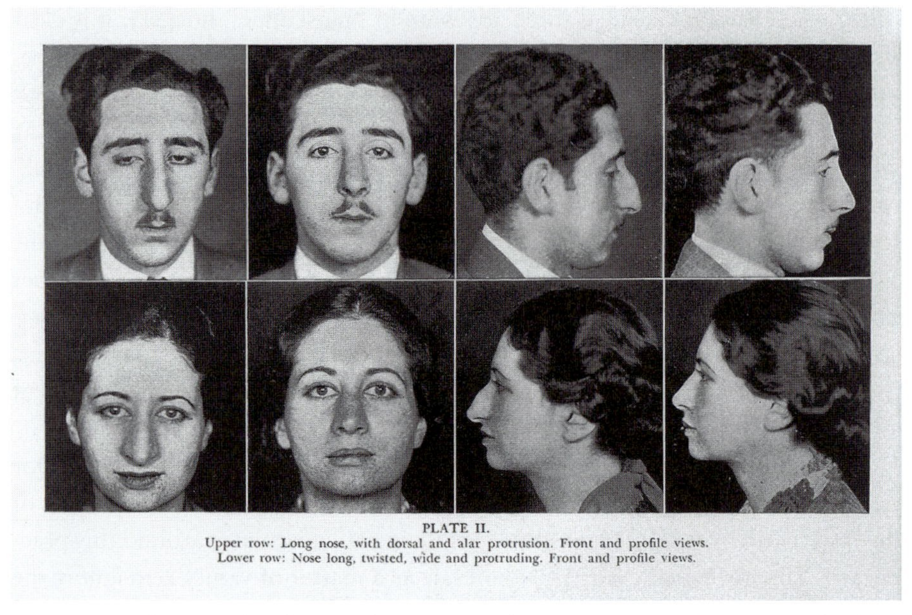

FIGURE 6.2: Photographs of noses, front and profile views. From J. J. Eastman Sheehan, *Plastic Surgery of the Nose* (New York and London: Paul Hobber, Harper and Brothers, 1936), p. 31, plate II. Wellcome Library, London.

SLENDERNESS AND THE IDEAL SHAPE

As the culmination of developments put into play decades earlier, the ideal body shape for men and women had changed considerably by the 1920s. Among women, the Victorian hourglass shape that (with the aid of corsets) emphasized large breasts and buttocks was replaced by a shape suggesting a more youthful and exercised body with no apparent need for supports of any kind. Aware of this shift, manufacturers redesigned their products to be less conspicuous, thus creating a corsetless figure that kept pace with current fashions. This purportedly more natural shape emphasized straight lines and a tubular figure that were incompatible with large or even medium-sized breasts, which needed to be concealed through bandeau brassieres or tubular one-piece undergarments. (Indeed, even though the traditional whalebone corset had largely disappeared by the 1920s, some form of support undergarment was worn by most women until the 1960s.)[10] In addition to being considered beautiful, slenderness conveyed moral qualities that suggested a conquest of the self through willpower alone, at least as it pertained to exercise and dietary discretion. The tight grip on the body that was once exercised by the corset was now replaced by what health reformer Bess Mensendieck called a "muscle corset," a purely "natural" accessory crafted and applied by the woman herself.[11] Aside from this growing emphasis on personal responsibility for one's body shape, the more slender female ideal was accompanied by a growing desire to reveal flesh that had hitherto been concealed beneath clothes. As hemlines rose during the 1920s, women's legs were increasingly exposed to the admiration and scrutiny of others. The perfect leg, explained a French beauty guide, should be "long, plump, supple, sleek, white and in proportion to the rest of the body. . . . Fatness or thinness of the leg is contrary to beauty."[12] More flesh would be revealed in the decades to come as bathing suits gradually became smaller to reveal backs, tummies, and cleavage, thus showing off the body while leaving very few places to conceal imperfections. Since the 1960s topless and even nude sunbathing has become more acceptable on European beaches, but here, too, norms based on age, aesthetics, and comportment are typically observed. In France it is acceptable for women to be topless if they are under forty-five, do not possess overly large or pendulous breasts, and remain prone while their breasts are exposed.[13]

In an era that promoted consumer hedonism more than ever before, balancing self-indulgence with self-restraint was a difficult feat to achieve, and many understandably fell short of the ideal. Moreover, the emergence of a slender and more sensual female body was not an uncontested development, and at least through the 1950s modern women sometimes found themselves caught between the sexual emancipation that validated greater attention to beauty and natalist associations of female bodies with motherhood. Objections to the slimming down of women had been voiced since the 1880s, when novelists and artists deplored how the New Woman seemed to be forsaking her reproductive "destiny" for selfish pleasures.[14] During the 1920s and 1930s, French women were encouraged to exercise moderately but not to the extent that it would jeopardize their roles as mothers, an insistence that was even more strongly posed by the stridently pronatalist Vichy regime of the 1940s.[15] Opposition to the cult of slenderness was also voiced in fascist Italy, where a slim figure was associated with the dangerous "crisis woman," usually depicted as a masculinized slave to the fashions inspired by Paris and Hollywood. During the 1930s the Italian state tried to redefine canons of female beauty by encouraging purportedly more authentic female bodies: Like earlier ideals, these were peasant images of rounded flesh and fecundity that bypassed modernity to reinforce the link between woman and maternity. On this point official ideology foundered as more influential ideals were promoted in commercial culture. Voluptuous, slender, and sexually emancipated bodies were the stuff of Italian beauty contests and magazine covers well into the 1940s.[16]

Over the next few decades the female body would undergo further transformations in terms of size and shape. Large breasts and an hourglass figure became prominent once more in the late 1940s and 1950s, and during this time manufacturers identified teenagers as a market ripe for training bras and other products to enhance their appearance. This development raised concerns about the negative effects of instilling in children potentially unattainable ideals, and some critics pointed to the feelings of inferiority and low self-esteem that the psychologist Alfred Adler had warned about years earlier. One child psychologist took aim at the images offered to children in comic books. In numerous articles and his book-length indictment of the genre, *Seduction of the Innocent* (1953), Fredric Wertham expressed concern about the effect that these images, as well as advertisements, had on the body images and self-esteem of young readers. "Comic-book advertisements give children the idea of scrutinizing themselves in a mirror, to look for anything they should worry about." A definite sense of inferiority was being instilled in girls, Wertham argued, who would inevitably compare their own developing breasts to the fantastic conical protrusions of the typical comic book female. But breasts were just the tip of the iceberg when it came to exploiting adolescents' self-consciousness about their own bodies. A clear complexion marred by pimples or blackheads, it was suggested, could be remedied through any number of products, as could weight gain or weight loss, and just about any deviation from the ideals presented in the cartoons themselves.[17] Depictions of the bodily ideal (as well as the products offering to facilitate it) were reflected in many quarters of popular culture, making representations of the ideal body virtually inescapable.

Faced with the daunting prospect of changing social attitudes or changing themselves, many women opted for the latter. American surgeons identified *hypomastia* and even *micromastia* as "deformities" that brought discomfort that was more psychological than physical and that could be addressed by reshaping the breasts themselves through fat grafts and sponge implantations. Just on the horizon were silicone injections, which were increasingly offered and demanded in the 1950s and 1960s, and the silicone gel implant, first performed in 1962 and, with frequent modifications, a consistent favorite to the present.[18] By the 1960s and 1970s psychological objections to the cult of female beauty had become politicized as the women's movement condemned how female bodies had been perennially subjected to dominant ideals of beauty and slenderness while being marginalized in most other areas of society. Defiantly exposing breasts and burning bras became some of the most frequently cited expressions of female protest against the weight of beauty ideals. Fat, too, became a feminist issue during this time as the seeds were sown for the size-acceptance movement that would later rally overweight women (as well as many men) against discrimination based on size and appearance (often while dismissing medical warnings about the health dangers of obesity).[19]

Despite these objections, the cultural demand for thin yet shapely female bodies has continued. Changes in the ideal female body are quantifiable as well as cultural. If beauty contests are any indication of changing ideals, the winners of the Miss America pageants of the 1920s were heavier and had smaller breasts than those of today, though their waist-to-hip ratios were largely the same.[20] Nevertheless, it is also clear that the female ideal has become significantly slimmer and more androgynous since the 1960s, with the waiflike fashion model Twiggy (Leslie Hornby) usually cited as the most extreme example: She weighed a mere forty-one kilograms (ninety pounds) in the mid-1960s.[21] A recent study of *Playboy* centerfolds from 1953 to 2001 reveals how their body mass has descended "below corresponding population levels, whereas their typical waist:hip ratio now approaches population levels. In sum, centerfold models' once shapely body characteristics have given way to more androgynous ones."[22] Dismal testimonies to the pervasiveness of this slender ideal are to be found in the bodies and psyches of anorexic and bulimic girls and women (and occasionally men), where the attempt to control their flesh often results in emaciated bodies that fulfill neither female nor male aesthetic ideals. In the early twenty-first century, dieting and exercise continue to be essential attributes of the project of bodily "fitness," for both men and women.

MUSCULARITY AND FITNESS

Although manhood and muscularity have often been linked in the West, the emergence of a more muscular physical ideal is traceable to social and economic transformations that began in the late nineteenth century and accelerated from the 1920s on. Anxieties about the degeneration of white male bodies during the 1890s were closely linked to concerns that men were becoming "overcivilized" through modern society's emphasis on education, manners, material comforts, and sedentary lifestyles. These "softening" elements of modern civilization were exacerbated by the increasing visibility of women in the modern workplace and the strengthening of suffragist movements, all of which encouraged many men to reinforce their sense of masculinity by exaggerating certain bodily features. The compensatory function of muscles was especially evident among the educated sons of working-class parents, young men who often sought access to middle-class culture through low-paying clerical jobs that carried little status or potential for advancement. Among such disaffected men, both the spectacle of muscularity as well as actual bodybuilding provided means of celebrating the muscular robustness of their fathers while ridiculing the physical weakness that seemed to characterize solidly middle-class manhood. This tension between middle-class and lower-middle-class approaches to muscularity is evident in Germany, where the latter arguably turned to muscularity as a means of compensating for dull white-collar lives while taking a swipe at the more culturally privileged educated middle class. Making a virtue out of a necessity, these men found in the developed body an alternative form of social distinction that condemned the muscular atrophy that sedentariness and excessive study seemed to generate. To some extent, then, idealizing the muscular male physique was a euphemized form of class struggle.[23] Among others, as was the case with Jewish men, the muscular body was also a means of collective self-assertion: In a culture that dismissed them as overly cerebral, weak, and nervous, physical development allowed Jewish men to "pass" at a time that increasingly focused on more robust models of masculinity.[24] As physical labor became less central to work in the twentieth century, the muscles that signified strength were forged out of feelings of weakness and exclusion. As the usefulness of muscles decreased throughout the century, their role as spectacle came to the fore.

Despite these developments, the extraordinarily well-developed body was not a universally accepted ideal. Instead, many men were content to slim down as the antiobesity campaign so often directed at women was increasingly waged against male fat, which was often associated with a "softness" that undermined masculine willpower and autonomy.

A chorus of opposition to fat men was voiced by health reformers, and women, too, increasingly demanded that men should take care of their appearance. "Men who are nearing middle age take heed," warned the *Australian Women's Weekly* in 1933, "if you have an attractive young wife and wish to keep her! Look out for that ever-spreading waist-line and double chin which usually accompanies it. . . . Mr. Husband should take stock of himself before a mirror—a full length one!"[25]

Some men turned from dieting and modest exercise to more ambitious bodily projects. Bodybuilding is a good example of a "sport" that experienced

FIGURE 6.3: Two paths to physical perfection. *Adam: Revue de l'homme*, no. 8 (November 1926).

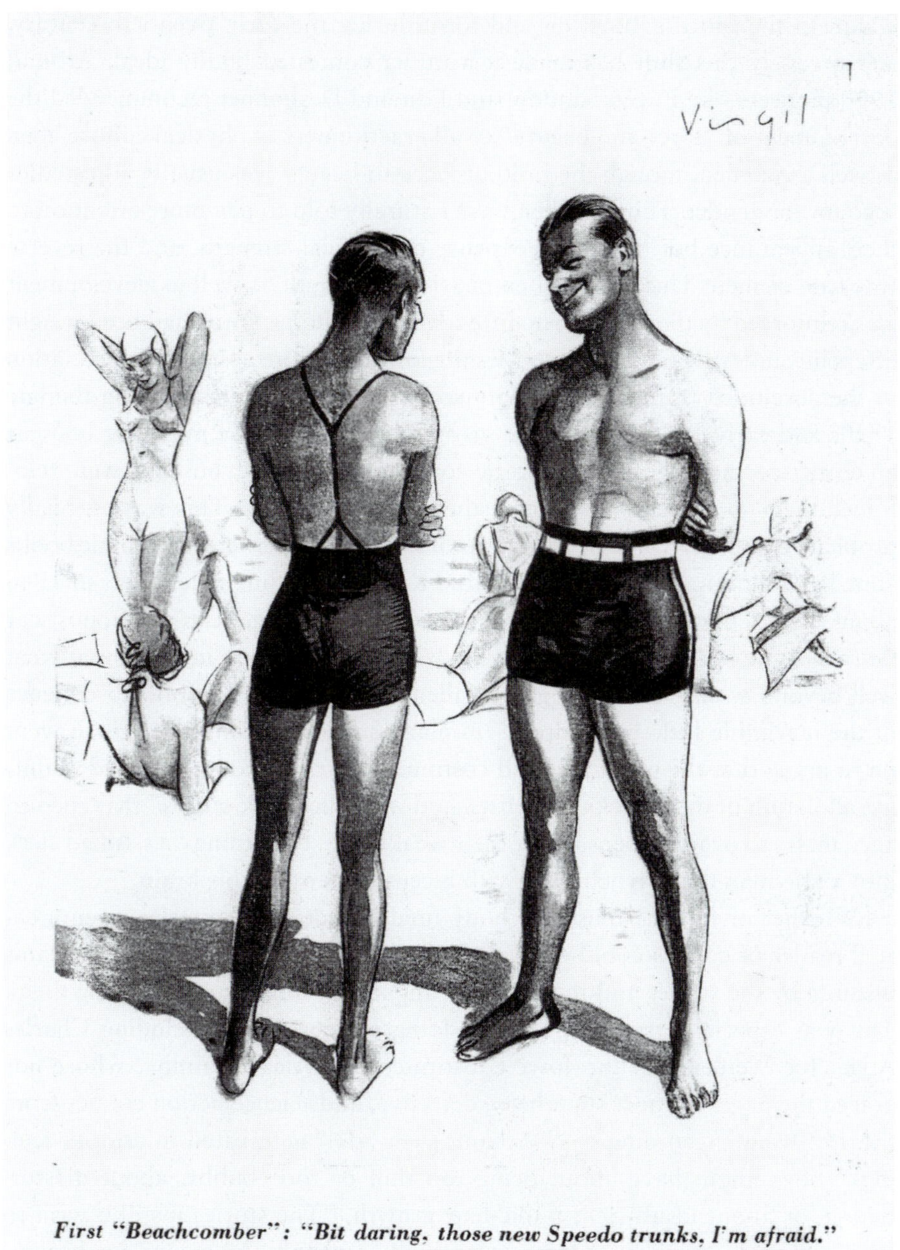

FIGURE 6.4: Speedos advertisement. *Man: A Magazine for Australian Men* 3 (December 1937): 106.

a shift in the ratio of function and form during the early twentieth century, and precisely this shift has made it a rather contested bodily ideal. Around 1900 pioneers like Eugen Sandow and Edmond Desbonnet recommended the development of "force and beauty" to all practitioners of physical culture, men as well as women, though the ratio of force-to-beauty was usually adjusted to account for gender. Thus, women were naturally told to pay more attention to their appearance but not at the expense of physical strength, and the reverse was true of men. This practical emphasis on strength as well as development was reinforced by the fact that many early bodybuilding gurus had been weight lifters in vaudeville or the theater. A shift toward a more aesthetic appreciation of the developed (rather than simply strong) body took place during the late 1920s and early 1930s and placed greater emphasis on the muscular body as an erotic spectacle.[26] The homoerotic connotations of bodybuilding were reinforced through this emphasis on aesthetics over function. This was especially problematic when it came to depictions of muscular superheroes in comic books aimed at children. In the wake of Alfred Kinsey's startling revelations in 1948 about the extent of same-sex experiences reported by American males, postwar moralists saw homosexual elements all over the comic book industry that went well beyond what Gershoy Legman called "the explicit Samurai subservience of the inevitable little-boy helpers" in many superhero comics. Legman went on to argue that the physiques and costumes of superheroes generated "fainting adulation of thick necks, ham fists, and well-filled jock-straps" that encouraged males to overcompensate for their weaknesses by turning "a sissified clerk into a one-man flying lynch-mob with biceps bigger than his brain."[27]

Whether or not the muscular body fired homoerotic fantasies, even a casual reader of comics could not miss the connections between the daydreams inspired by the stories and the products apparently aimed at actualizing them. This was especially true with bodybuilding entrepreneurs, including Charles Atlas, Joe Weider, and the Jowett Institute of Physical Training, whose ads graced the pages of most superhero, detective, and science fiction comics from the 1920s on. According to Wertham, such ads "aggravated inferiority feelings" boys might have about being too thin or too chubby, about masturbating, or about insufficient pubic-hair growth. "The stories instill a wish to be a superman, the advertisements promise to supply the means for becoming one."[28] Others in the immediate postwar era were concerned about the association of hypermuscular bodies with the fascist thuggery that had been only recently quelled, as is evidenced by the controversy that swirled in France around the very popular *Tarzan* comic books. Here, the nearly nude hero was condemned in part for his brutal manner of resolving problems with his fists, a policy that seemed far too redolent of the tough tactics associated with fascists less than a decade before. Memories of the fascist era were pushed under the carpet in Italy, where American bodybuilders often starred in the more than

300 sword-and-sandal or *peplum* films of the late 1950s. For the *peplum*'s largely southern, peasant and proletarian male clientele, who had witnessed the migration of muscular labor from the rural south to machine-based jobs in the industrial north, the spectacle of muscles was "an affirmation of the value of strength to an audience who was finding that it no longer had such value."[29] The erotic potential of these spectacles was also part of their wider popularity: While *peplum* films deliberately catered to female as well as male viewers, their unintended appeal surely extended as well to homosexuals and bisexuals.[30]

Despite bodybuilding's ambiguous relationship with heterosexual manhood, the last quarter of the twentieth century witnessed a definite bulking-up of the ideal male body and greater acceptance of exercise (and anabolic steroids) as a way to achieve it. In the 1970s champion bodybuilder Arnold Schwarzenegger was instrumental in bolstering the virile and heterosexual image of muscular development and helped make the practice more respectable as a mainstream male ideal (though this did not diminish the appeal of bodybuilding for gay men, many of whom found in physical development a way of countering associations of homosexuality with bodily weakness). Since that time muscular male ideals have become more pronounced in commercialized boy culture. American action figures like G.I. Joe and the male heroes from *Star Wars* have clearly been "juicing" since they first appeared in 1960s and 1970s, so that whereas decades ago their bodies might have more closely approximated those of fit (but otherwise ordinary) men, today one is more likely to find toys with unrealistically slender waists and equally implausible chests and biceps (a tendency that parallels the fantastic proportions of girls' toys, notably Barbie). Similarly, a recent study of male centerfolds in *Playgirl* magazine reveals a significant increase in muscularity (about twenty-seven pounds worth!) from 1973 to 1997. The authors of this study speculate that, along with the increasing availability of anabolic steroids, the expanding role of women in society has played a significant role in this development, "leaving men with only their bodies as a distinguishing source of masculinity."[31] Others suggest that such obsessive cultural interest in hypermuscularity reflects anxieties about the perceived "emasculation" of nations like the United States, particularly after the Vietnam War, as the popularity of the *Rambo* trilogy of films (1982–1988) may suggest. To these sociohistorical explanations one might add that the desire for muscles need not spring only from matters of self-presentation, for in psychological terms muscles also serve an insulating and defensive function for a potentially frail sense of self, a point vividly illustrated in Sam Fussell's *Muscle: Confessions of an Unlikely Bodybuilder* (1991). In technologically advanced societies, where extreme muscular development has scant functional value in everyday life, the look and feel of strength may be attractive for the aura of control and invulnerability they seem to confer.[32] As a bodily ideal, then, the muscular physique surely conceals more than it reveals.

THE PERFORMANCE PRINCIPLE

"Faster, higher, stronger": Pierre de Coubertin's influential motto for the burgeoning Olympic movement remains a useful corrective to views of the twentieth-century bodily ideal as being primarily a matter of surfaces and appearances. When it was first expressed in the late 1880s, this vision of a body whose capacities seemed to defy physiological limits was controversial in the eyes of many physicians and health reformers, most of whom were more concerned with the functioning of the "normal" body under ordinary circumstances. But theirs was a world in the throes of change, and "ordinary" circumstances were clearly in a state of flux. As most commentators noted, often with anxiety, the pace of modern life was accelerating at a dizzying rate, at least for those who could recall more relaxed times. The twentieth century was ushered in by a popular mania for record-breaking performances, by both machines and men, that in many respects mirrored the quickening pace of modern life generally. Futurist aesthetes like F. T. Marinetti were among the more memorable proponents of faster and more efficient bodies to go with the new times: "We believe in the possibility of an incalculable number of human transformations, and without a smile we declare that wings are asleep in the flesh of man."[33] Whether exhibited in academic, military, musical, sexual, or sporting contexts, the twentieth century gave birth to what John M. Hoberman aptly calls "an ideology of uninhibited performance" reflecting "the sheer ambition to improve performance in the absence of any restraints upon this ambition."[34] The modern culture of machinelike efficiency thus promotes a bodily ideal of boundless energy and expert performance in most spheres of life.

The notion of performance assumed a double valence in the 1920s, suggesting at once role-playing in the service of dating and seduction as well as the carrying out of specific bodily tasks, in this case a spectrum of sexual acts. Here again Hollywood provided a range of models for potential emulation: Just as many women copied the makeup and hairstyles of starlets, young men actively studied the physiques and methods of stars like Rudolf Valentino and Douglas Fairbanks to enhance their attractiveness and dating techniques.[35] "Understudy" of this sort was picked up by men and women throughout the West, and many learned their lessons well. Some British women who frequented films in the 1930s even agreed that they now had a more discriminating view of dating and lovemaking, which one had come to see as "more of a technique than as an outcome of emotions."[36] Thus, women too were in a position to place certain demands on men, though it is likely that many of them would have also experienced pressure to comply with new standards of acceptable pleasure.[37] Advances in the science of sexuality resulted in a greater rationalization of coitus, both as a pleasurable sensation and as a potential act of reproduction. The growing acknowledgment of female pleasure on the part of sexologists like Havelock Ellis, Marie Stopes, and Theodoor van de Velde also altered how men approached their performance of sexuality, which now called for greater attention to the techniques of foreplay and their partners' pleasure, though usually within the context of marriage. If a century earlier self-control might have compelled men to refrain from sex, by the early twentieth century the site of restraint seems to have been relocated within the act itself, with a man's ejaculation ideally postponed until his partner achieved orgasm (or pretended to, as the case may be—another aspect of erotic performance).[38] In Germany sex therapists like Dr. Ludwig Levy-Lenz encouraged men to employ both tongue and finger in their efforts to overcome female frigidity and to achieve the ultimate goal of simultaneous orgasm. "The sexual personality of a woman," declared another German therapist, "unfolds only under the hands of the man."[39]

Alongside the expert play of fingers and tongues, a growing concern with penis size and performance seems to have trickled upward from its traditional association with the lower orders and "savages" to become the preoccupation of the middle-class males who eagerly embraced aspects of what Kevin White calls "underworld primitivism."[40] If many married men's inability to stave off climax for the full twenty minutes recommended by sex reformers like Stopes generated considerable feelings of ineffectiveness, so too did the problem of impotence, complaints about which seemed to be on the rise throughout the century.[41] As the allure of youthfulness pertained to energy as well as beauty, the aging body was disadvantaged in terms of performance as well as devalued as an image. In the 1920s and 1930s, rejuvenation through a variety of techniques was promised for middle-aged and elderly men, as well as menopausal women, seeking to recapture their youthful vitality. Serge Voronoff's famous "monkey-gland" procedure attracted many men to Paris to have the testicular extract of monkeys grafted onto their own testes, just as Eugen Steinach later lured others (like Sigmund Freud and W. B. Yeats) to his Viennese clinic with the promise of bodily rejuvenation through a simple vasectomy.[42] At the same time, the emerging fields of gerontology and geriatrics discarded traditional questions about why one grew old to focus on the process of aging itself, and by the 1940s the scientific study of aging proceeded alongside social services for the elderly.[43] Midcentury claims that male sexual impotence was primarily the result of psychological causes gave way during the late 1980s to the urological concept of "erectile dysfunction," a thoroughly organic "disease" whose frequency among aging men was thought to constitute an "epidemic" and thus a major public issue. Heralded as a solution to nearly all male sexual problems, Viagra (sildenafil citrate) was marketed from the late 1990s on through the language of machinelike power and performance, a prospect so sexy that Viagra quickly transcended its therapeutic purpose to become a party drug in certain circles.[44]

However often sexual performance has revolved around the production of pleasure, in many cases this apparently hedonistic aim has been enlisted for the preservation of marriage and the reproduction of citizens. As Western countries addressed the decline of their populations after World War I, greater emphasis was placed on pregnancy and childbirth, and many states used welfare and other incentives to encourage women to have more babies. Having already been colonized by male medical expertise early in the nineteenth century, the experiences of conception and childbirth continued to be medicalized throughout the modern era as the cultural tendency to downplay female agency in these domains persisted. Hospitals were increasingly presented as the only legitimate and safe place to deliver babies, and within the hospital walls the experience of childbirth was broken down into carefully monitored stages and specialized rooms through which women moved, as if on an assembly line, from the time they arrived at the reception desk through the delivery room and finally the postdelivery ward.[45] Thus, if sexual emancipation encouraged women to seek pleasure, many contended that procreation should still be the ultimate aim. In other contexts, women's pleasure could be downplayed altogether. During the 1940s and 1950s, marital sex manuals even advised American women to approach their amorous behavior as they would their cooking: as a means of selflessly offering comfort to their men, whether or not they achieved climax themselves.[46] It is therefore significant that when, in the 1970s, the women's movement promoted greater control by women over their own bodies, it took careful aim at the subordination of female pleasure to the reproductive demands of male-dominated states.

In the twentieth century, the model of the efficiently performing body was not merely that of a sexual athlete or baby maker but a tireless worker who placed efficiency and productivity above other goals. Conceptualized in the nineteenth century as a thermodynamic motor whose stock of energy was always capable of being dissipated through fatigue and other expenditures, the working body remained the target of reform efforts well into the twentieth century. Social theorists and medical specialists tried to solve the riddle of maximizing productivity without causing exhaustion. The machine provided a perceptual model for a range of human activities, and in American society of the 1910s and 1920s "efficiency" was repeatedly contrasted to "waste" as reformers sought to streamline social and industrial life along mechanical lines. Drawing on the time-and-motion studies of earlier decades, the father of scientific management, Frederick Winslow Taylor, revolutionized factory production through his careful scrutiny of how workers performed tasks and insistence on the elimination of extraneous movements that impeded efficiency and diminished productivity. Taylor's dream was for a society in which all personal, social, and professional contexts were subjected to scientific management.[47]

Along with the assembly-line innovations of Henry Ford, Taylorism was quickly exported to countries like Germany, where the drive toward American-style industrial methods gained popularity from the 1920s on. The search for the optimal worker was undertaken with particular fervor in the Soviet Union, where "Stakhanovite" superworkers comprised those unskilled or semiskilled workers who, with little education or political clout, had once counted among the "little men" of an earlier period. Among such heroes physical strength was complemented by courage and daring to push themselves beyond normal production quotas, whether this daring was said to flow spontaneously from within or, as many Stakhanovite biographies claimed, to have been inspired by one of Joseph Stalin's speeches. Machinelike in their awe-inspiring disregard for limits, Stakhanovite superworkers were not meant to be models of brute strength only but of the harmonious blend of physicality and intellect.[48] Despite this celebration of worker productivity, since the 1970s technological advances have threatened to render the traditional laboring body obsolete, a prospect that seems increasingly plausible as cerebral work and the production of "information" eclipse the need for muscular exertion.

An emphasis on efficiency continues to govern the way in which individuals are meant to manage their bodies and emotions. Advances in the management of pain, not least the development of aspirin in Germany, were bound up with the overriding need for efficiency. "Pain is costly," declared the American youth worker Luther Gulick, Jr., in 1920. "It unfits us for giving attention to other things. It keeps us on a constant strain. It destroys efficiency."[49] Variations on this theme have been the mantra of pain-relief advertising ever since, and today purveyors of aspirin, acetaminophen, and other medications insist that our busy lives leave "no time for a headache." A similar approach has been adopted toward mental illness, which, like impotence, has come to be more firmly situated in biology rather than in the psyche. Psychopharmacology has come to the rescue of people suffering from depression, and drugs like Prozac have often been promoted as means of enhancing individual performance, whether at home or in the office, without necessarily addressing underlying personal or societal factors.[50] Nikolas Rose even suggests that this new emphasis on the somatic or "neurochemical self" targets change at the molecular level as correctable problems rather than problems of the "abnormal" individual, thus emphasizing performance without recourse to the connection of normality with morality.[51]

The world of professional sports is by far the most spectacular location for imagery of the performing body constantly seeking to transcend conventional limits. From the 1920s on, mechanistic models of the body and behaviorist theories of mind were combined in sports medicine's approach to athletic bodies, the rigorous training of which would unlock automatic reflexes and promote the efficient expenditure of energy.[52] Yet if the performance principle often approves of bodies that transcend physiological limits, its inherent masculinism often frowns on the training and competition of women, for whom transcending the limits of the body can also entail a transgression of conventional ideals of womanhood. Early in the twentieth century, women's physical education was delimited by an emphasis on "moderation" in the name of health rather than record-breaking performances; in this way exercise would encourage the slenderness of the female body while promoting the overall health considered essential to childbearing. Through the 1950s and 1960s, advances in women's competitive sports were dogged by accusations that such "mannish" pursuits rendered women unattractive and even made them lesbians, though certain sports (like rugby and boxing) have always been viewed as more intrinsically "masculine" than others. A greater acceptance of female athletic performance has occurred since the 1960s and 1970s, though (among other inequities) gender stereotypes continue to intrude into women's sports. However much the muscular female athlete may fulfill reigning ideals of maximum performance and efficiency, by challenging prevailing gender stereotypes about how "feminine" bodies should seem and act, her success risks being undermined through the violation of competing ideals of female beauty. Different attitudes toward doping during the Olympics reveal this tension: Whereas men who test positive for steroids are denounced as cheaters, women who are caught are more often attacked for having violated gender norms.[53]

Each of the areas outlined in the preceding reflects the modern tendency to fashion bodies in accordance with prevailing ideals and to push bodies beyond what had hitherto been considered their "normal" capacities. This reflects what Zygmunt Bauman sees as the two central features of the modern spirit: "the urge to *transcend*—to *make* things different from what they are—and the concern with the *ability* to act—the *ability* to *make* things different."[54] To be sure, techniques have changed since the dandies of the 1830s resorted to calf pads to accentuate the beauty of their legs or women donned whalebone corsets to achieve the desired shape, but there is a continuity between such practices and those of today, where one's calves (along with many other body parts) may now be surgically enhanced through the use of implants, and twisted or smoothed into more pleasing shapes and textures, especially through new exercise technologies that aim at developing specific muscle groups. Perhaps what most clearly separates twentieth-century bodily ideals from those of earlier periods is not the strong social emphasis that is placed on appearance and performance but the ever-expanding technologies that allow one to alter the body

FIGURE 6.5: Oliver Burston, *Illustration of a Man Running on a Treadmill*. Digital artwork/Computer graphic. Wellcome Images, London.

in hitherto-unexpected ways and that, in turn, further fuel the drive to alter oneself. Accepting the "natural" body as a given is clearly less legitimate in the twenty-first century, where cosmetic surgery is an increasingly attractive option and where elite athletes still subject themselves to extreme dietary and exercise regimens as well as a range of performance-enhancing drugs.

CHAPTER SEVEN

Re-markable Bodies

ANNA COLE AND ANNA HAEBICH

> The body that can be told is not the lived body; the names we give to embodiment are not the experiences themselves.[1]

In the recent historical period in the West (from about the 1920s to the present), a defining characteristic of the cultural experience of the human body is the ubiquitous idea of its mutability. That the body, rather than being destiny, is ours to write, rewrite, and perform—however consciously or unconsciously that performance takes place—is an important feature of both late twentieth-century popular culture and recent feminist theoretical work on corporeality. In both popular culture and academic theory, the body is no longer taken as an immutable fact of nature but more a site of potentialities.[2] Feminist theories of the body since the early 1990s have noted that the "biological" body can no longer be perceived as a closed, monolithic, objective entity, and thus the body shifts from being a complete "object" to an unfinished, "becoming" agent.[3] At the level of the everyday, this can be seen in a widespread gym and fitness culture, including the use of anabolic steroids; in plastic surgery, including genital surgery; in drag acts and transsexuality; in the emergence of a "queer" identity; and in the late twentieth-century efflorescence of tattooing. These are just some of the ways that the body we are born with can be lived as changeable and transformable. In interesting ways, as we explore in the following, this contemporary theme in Western popular culture and academic writing on corporeality reflects cross-cultural understandings and uses of the body that predate it.

Over the past twenty years or so, a new canon of work on the body, driven by feminist theoreticians, has argued against the taken-for-granted assumptions of the nature/culture divide. The nature/culture divide as a binary opposition is implicated with other binarisms that inform an everyday understanding of the world: man/woman, mind/body, sex/gender, sign/referent, West/rest. Given the association of woman and cultural "others" with nature, feminist theoreticians argue that an uncritical acceptance of the culture/nature divide would have critical ethical and political consequences. The theoretical insights of now-canonical writers in this field such as Judith Butler and Elizabeth Grosz were to a large extent driven by a political consciousness formed in the 1960s and 1970s that remained unconvinced, as Vicki Kirby puts it, "that the roles and expectations inherited by men and women and by other socially differentiated groups were natural determinations."[4] In this theoretical work the body is conceived of not as a stable, fixed object or "thing" but as historical, cultural, and intersubjective, endowed with intentionality, as well as a proliferation of unintentional meanings. The idea that so-called natural prescriptions are political designations, historical and cultural values that can be transformed, has had a significant impact on studies of the body and embodiment. There are no "natural bodies," as Elizabeth Grosz writes, only those marked by the "history and specificity of their existence."[5] This chapter on marked bodies of the twentieth and twenty-first centuries is situated within this trajectory of theoretical work on corporeality. Human bodies from all cultures are marked with inscriptions of cultural and social meaning and significance. Corporeal surfaces may be deliberately and irreversibly altered through whipping, scarification, cicatrization, piercing, branding, and tattooing. Other marks, blemishes, lines, wrinkles, age spots, stretch marks, lesions, blisters, wounds, scars, and abrasions are the unintended by-products of the lived body. There are beliefs, symbols, codes, and rules about these markings that can vary through time and across territory, language, race, class, gender, and age. These marks and associated technologies are part of wider cultural systems of bodily care and presentation, linked, in turn, to larger sociopolitical institutions and cosmologies.

While bodies can be indelibly marked by acts such as tattooing and by processes such as aging, it is equally true that the cultural meaning of these marks shifts and changes dramatically depending on the spatial and temporal context. In times of war, for example, bodies that have been marked by violence can have symbolically powerful but highly contested meanings. Ali Ismaeel Abbas, age thirteen, the victim of a U.S. bombing raid on Baghdad that blew both his arms off (and killed his father, pregnant mother, brother, aunt, three cousins, and three other relatives), had his photograph splashed across the world's newspapers in April 2003 as he lay in a Baghdad hospital. To the millions across the world who marched against the war, holding posters showing his limbless body, this young boy's marked body represented one of the most potent and terrible symbols of First World mass terror and the murder of innocent civilians. His body came to symbolize the massive inequities between the Iraqi state and the U.S. and U.K. military-industrial technological might that was responsible for the bombardment of Iraqi civilians. It is unlikely that the same marked body had a similar symbolism for the U.S. and British governments that authorized the bombings. If Ali Ismaeel Abbas's body meant anything to these governments, he more likely symbolized an unfortunate but necessary by-product of a war against terror.[6]

As bodies can be marked by deliberate—or unintended—actions, so are they marked *out* by the apparently natural properties of gender, race, class, age, disability, or disease. Attributes seen as occurring naturally because of gender, class, race, disability, age, and disease have all been used historically, and in the present day, to mark individuals and groups in particular, often pejorative and limiting ways. A historical link between the "disease" of hysteria and the womb (called *hystera* in Greek) meant that women's bodies, in this instance, were marked by an apparently biologically determined irrationality and instability. That women, in general, menstruate, have the ability to develop another body unseen within their own, and can give birth and lactate, has, in Western society at least, historically suggested a potentially dangerous volatility that can mark the female body as out of control, beyond and even against reason.

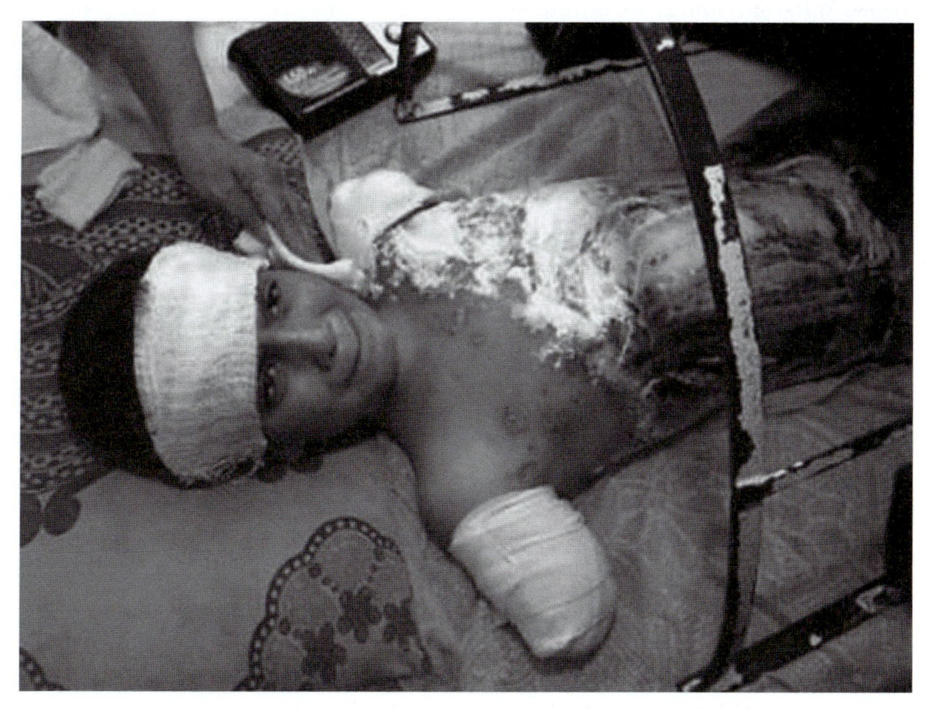

FIGURE 7.1: Ali Ismaeel Abbas after U.K./U.S. bombing raid on Iraq, April 2005. Photo supplied by authors.

Other bodies not considered normative are marked out to mean things that do not tally with the experience of those who live that embodiment. Alison Lapper, the award-winning U.K.-based artist, born without arms and with shortened legs due to a condition called phocomelia, describes how her body, in 1960s Britain, marked her out, wrongly, as having a mental deficiency. Her experiences of living in a body different from the norm, in a "disabled" body, show the dissonance between the experience and meaning of a body considered "marked."[7] In Lapper's case, as her autobiography demonstrates, her body for most of her life marked her out as mentally deficient, asexual, and even grotesque.[8] However, this same body became in later life literally a statuesque symbol of courage and beauty, depicted when she was seven months pregnant in a thirteen-ton, fifteen-foot-high marble sculpture on Trafalgar Square's Fourth Plinth.[9] As one reviewer wrote in an article entitled "Bold, Brave, Beautiful," "Alison Lapper Pregnant . . . brings to mind the classical statues that grace our greatest museums, other sculptures from other times which also have, whether by accident or design, missing arms and legs."[10] The artist who created the sculpture, Marc Quinn (whose previous work is mostly preoccupied with the ever-changing physical states of the body), described his sculpture of Lapper as representing a "new model of female heroism" and stated that her pregnancy "makes this a monument to the possibilities of the future."[11]

It is clear, then, that a marked body, no matter how indelible that mark, means different things in different contexts. In this chapter, we explore the ways bodies can be marked and re-marked across the categories that are sometimes seen to define and confine them. They are the re-markable bodies of this chapter's title. In what follows we employ a series of embodied examples, some controversial, some more commonplace, to constellate questions about how the marks of gender, race, class, age, and disability can shift across culture and time; in so doing, we suggest some alternative readings of marked bodies and of body markings.

CROSS-CULTURAL MARKED BODIES

It is an "anthropological commonplace" that every culture's ideas about the body both reflect and sustain ideas about the broader social and cultural universe in which those bodies are located.[12] Any serious consideration of the rich material on body modification in diverse cultures draws attention to the many different ways bodies are understood, used, and experienced culturally and how, in turn, the different ways in which bodies are conceived culturally affects the "relationship between corporeality and the self."[13] There are a wide array of cross-cultural techniques to mark and reshape the body. Techniques of scarring, piercing, tattooing, distorting, mutilating, smearing, and painting are used on body surfaces and parts including the feet, earlobes, lips, nasal septum, skull, genitalia, fingers, and teeth. This reworking of the body creates rich surfaces and depths that carry social and cultural signification. Recent anthropological research, when used in combination with the insights and analysis of new body theory, offers some profound examples of the different ways that body and culture interact and influence one another. The best of this research suggests, as do the theoretical arguments of corporeal feminism, that the very categories of nature/culture/mind/body need to be rethought.

Kirby's theoretical work on corporeal categories analyzes the Hindu ritual festival of *thaipusam* to illustrate vividly her questions about how the cultural context that surrounds a body can, as she puts it, "also come to inhabit it."[14] As Kirby explains, the *thaipusam* devotee is impaled by metal spikes inserted through the skin and into the organs of the body. The hands, face, lips, neck, belly, and cheeks are all skewered, and the *thaipusam* then walks a considerable distance as part of a religious procession, with many others who are similarly impaled, within an elaborate metal scaffolding that is held up by the many deep piercings and the devotee's own body. "To be skewered by any of these metal prongs would prove at least painful for most of us, and conceivably lethal. Bleeding, scarring and internal injury would be the inevitable results . . . yet for the serious *thaipusam* devotee, none of these effects are realized. The man does not bleed, nor does he scar."[15] How then, asks Kirby, do we understand the impact of the *thaipusam* devotee's culture on the physical act of the ritual? As Kirby suggests, contemplating this remarkable act of marking without scarring, impaling without fatally wounding, confounds everyday assumptions about the static nature of the biological body and its sensations. This example asks us to consider the situational and relational aspects of seemingly natural, biological phenomena such as wounds on the body.

SCARIFICATION

The practice of cicatrization, or scarification, among Aboriginals in Australia raises similar questions and challenges to normative ideas about the body. Early twentieth-century commentators on the Aboriginal practice of scarification expressed amazement at the seeming lack of pain felt by those undergoing this ritual act. Their limited observations inadvertently suggest similar processes of culture "inhabiting the body" to those Kirby describes in the ritual of *thaipusam* and the reception of pain. One contemporaneous source on scarification in Australia noted a "comparative insensitivity to pain" in those undergoing the process.[16] Clearly, such observations need to be placed in the context of frontier colonialism, which preached the insensitivity of Aboriginal bodies to pain as a justification for the violence wrought against them by colonists. Nevertheless, we can speculate today on how the ritual experience of pain, which left the scars of cicatrization, compared, for example, with the pain felt in the acquisition of scars left on indigenous and convict bodies by the widespread use of the lash in colonial Australia.[17] As contemporary analyses of pain, including acute and chronic pain, argue "there is simply no pain that is strictly biological; pain always has meaning, always is 'socially informed.' "[18]

In the case of scarification, the raised scars of cicatrization are made when substances such as plant sap, blood, or charcoal are rubbed into incisions made by cutting the skin deeply with a piece of flint, quartz, or glass or by burning the skin with charcoal. In the Western and Central Desert region of Australia, as Jennifer Biddle has documented, Warlpiri cicatrices (also called *kuruwarri*) are made by packing open cuts with site-specific country (land, ground) to create ridged visual and tactile surfaces that are said to embody and enact ancestral life force as well as creating bodily links between people and named ancestors.[19] Anthropologists working with Australian Aboriginal communities whose members practice ritual scarification, and other forms of body marking, have these in the context of broader social and cultural institutions and cosmologies. Deborah Bird-Rose's work, for example, with Yarralin people in the

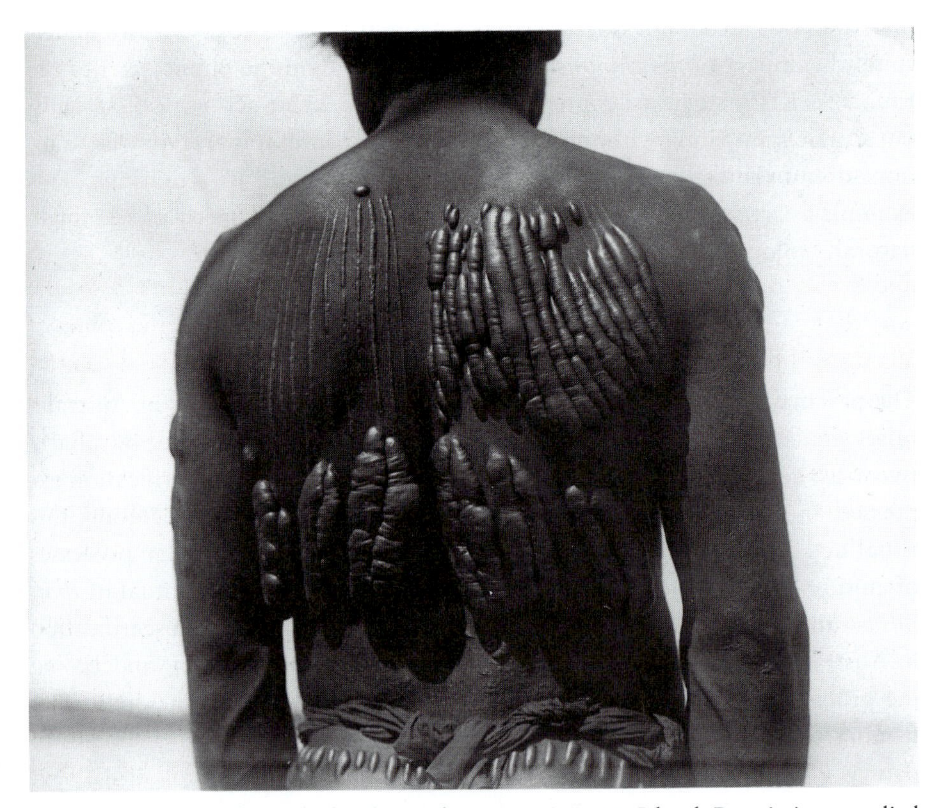

FIGURE 7.2: Australian Aboriginal man from Montgomery Island. Permission supplied by Mowajam community, Western Australia.

Northern Territory, highlights the significance of social relationships between the cutter, the person being cut, and family members:

> Kin are mapped on the human body through sign language; my brother is equivalent to my right calf; my sister to my left calf. . . . Formerly when a boy was made into a young man his sister had narrow lines incised on her calves. Evidence of the brother's states were borne on the sister's body. When a woman's son was made into a young man, scars were incised on her belly, and, scars on her arms indicated her husband's state.[20]

Christine Watson, working with Kutjungka women in the Northeast Kimberley, similarly documents how systems of corporeal inscriptions correspond with sand drawing, ceremony, song, and sign and body language to create complex visual and haptic sensory fields linking Aboriginal bodies with country, kin, and ancestors. Watson emphasizes the body and its surface, the skin, as a site where "the identity of individual Kutjungka people is linked with that of human relatives, ancestors, ceremony and land. . . . A person's relatives all have a place within or on the body."[21] As women mark their bodies with signs of initiation, kinship, marriage, and grief, they recreate geographies of their bodies and create social maps of their world. The cutting of the head and other forms of scarification in mourning and bereavement ceremonies, as well as the male practice of circumcision and subincision, also indicate the degree to which the surface of the body—the skin—is made, like country itself, a text for ancestral imprinting.[22]

Work on the intercorporeal nature of the art of Warlpiri artists from Aboriginal Australia provides another perspective on the work of cicatrization and breaks down familiar categorizations of body/culture. Biddle explores the way that country, skin, and, and in this case, canvas, are made "the same."[23] Via an embodied analysis of the work of contemporary Western and Central Desert artists, she demonstrates the ways that body marking, marks on the land, and marks on the canvas interact in a profound exchange between the individual and the society that surrounds and inhabits him or her. As Biddle explains, "What these paintings demonstrate is a procedural enactment of how it is that canvas, country and skin are knowable—mark-able, make-able—as the same cultural stuff."[24]

Central and Western Desert marks, as she explains, are "indexes" (signs that remain existentially tied to what they "represent") rather than "icons" (signs that look like what they represent). They are indexes in that they "embody original Ancestral potency."[25] This potency does not simply arise. It must be enacted by precise, repetitious, and regulatory operations, what might otherwise be called law.[26] Country (land) is itself understood to be indexical—the product of ancestral imprintation. Sites and places hold precise affiliations and identifications as well as powerful and potentially dangerous forces. "Even if disengaged from the body of the Ancestor these forms, features, marks, places do not cease to retain Ancestral presence. Hence the constitutive force, power and effects associated with marks and mark meaning: to rejuvenate country, species; to control fertility; to regulate social relations and relatedness; to cause illness and to heal and other such effects."[27]

Warlpiri call these marks *kuruwarri* (as they do cicatrices), a complex term meaning "mark, trace, Ancestral presence and/or essence, birthmark and/or freckle." It is not only in country itself that Ancestral visceral presence resides, but these presences (located in certain sites and affiliated with certain species) can enter women's wombs, cause conception, and in turn leave birthmarks, freckles, and other identifying traits of specific kinds of subjectivity on individuated people.[28] Biddle explains,

> The fleshy traces of birthmarks and freckles are indicative of how "skin" is literally and materially the same "substance" as country, in that it is equally a medium in which Ancestral traces reside. Embodied originally by Ancestors, these marks have visceral effects on contemporary humans because they are visceral remains of Ancestors. In turn, they provide a necessary material, intercorporeal means for linking Ancestral bodies to contemporary bodies.[29]

Her research provides a vivid example of alternate ways of conceiving the body/mind/culture/nature "split." Through the work of Warlpiri artists we can see how the "split" does not exist in any taken-for-granted Western sense. The process of mark making involved in the creation of the artworks, and similarly in the creation of scarification, expands our sense of the boundaries and borders of the body and the world outside it.

This embodied example of Aboriginal scarification grounds some of the theoretical insights of the philosophers of embodiment who question any limited definition of the body. The painful process of the marking of scarification returns us to our earlier questions, as in the ritual of *thaipusam*, about how the culture that surrounds a bodily intervention comes to inhabit the very body itself. Or, to put it simply, how is it that the extreme physical sensations involved in the process of cicatrization are borne with an apparent lack of abjection and suffering? Unless we imagine, as those early Australian colonists did, that Aboriginal bodies simply feel less pain than European bodies, we must concede the powerful impact of culture in the experience of wounding, rewounding, and indelibly marking the body.

Pausing for a moment with those imaginings of the early Australian colonists, we can also consider scarification as an embodied example of the way a marked body means different things in different contexts. Scarification, as we have seen, holds a complex of meanings within Aboriginal culture. Permanent scars and keloid cicatrices, deliberately cut into the skin in ritualized processes, engrave social and cultural meaning into Aboriginal bodies and mark them as inextricably part of the social collective. As Elizabeth Grosz theorizes, "Unlike messages to be deciphered, they are like a map correlating social positions within corporeal intensities."[30] From the limited colonial descriptions and contemporaneous anthropological research into scarification, it would seem that this was a map that few from the colonizing society sought to learn to read or understand.

The limited colonial records available suggest that observers were generally not concerned to understand the meaning of the scars, seeing them as largely decorative and peripheral. In part, this reflected Western elite criteria that largely condemned deliberate body marking as both pagan and narcissistic and therefore unworthy of serious study. As we have seen, cicatrization marked the body as inextricably social and as profoundly, and reciprocally, steeped in culture. In a process of colonial reversal, this intense corporeal mark of sociality was interpreted most often by the newcomers as a sign of these bodies' *lack* of "civilization," as a mark of brutality. A scientific opinion of the body marking of Arunta and Luritcha men, for example, held that it was merely "ostentatious display" for "successful courtship of the opposite sex" and for the "charm of novelty."[31] In a familiar trope from colonial contexts around the world, these scientists feminized the colonized men, commenting that while women adorned themselves to attract the opposite sex in the "civilized" world, the reverse was the case in "primitive" Aboriginal society. The pioneer psychologist Stanley Porteus concluded dismissively that Aborigines' "readiness" to submit to the pain of ritual scarring was driven only by the "need of social approbation":

> If hardihood is estimable, then the young man will go to extremes to gain that esteem. Similarly, since it is proper to mourn the dead, the women will go to undue lengths to conform to custom and will cut and gash themselves most terribly to mark their grief.[32]

Their comments reflect widely divergent ways of seeing and reading the marks of scarification depending on cultural context.

SEXUALITY AND THE MARKED BODY

> Is it, after all, unreasonable to be suspicious of Westerners who are exercised over female circumcision, but whose eyes glaze over when the same women are merely facing starvation?[33]

We began the section on cross-cultural bodies by arguing that any serious consideration of the rich material on body modification in diverse cultures draws attention to the many different ways bodies are understood, used, and experienced culturally, and this in turn shows how the different ways in which bodies are conceived culturally affects the relationship between corporeality and the self. The permanent body marking known as female circumcision, clitoridectomy, or pejoratively as female genital mutilation has, over the past twenty years at least, become a controversial area in feminism and "development" studies and holds important theoretical implications for the corporeal feminisms outlined at the start of the chapter.

Western opposition to female genital cutting has a long history, extending back at least to the colonial era. In the 1970s, with the emergence of "second-wave" feminism, it became a contested issue in the United States and Europe. Influential articles by Gloria Steinem, Mary Daly, and others, including Third World activists such as Nawal El Saadawi, condemned the practices.[34] Frances Althaus notes that prior to this concern within feminism, African activists and medical practitioners, in the 1950s and 1960s, had brought the health consequences of female circumcision to the attention of international organizations such as the United Nations and the World Health Organization.[35]

In these depictions of female genital cutting for an international audience, practices became largely severed from their sociocultural, symbolic, and economic context. During the United Nations Decade for Women (1975–1985), female genital "mutilation," as it became known, was a prominent and controversial issue. However, as Christine Walley notes, the response to the ensuing publicity was not what many First World feminists might have expected: "Instead of being congratulated for their opposition . . . they were called to task by some African and Third World women, including a group who threatened to walk out of the mid-decade international women's conference in Copenhagen in 1980."[36] While some of these women themselves opposed female genital "operations" (as Walley calls them), they objected to the way the issue was being publicized, and they called attention to the complex and unequal power dynamics between the so-called First and Third Worlds and First and Third World women. They argued that presentation of female genital cutting fed into value-laden understandings of differences between Africans and Euro-Americans. Such understandings presumed a difference between First and Third worlds that was built on the historical belief in a chasm between "modern" Euro-Americans and "native" colonized others, in this case Africans.

One common trope in much of the Euro-American–oriented literature opposing female genital "mutilation" is the characterization of African women as oppressed victims of patriarchy, ignorance, or both, not as social actors in their own right. As one authoritative "overview" of female circumcision in Africa states, the "reasons given for the circumcision of females, other than 'it is the custom', seem to be consistent in most African societies and are, for the most part, based on myths, an ignorance of biological and medical facts, and religion."[37] As Penelope Hetherington writes, "In much of the [anti–female circumcision] literature . . . women are almost always represented as victims of patriarchy in backward societies yet, at the same time, it is made clear that men are not permitted to take part in the ceremonies and that the practice is controlled entirely by women."[38] As both Hetherington and Kirby point out, this position inevitably leads to claims of "false consciousness."[39] These women are illiterate, ignorant, backward or, in other words, non-Western.[40] Walley argues that this position is "particularly belied by African women's groups, on and off the continent, in relation to female genital operations."[41] And, as Hetherington succinctly writes, "it makes no sense to accuse African women of 'false consciousness' if their personal survival depends on their acceptance of the inevitable and 'natural' links between circumcision, eroticism, marriage and reproduction."[42] In this respect, debates over female circumcision in Africa resonate with colonial discourse on other practices such as *sati* (widow burning), foot binding in China, and the wearing of the veil in Muslim societies, where representations of non-Western women's domination by non-Western men were used to justify British and French colonialism.[43]

As critiques of the literature opposing female genital cutting argue, much of it depends on static and reified notions of "culture" and "tradition." Rather than seeing culture as historically changeable, the predominant discourse on genital cutting understands culture as ahistorical, as a set of customs or traditions. Such an approach remains, it seems, willfully ignorant of the complicated ways that colonialism and modernity have interacted with the "tradition" of female circumcision. In Kenya, for example, in the late 1920 and 1930s, colonial attempts to stop the practice of clitoridectomy among the Kikuyu who lived in the area surrounding Nairobi revitalized what, arguably, was a practice on the wane and gave new meaning to it, not to "traditionalists" but to the younger nationalists of the Kikuyu Central Association.[44] Alternately, as Hetherington points out, the history of female circumcision in Britain and America from the mid-nineteenth to the twentieth century, ostensibly to control the evils of masturbation, and the various contemporary forms of Western mutilation of the female body problematize static notions of the West and the "rest."[45]

As Hetherington argues, the issue of female circumcision became the ideal "site" for the expression of deeply held but culturally and historically bound convictions because of a number of intersecting discourses produced mainly in Western thought. Central to this set of meanings, she argues, were ideas about sexuality and the specific rejection of Freud's notion of the vaginal orgasm. A large body of literature opposing the practice of female circumcision has emerged from the "commonsense" position that the clitoris is the seat of sexual pleasure in women, and, as such, the cutting of this part of the body is

FIGURE 7.3: An elder Mbakwa-Manja woman of Ubango Shari, Central African Republic, holding knife used for female genital excision (clitoridectomy). Source A. M. Vergiat, ca. 1930. Wellcome Library, London.

tantamount to torture for women who undergo this process. As Hetherington writes, "every woman in the West now 'knows' that the uncircumcised clitoris plays an important part in her enjoyment of sexuality. The removal of the clitoris has therefore now become one of the most patriarchal and sexist acts which can be imagined."[46] However, as Hetherington continues, this "knowledge" about the clitoris may have no power in African societies where women "know" that the proper expression of sexuality involves the removal of the clitoris.[47]

This sense that what we "know" about the clitoris is socially and culturally constructed, and therefore limits the meaning we can give to the cutting of it, seems to be borne out in some anthropological texts on the practice. Walley's ambivalent and complex account of female circumcision in Kikhome, in the western province of Kenya, in 1988 notes that, by undergoing the "painful public ordeal, of initiation," the young women "not only developed a personal sense of self-confidence and pride that made them feel like adults, they were awarded considerable public respect."[48] Similarly, Walter Goldschmidt's anthropological observations of the practice among the Sebei, a Ugandan subgroup, suggest that it was less about controlling women's sexual desire than about creating a ritualized site for Sebei women to celebrate their collective strength.[49] Janice Boddy, based on her work in the northern Sudanese village of Hofriyat in the early 1980s, concludes that the practice of female circumcision (including the practice of excision and infibulation) is an "assertive symbolic act that serves to emphasize a type of femininity that focuses on fertility while deemphasising their sexuality."[50] Boddy notes that in a sex-segregated society, such as the Hofriyati, women achieved social recognition not by becoming like men, but by emphasizing their differences. The idea that sexual fulfillment and the clitoris are not aligned in any "commonsense" understanding is illustrated in the work of anthropologist Hanny Lightfoot-Klein, herself an outspoken critic of the practice, who noted the sexually engaged behavior and conversation of older North African women who, to her surprise, despite being circumcised, gave detailed descriptions of their sexual pleasure, including orgasms, presumably because other areas of the body had become intensified erogenous zones.[51] Nahid Toubia, a Sudanese doctor, argues that little is known to medical practitioners about the sexual functioning of infibulated women or the processes by which women might compensate for loss of the clitoris through other sensory areas, emotion, or fantasy.[52] As Hetherington observes,

> It is clearly the case that a survey of Western uncircumcised women would reveal a high percentage who do not regularly achieve orgasm and a high percentage who suffer from a range of the medical and gynaecological problems mentioned [in the medical, paramedical, and news literature on female circumcision].[53]

A discussion of clitoridectomy at first glance seems to return us starkly to the key questions at the core of theoretical debates around corporeality. Just how powerful is culture in reshaping "nature" in the face of this procedure? Isn't the biological implication of female circumcision inevitably monolithic in its meaning? Can the removal of the clitoris be rewritten culturally, or does the violent effect of the procedure defy the demands of contemporary theory to assert the permeability of the categories of nature/culture? However, as the more complex and historically informed literature on this subject argues, this act of marking the body is as infused with meaning and culture as any of the other acts described in this chapter on the marked body, and it takes place within complex and historically specific contexts, as well as complex contemporary politics.

TATTOOING

The efflorescence of tattooing in the past twenty years in the United Kingdom, Europe, and America, and its concomitant resurgence in the Pacific, traditionally seen as the cradle of tattooing, raises related questions about embodiment. Theoretical work on tattooing suggests that the increasing rates of acquisition of tattoos, as an acceptable and desirable form of cosmetic enhancement or body modification, has been precipitated by changing understandings of "the body" and what one can "do" with bodies. In some approaches, the current popularity of tattoos, or the so-called tattoo renaissance, is understood as part of late-modern or postmodern identity projects or "body projects," in which individuals reflexively construct their biographies and identities.[54] The concern with the body and embodiment through practices such as tattooing can be thought of, at some level, as a response to a modernist preoccupation with, and experience of, the rationalization and bureaucratization of everyday life. For some, as Anthony Giddens argues, this has resulted in an increase in the experience of "ontological insecurity."[55] The current appeal of many body-modification practices such as tattooing may be attributed to this effect, as they are taken to represent a more bodily, affectual, and social way of being.

Recent work on tattooing challenges the implicitly passive or docile body found in the writings of Kafka, Nietzsche, Foucault, and to an extent Deleuze, which is "marked, scarred, transformed and written upon or constructed."[56] Rather, the tattooed body is seen as a lived body, an agent that is active in its own self-transformation and the transformation of the social spaces it inhabits and produces. The tattooed body, as an increasingly common part of popular culture, can, to a certain extent, be seen to challenge the idea of the body as a "natural" entity. Instead, the body is seen as a site of performativity.

The historical association of the tattooed body with the ethnic or class "other" in Euro-American contexts has resulted in the deployment of tattoos and tattoo-derived imagery in many contemporary films and marketing campaigns as a means of denoting the wild, natural, instinctive, and virile. Such representations of tattoos and the tattooed body depend on a cultural tradition in which the tattoo is represented as being intimately bound with "difference."[57] This can be traced both to representations of exotic, tattooed tribal "others," beginning with the voyages of Captain James Cook, and to an outmoded tradition of criminology that drew on the theories of phrenology and physiognomy in which the "irresistible disposition to become tattooed" was attributed to a pathological (criminal) disposition.[58] Despite the increasing popularity of tattooing in contemporary Euro-American cultures, a link can be made to an ongoing tradition in which the tattooed body is linked with the criminal in popular representations of urban gang culture.

In contrast to this approach, one persuasive argument in recent studies of tattooing argues that it has been transformed from a working-class practice to a middle-class one as the practice has become significantly mainstream. Margo DeMello's *Bodies of Inscription*, based on ethnographic research at the increasingly widespread tattoo conventions in North America, argues that tattooing has become increasingly accepted and respected and has, in fact, achieved the status of an art form.[59] While DeMello acknowledges that the working class and middle class are not homogeneous groups, for her the working class is framed in terms of aesthetic, cultural and spatial practice, and little is said about class as a lived experience, nor is there a sense of her informants' everyday conception of the term *class* in relation to the tattooed body. The primary concern of DeMello's book is to consider how tattoo practice has become incorporated into a middle-class discourse equivalent to that of other art forms, and she therefore does not explore how tattoos are experienced in relationship to the different social positions and cultural representations people inhabit.

In contrast to this position, despite its increased popularity in recent years, it seems that tattooing continues to possess very diverse uses and meanings for a range of different individuals and groups from different sociocultural and economic backgrounds, including the working classes, youth, and alternative cultures. The most recent work on tattooing suggests that it is used, intentionally or unintentionally, in such a variety of contexts—for a variety of different purposes, with a multiplicity of sociocultural functions and meanings—that it illustrates how the body, even when marked with the painful and permanent lines of tattoo, shifts depending on context. Recent work on tattoo shows how, to understand the sociocultural significance of tattooing, one needs to pay attention both to tattoo's necessary corporeality and to the sociocultural and historical contexts in which it is deployed.

Karl Broome's recent ethnographic research among London's youth tattoo milieus involved a year observing his brother's apprenticeship as a tattoo artist in his southeast London kitchen.[60] His findings suggest that the appropriation of tattoo as a sanitized, middle-class representational symbol is only part of the story of these marked bodies. Broome takes a phenomenological approach to his own intimate and visceral involvement with the youth tattoo milieus and suggests that many recent sociological and cultural studies approaches to contemporary tattooing tend to be subjected to a form of textual analysis that looks only, or mainly, at the "symbolic" aspects of tattoo. Broome argues that when tattoos are understood as part of a postmodern "cultural script" in which the focus is on the tattoo's increasing self-referentiality, they become further removed from what is understood as their "social processes." The tattoo here is seen as a superficial mark, indexing little outside of the fashion system by which it is constituted. Tattoos are treated as "floating signifiers" that refer to nothing but themselves.[61]

Broome suggests that such accounts fail to acknowledge a reemergence of forms of sociality based on sensuous rather than cognitive criteria. Rather than being preoccupied with the tattooed body as a "symbolic construct," his ethnography focuses on the corporeal process of tattooing and considers the tattooed body as a lived reality. For the people described in Broome's contemporary ethnography, far from being a middle-class fashion statement, tattooing was a profoundly meaningful "sociosensual" experience.[62] The relationship to the tattooist and to one's body extended beyond that between the person tattooing and the person tattooed to an intercorporeal network of friends and associations.

GENDER AND TATTOO

Les Back argues that, until recently, young women who were tattooed engendered accusations of involvement in sexual promiscuity, or prostitution, and of being a "slag" (loose woman) or "sluttish."[63] He argues that the tattoo renaissance of the last ten years or so has changed this situation to some degree, as more and more women have begun to wear tattoos and the stigma associated with them has lessened. However, as recent research shows, it is not simply the case that the stigma attached to tattoos on women has lessened in Western culture. This depends on several interrelated factors, such as the tattoo design itself, the quality of the work, who "did" the tattoo, and, more important, whose body it is on. Following from this, recent studies of body-modification practices suggest that, far from becoming ironic postmodern statements—little more than aesthetic enhancement, as some commentators argue—they still hold important and diverse meanings.

Victoria Pitts's study of the cultural politics of body modification in North America focuses on radical body-modification practices such as tattooing, piercing, implants, suspension by hooks, branding, and scarification. She argues that such practices are the effects of a cultural politics enacted through the body and are forms of signifying practice that enact gestures of social bonding, community building, personal healing, and cultural dissent. Her study considers in particular the reclaimative discourses for women involved in such body-modification practices, arguing that such practices give "expression to the tensions and fissures of gender and sexuality."[64] Her analysis, which includes but is not limited to tattooing, suggests that many women located on the "U.S. cultural margins" used body modifications as part of an "agency discourse of reclaiming the body."[65] Body-modificatory practices are treated as forms of self-intervention and self-control over bodies that are normally taken as being inscribed by powers and forces over which the modifiers may otherwise feel they have little control.

Pitts argues that some of the approaches that women deployed to challenge essentializing discourses and exploitative treatments of their bodies involve

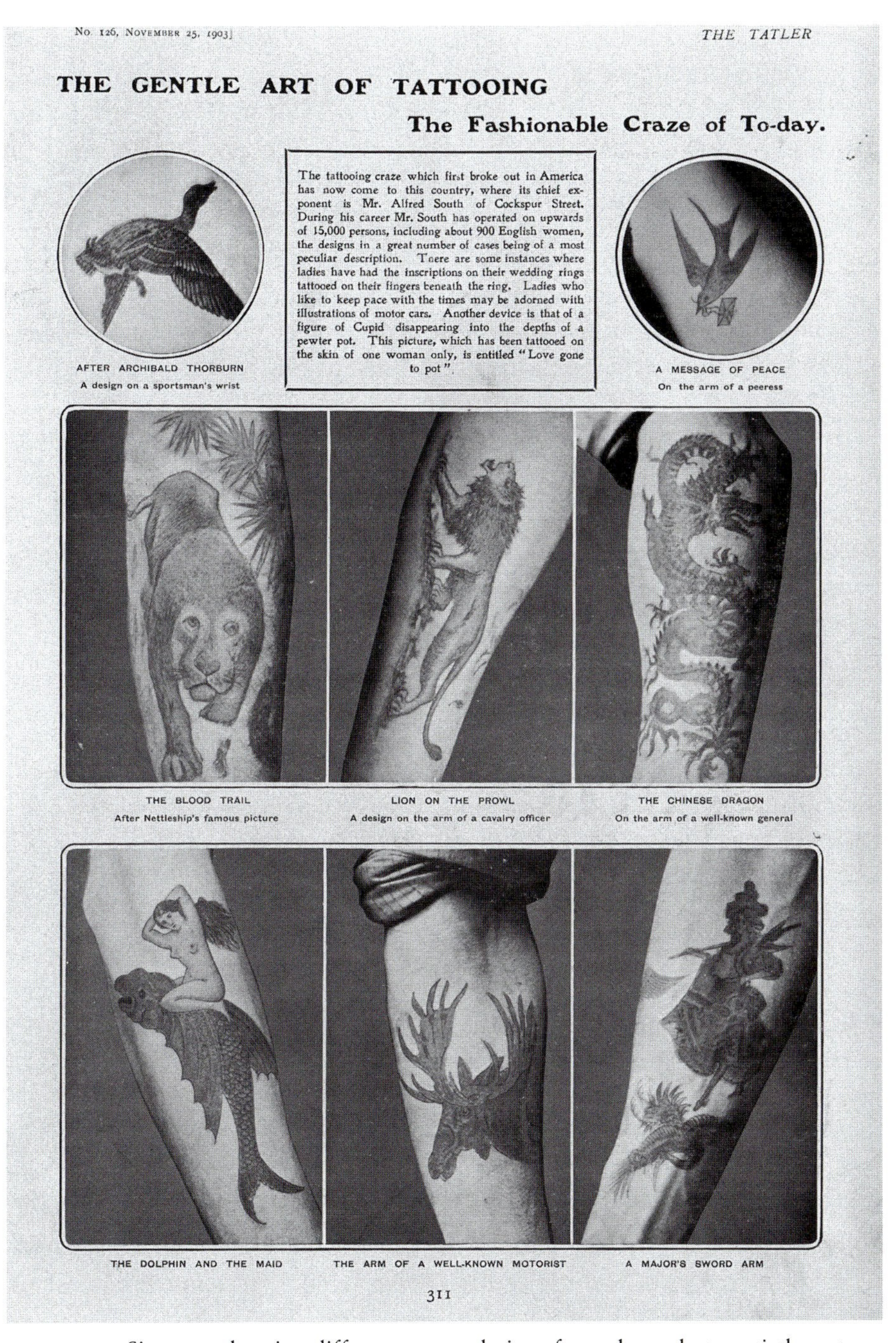

No. 126, NOVEMBER 25, 1903] THE TATLER

THE GENTLE ART OF TATTOOING

The Fashionable Craze of To-day.

The tattooing craze which first broke out in America has now come to this country, where its chief exponent is Mr. Alfred South of Cockspur Street. During his career Mr. South has operated on upwards of 15,000 persons, including about 900 English women, the designs in a great number of cases being of a most peculiar description. Tnere are some instances where ladies have had the inscriptions on their wedding rings tattooed on their fingers beneath the ring. Ladies who like to keep pace with the times may be adorned with illustrations of motor cars. Another device is that of a figure of Cupid disappearing into the depths of a pewter pot. This picture, which has been tattooed on the skin of one woman only, is entitled "Love gone to pot".

AFTER ARCHIBALD THORBURN
A design on a sportsman's wrist

A MESSAGE OF PEACE
On the arm of a peeress

THE BLOOD TRAIL
After Nettleship's famous picture

LION ON THE PROWL
A design on the arm of a cavalry officer

THE CHINESE DRAGON
On the arm of a well-known general

THE DOLPHIN AND THE MAID

THE ARM OF A WELL-KNOWN MOTORIST

A MAJOR'S SWORD ARM

311

FIGURE 7.4: Six arms showing different tattoo designs from the early twentieth century, showing the longevity of interest in female tattoos. From *The Tatler*, November 25, 1903. Wellcome Library, London.

various attempts at producing "subversive bodies" that defy what is perceived as normative, in appearance, practice, experience, and stylization, thereby challenging normative conceptions and categories of femininity and performativity. Tattooing is considered in her account as being counternormative in that it presents an alternative to how women's bodies have traditionally been marked culturally. From this perspective these so-called anomalous bodies are taken to resist the dominant symbolic order, resisting "orderliness and social control."[66] The subversive body presents contradictory images to those produced by normative conceptions of the female body and thus openly celebrates Kristeva's "abject borders of the body" by intentionally producing blood, pain, and permanent markings.[67]

Pitts suggests that the "liminality of the becoming" afforded by body modification allows for subversive "identity effects."[68] For Pitts, like Kristeva, liminality is achieved through the opening up of the borders and boundaries of the body. By doing so one produces what Mikhail Bakhtin calls a "grotesque" body, one that subverts what is normatively considered acceptable embodiment.[69] This is similar to the "abject" body celebrated by Judith Butler and Queer theorists more generally: a material body that evades categories and boundaries by transgressing and subverting them.[70] Pitts's analysis of the reclaimative discourses of women who may have suffered from serious illnesses (such as breast cancer) and/or have been subjected to various forms of physical and emotional abuse (such as rape) suggests that body-modification practices such as tattooing can enable the reappropriation of the body, challenging bodily victimization.[71]

Pitts's analysis of the liminality created by body-modification practices reflects Alfred Gell's seminal study of Polynesian tattooing practice, *Wrapping in Images*. For Gell, the process of tattooing has the capacity to open up and then reseal the boundaries of the body and thus, by extension, the boundaries of the self. Gell refers to tattoo's "invariant processual contour," the wounding, the scarring, and the healing involved in tattooing, which, until another painfree technology exists, indexes the experience of pain/discomfort involved while being tattooed.[72] The corporeal experience of being tattooed is significant here, as we can also understand it as a means by which people attempt to embody what they perceive as "cultural" experiences and affects. Such an analysis returns us to the work of the intercorporeal artists described in recent work on scarification in Australia.

In conclusion, despite the view that tattooing and other practices of body modification have become assimilated into the logic of postmodern consumer culture in which such practices refer to nothing except themselves, the efficacy of the tattoo, the painful process of tattooing, and what it indexes remain. In this way, tattoo mirrors the other practices of permanently marking the body—or permanently marked bodies—that are the subject of this chapter.

CHAPTER EIGHT

Body Marks (Bestial/Divine/Natural)

An Essay into the Social and Biotechnological Imaginaries, 1920–2005 and Bodies to Come

MICHAEL M. J. FISCHER

The soul . . . is produced permanently around, on, within the body.

Michel Foucault, *Discipline and Punish*[1]

From the spectacular semiosis of the Renaissance body—albeit a body in pain—modernity fashions a new body for its own labor-intensive and empirical epoch. Modernity creates the status of two new bodies . . . Concealed within the first body, [the second] is the diagrammatic, fibrous, structured, organized object of investigation . . . the object of the disciplinary interventions which will thenceforth sanitize it, train it, and prepare it for labor. . . . Throughout the culture of the modern period, the figure of the individual has been haunted by the figure of the artificial man . . . ghosted by the interference which that possibility offers to the task of sustaining the original sense of human authenticity itself.

Francis Barker, *The Tremulous Private Body*[2]

. . . as was made terribly clear by the last war, which was an attempt at new and unprecedented commingling with the cosmic powers. Human multitudes, gases, electrical forces were hurled into the open country,

> high-frequency currents coursed through the landscape, new constellations rose in the sky, aerial space and ocean depths thundered with propellers, and everywhere sacrificial shafts were dug in Mother Earth. This immense wooing of the cosmos was enacted for the first time on a planetary scale—that is, in the spirit of technology. . . . In the nights of the annihilation of the last war, the frame of mankind was shaken by a feeling that resembled the bliss of the epileptic. And the revolts that followed it were the first attempt of mankind to bring the new [techno-cosmic] body under its control.
>
> Walter Benjamin, "To the Planetarium"[3]

The more we tinker and experiment with the body, the more the nature of bestiality and divinity are redefined, the more bodily markings take on new connectivities, significations, intensities, and transductions. We have always been able to read the aging body for traces of experience, but increasingly we now enter the age of biological sensibility.

Four score and five years (1920–2005)—from the concussions and crutches of World War I through cancers produced by Cold War radioactive tests, to pesticides and industrial and military toxic runoffs producing endocrine disruptors, leukemias, and kidney disease; from *Wretched of the Earth* conditions producing *Black Face, White Masks*, and torture in North Africa and Latin America, "dirty" wars in the Middle East and Asia, posttraumatic stress syndrome from the Vietnam War, and Tamil Tiger and Islamic jihad suicide bombers' refunctioning of the body, to twenty-first-century First Gulf War syndrome and Second Gulf War enhanced prosthetic developments—these biologically traumatizing eighty-five years have left "new" signifying marks reverberating within and on, and projected through, the bestial, divine, natural body. Experimental sciences and engineering technologies attempt cosmic-cosmetic repair with lipstick and makeup as well as with new organs, prostheses, xenotransplants, autologous and regenerative techniques, pharmakons, filmic and networked gymnasia of the senses. The demiurgical filmic eye is one of a series of prosthetic or cyborgian eyes that allow passage between *Leib* ("body") and *Körper* ("corpus"), "the first oriented to humanity, the second to divinity."[4] The filmic and other prosthetic eyes attempt to repair the inaccessibility of the body to ourselves, the inability to see our face, our back, our whole head. The filmic eye, like our own senses, is productive of phantasms as well as illusionary realisms by way of the filmmakers' artifactual editing, cutting, and suturing.

If time ("since 1920") suggests questions of the "new," the sequence "bestial, natural, divine" suggests questions of evolution, emergence, and illness/healing, if not older languages of the great chain of being and merit cycles of reincarnation. The "bestial" does boundary work on one side of the natural body's ambivalence, signaling internal disease processes (the wolf of lupus, the crab of cancer), viral/retroviral species crossings (the fevers of mosquito-transmitted malaria, HIV/AIDS from hybrid monkey immunodeficiency viruses,[5] avian influenza A [H5N1], fears of porcine retroviruses should we pursue xenotransplantation), and parasite-host and prion relations out of control (mad cow disease). The "divine" does boundary work on the other side: the miracles of healing, the touch of ecstasy, empowerment, feelings of transcendence (illustrated in Michelangelo's divine hand of God touching Adam's outstretched finger and Eric Avery's photograph of the "space of healing" in reattaching a hand), the transports and transformations of drugs (oscillating between the demonic and the divine), and the extensions of electromechanical media (cochlear implants, lenses, the body electric, the cyborgian body). The natural body is an ambivalent term, meaning both our Other and ourselves. Even as ourselves (my character, my body, my selfhood), our nature is often Other, that which we attempt to separate ourselves from and which we are dependent on, which we attempt to control but which always escapes our reach.

In matters of cultural imaginaries, the dance between science and the arts can provide a dialectical cultural critique: science and engineering interrogating the current conditions of possibility; art exploring techniques, tools, and concepts in contexts unjustified beyond the aesthetic. In this dance, the arts are often early adopters of technique yet often lag behind the frontiers of what can be done and of what scientists themselves plan and dream. They can serve as

FIGURE 8.1: Eric Avery, "Hands Healing: A Photographic Essay" (5/6 images), 1977. Reproduced with permission of the artist.

advance publicity to ready the public imagination as not only a *fort-da* mechanism of anxiety reduction, nor just a carnival of the technoscientific, but also a reminder to scientists of the real-world constraints and difficulties of implementation by exploring embedded demonic fears and anxieties of technologies going wrong, being misused ("in the wrong hands," exacerbating inequalities and discrimination), and causing unintended social consequences.

How "new" either the marks or their modes of inscription are is a minor-key harmonic of this essay, to which we return in the conclusion, "GATTACA: Back to the Future Body," along with its twin, the much-contested philosophical question of what "emergent forms of life" might mean.[6] For body marks since 1920, the "new" might involve at least three forms: First, body marks, at a minimum, participate in their historical cultural contexts ("new" as marks of cultural historicities, as in Michel Foucault's and Francis Barker's contrast of Renaissance versus modern French epistemes). Second, there is the "new" in a structural, ritual, or trope sense.[7] How different, for instance, is the embodiment claimed by Flannery O'Connor (drawing on Catholic corporeal theology while suffering from the wolf/lupus within) from that of "reclaiming the body" by 1990s body-modification artists, as described by ethnographer Victoria Pitts or by any number of anthropologists of ritual process, whether Catholic or Ndembu, Jewish or Islamic, or Hindu-Buddhist?[8] This measure of the "new" might be gauged by reading against older anthropological accounts such as (a) Victor Turner's 1987 survey of "Bodily Marks" of growth and aging, clothing and headgear, masks and ornaments, permanent tattoos, scarification and cicatrization, circumcision and clitoridectomy, face and body painting, and the stigmata and signs of religious elevation;[9] (b) the 1925 *History of Tattooing* by Wilfred Hamby; and (c) semiotic-hermeneutical-cultural explications of the meanings of bodily marks in particular societies—for example, Andrew and Marilyn Strathern on Mt. Hagen in New Guinea; Terence Turner on the Chikris of Brazil; or Victor Turner's own account in *Drums of Affliction* of cicatrization in Ndembu girls' puberty rituals, including erotic Braille around the navel " 'to catch a man,' by giving him enhanced sexual pleasure when he runs his hands over them" and parallel cuts above the breast line "to deny the lover," one representing her premarital lover, the other her husband-to-be (the former never to be mentioned to the latter so that they will not fight). A third measure of the "new" is contained in Walter Benjamin's suggestion that while the ancients' discourse with the cosmos was through the bodily experiences of ecstatic trance, twentieth-century technologies are transforming the sensorium, the collective techno-body, biological exchanges (viruses, retroviruses, transfusions, organ transplants, grafts, genetic engineering, pharmakons, and biologicals), and affect and emotion—the "new" as biotechnologies.

To be sure, traditional body markings—natural, bestial, and divine—remain potent in many contemporary worlds as anyone dealing with the spirits that possess in Indonesia, the Persian Gulf, or Amazonia may attest, as they are potent also in snake bites as signs among snake handlers in the American Bible Belt, in glossalalia among new Christians in Kenya, or in snakelike dreadlocks among the ecstatic Buddhist-Hindu priests of Sri Lanka's new urban cults (whose psychoanalytic case histories are charted by Gananath Obeyesekere in his cross-cultural, contrastively-comparatively titled and argued *Medusa's Hair*[10]). Vincent Crapanzano, with respect to Hamadsha trance dancing, dismisses the label "masochism," suggested by a psychoanalyst, for the ecstatic "mortification of the flesh," not only because the conceptual category is misapplied from one culture to another but also phenomenologically because it is too crude and undifferentiated for the different dance steps and breathing rhythms that make the body feel different, invoking different spirits.

"Divine eye," "natural skin," and "bestial body" provide notational names for the parts of this essay. But the three forms of the "new," the recombinatorial modes of emergence, and the boundary work are active "shifters" (to use a linguistic term) in all three sections, as are imaginaries, physical bodies, and technological and social connectivities. Chart enthusiasts might thus play with these as 3 × 3 × 3 matrices; but perhaps the torques and twists of topological forms (Möbius strips, Klein bottles) and knot theory would be better analogies for the markings on and by the body, its subjectivations and subjectivities.

We begin with the surrealist and biotechnological eye.

THE DIVINE EYE: EYE(I)ING THE BESTIAL, NATURAL, DIVINE

> The *eye*, acquiring cyclopean proportions when seen reversely through the magnifying glass before it, is not still . . . It might be said that the eye quivers . . . The gray-blue *iris* . . . also moves, though with a more calculated exactitude than the greater ocular structure. It moves economically and without caprice, as does the geometer. The circle at the center of the iris, the dark void known as the *pupil*, appears to be moving, but, in fact, the inside edges of the iris's striated membrane are contracting and relaxing, varying the magnitude of light information allowed to stream through the eye's lens and into the posterior chamber, to disperse through and illumine the thick jelly of the vitreous body, to bombard the retina's millions of photoreceptors.
>
> Michael Mejia, *Forgetfulness*[11]

Never still, marking and being marked, the eye is an organ both of *transfer* (Greek: μεταφορά, or metaphorá) and *interaction* (affect, emotion, recognition, eye contact/avoidance, eye play, winks and other signals). From the ancient mythic third eye of wisdom, the blind seer's insight, and cyclopean monstrosity to contemporary posthuman and cyborgian prostheses (colored contacts; shades; jewelers' loupes; dentists' binocular microscopes; biologists' atomic tunneling microscopes and scanning tunneling microscopes; medical positron emission tomography, computed tomography, and magnetic resonance imaging [MRI]; soldiers' night-vision gun sights and goggles; surgeons' and game players' virtual-reality headsets), both the evil eye and the "postnatural" techno-eye are never still, marking, noting, inscribing, glancing; becoming or appearing hidden, opaque, glassy, jealous, envious, lecherous, inquisitive, probing, shifty; reconstructing rather than directly seeing; transferring, projecting, hallucinating; serving as a vehicle of psychic transference, missing the mark.[12]

The postmodern Islamic eye has come to the fore—posting back and forth between the modernist drive for women's equality and the modernist-fundamentalist reaction of re-veiling—peeking out from its hijab seductively, marking gendered spaces, and protesting artfully. The artist Shirin has put her mark on the Islamic eye, filling it with calligraphic verses of martyrdom and eroticism, divine and physical love, witnessing (*shahid*, "witness, martyr") to women's agency, resistance, oppression, female power under and out of control, marking contested moralities.

The Islamic eye literalizes gendered spaces and interactions that elsewhere are more veiled: A woman makes eye contact with passing males but does not hold it lest it be taken as an invitation—a natural cue, say apologists; a bestial assertion of power, say some feminists. Black-veiled female traffic becomes a calligraphic form, ambiguously (un)surveillable, (un)readable by male eyes. In Elia Suleiman's satirical film *Divine Intervention* (2003), shot at the Qalandiya checkpoint between Jerusalem and Ramallah, a Palestinian young woman—stylishly and sexily dressed, perhaps Muslim or maybe Christian, with determined eyes set straight ahead—becomes a femme fatale upsetting the conventions of Israeli border guards, who, gazing open-jawed at her boldness and beauty, allow her to pass unchallenged, as two visual puns underscore the point: Their guard tower collapses, and she morphs into a Hong Kong–style aerobatic or ninja woman-warrior as balloons carrying Arafat's face float across the border. Traces of divine intervention (undoing the best-laid plans and boundary work of men) linger in the drifting balloons, distracting beauty, and the filmic eye panning from up-close and intimate to long shots above the fray. *Memoirs of a Geisha* (2005) produces the same results when the novice practices until a male eye is so distracted by her beauty that he crashes his vegetable cart.

The eye is already marked in the 1920s by shock, the shock of the transgressive, war, the psychosexually disturbed, the modern, the urban, and the cinematic. The shock of the eye is nowhere more sharply inscribed as a cultural icon than in Luis Buñuel and Salvador Dali's 1928 film *Un Chien Andalou*,

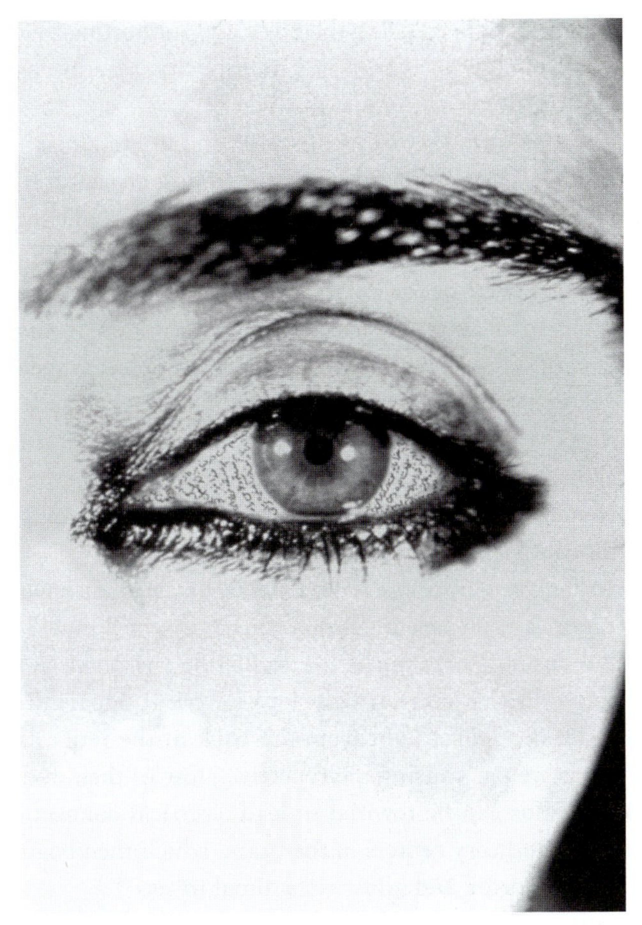

FIGURE 8.2: Shirin Neshat, *Offered Eyes*, 1993. Reproduced with permission of the artist.

with the terrifying slicing of an eyeball with a razor. Georges Bataille's *Story of the Eye*, published the same year, further explored the disturbed/disturbing specular zones of indistinction of eros/pornography/sexual disorder, life/death, morality/amorality, sacred/profane, and emotions "beyond good and evil" in registers of both illness/dementia and "post-death-of-God" sociality. Produced as part of his analysis with Adrien Borel, drawing on childhood obsessions with his syphilitic (blind, paralytic) father and the amorality of dementia, as well as debauchery, its authorship remained pseudonymic during his lifetime, functioning amphibiously across the divides of what is secret and public, noir and exposed, in surrealist and antisurrealist modalities. In his 1929 entry for "eye" in the journal *Documents*, Bataille links the eye to horror, cutting, seduction, cannibalism, appropriation, and the uncanny. Bataille's Nietzschean eye, in the 1930s, eventually recognizes and agitates to counter the rising choreographies of Fascist mass cults that claimed heightened superman (*Übermensch*) experience through battle-hardening the body and submitting to a single will.[13] He anticipates the biology of excess and multiplicity, reacting against economistic scarcity as the dominant model for life.

In efforts to probe different forms of seeing and interaction that can repair or extend this traumatized, diseased, and limited human (natural) eye, the terrain of existential, philosophical, and anthropological challenge moves ever more toward prosthetic cyborgian eyes and biological research on different kinds of eyes in the evolutionary palate. Snakes, for instance, see visually and with infrared sensors that create images of heat emitted from objects. The receptor cells on the sides of a rattlesnake head can detect wavelengths of ten micrometers (ten times the power of the best current human-made sensors) and are self-repairing. Might such "eyes" help detect the temperature difference between cancerous hot tumors and healthy tissue?[14] Or might the vibrations of tumor cells provide such heat signatures *sonically* to a "listening eye," transducing vibrations into sound and then into images?[15]

The mammalian vertebrate eye has a camera-like structure with the spherical vitreous humor (a transparent gel that the cut eye in Buñuel's image seems to spill) as the focusing lens, the iris for regulating the intensity of light, and a panel of retinal cells that convert light into electrical signals. Cats can focus the lens into a slit and reflect light from the back of the retina, allowing better night vision, as well as causing the reflective glow of their eyes in the dark. Newborn ferret brains can be rewired to lead electrical connections from the back of the eyes to auditory centers in the brain, which then begin to resemble visual centers in the cortex and allow the animal to see.[16]

A technology developed by Peter Meijer at Philips Laboratories, called vOICe, mounts a small webcam on the bridge of sunglasses that can scan the field of vision from right to left and convert height to pitch and brightness to loudness, producing a soundscape (through stereo earbuds) that already allows several blind users to gain a rudimentary sense of sight that improves with habituation.[17] This "rerouting of the senses," Meijer claims, works better than retinal implants also currently under development.[18] Steve Mann's earlier cyborg WearCom (EyeTap camera and reality mediator), one of a series of experiments in WearCom (wearable computers), was a sunglasses-like device that recorded people's faces and matched them to biographical information from previous interactions so as to function as a memory device for those of us who forget or want faster recall.[19] (It, and more recent EyeTaps, are also linked to the Web to allow others to share in the wearer's vision.) Steve Furness's Virtual Retinal Display is a somewhat similar device that casts images onto the retina so that they float over normal vision, providing neurosurgeons, for instance, with brain scans over the surgical area on which they are working, or pilots with additional virtual flight scenes over their cockpit view.[20] In a sense, Meijer draws on the body's resources, while Mann and Furness place the body within a distributed, neural-like, electronic network, paralleling in more medical or practical modalities the more bestial body arts of ORLAN and Stelarc (see later sections of this chapter).

Multiple functionality, rewiring, and different modes of seeing promise prosthetic repair, supplementation, and extension of the eye but also potential directed mutation. *Drosophila* genetics allow eyes to be multiplied or placed in novel positions; Mann's composite "bug eyes" assemble multiple fragmented views.

The focusing lens of the mammalian eye functions like a camera, the modernist technology par excellence of the early twentieth century, and of the 1920s technology of the *Kino-eye* (1924) and *Kino-Pravda* (a series of 1920s shorts that collaged fragments of reality to make visible truths not normally visible to the natural eye) of Dziga Vertov (né Denis Abramovich Kaufman). Filmic technologies—microcinematography of cell movement,[21] telescope and microscope lenses, later atomic tunneling microscopes, nuclear magnetic resonance, positron emission telemetry, and other visual technologies—allowed the human eye-brain to experience what could never have been seen before: beyond, below, slower, faster, further than the naked capacities of the human eye; cells, atoms, interiors of bodies and organs, distant planets and galaxies; daily traffic for crime control in urban spaces through networked video cameras; the night vision of guns, helicopter gunships, goggles, and remote-control targeting.

This eye, the filmic eye, opens new aesthetic possibilities (montage, speed, derangement), including the simulation of science fiction's bestial fears, from Dr. Moreau's experiments with organ transplantation,[22] to urban noir films (the detective eye), to the varied latter-day expressions of Friedrich Nietzsche's vision of the hypergrowth of eyes and brain, perched on spindly remnant limbs and torso. This is an image of cognition gone wild, bestial, and even, in a latter-day version, downloadable to a silicon chip,[23] which through a fantasy of escape from the body is a peculiarly perverse vision of the divine, a soulless variant of neo-Platonic, Illuminationist, and Enlightenment emanations, digitally purified.

The natural eye marked by and marking evil is revealed and embodied in John Howard Griffin's midcentury observation: "A 'hate stare' drew my attention like a magnet. . . . You see a kind of insanity, something so obscene the very obscenity of it (rather than its threat) terrifies you. . . . I felt like saying: 'What in God's name are you doing to yourself?' "[24]

In God's name: the divine eye, of course, is a shifter and a metaphor for insight transcending and immanent in the everyday. When it becomes hypostasized in "god-tricks," as Haraway puts it (the view from nowhere, or from Heaven), in the fantasies of omniscient surveillance and control, it goes awry.[25] The eye as a biological nexus, a complex of signal transducers, a space of inquiry, a door of perception can be divinely open and can see such things as "a kind of insanity, something so obscene the very obscenity of it (rather than its threat) terrifies you."[26]

THE NATURAL SKIN: EPIDERMAL SCARS, TATTOOS, PIGMENTING THE BODY

> I have carried this number all my life, that I and others should never forget. Now before I die, I wish to expunge it. Not through any technique that would eliminate it, but through a superabundance in which it will be lost. It will become just one grain of sand upon a beach. I wish to be covered, from head to toe, in one mighty number. *This* number [indicating the number on his arm], in becoming just six consecutive digits within the greater number will be relieved of the weight of evil it has borne until now. Yet [simultaneously] the greater number will be what it is through the presence of that part within its whole. That is inescapable.
>
> Michael Westlake, *Imaginary Women*[27]

> In body modification, you can take control of what you otherwise could not.
>
> Andrew, East Coast body modifier, quoted in Victoria Pitts, *In the Flesh*[28]

In 1959 John Howard Griffin—photographer, novelist, physician, musician, sufferer of a twelve-year temporary blindness, rescuer of Austrian Jews—chemically altered his skin color and visited racially segregated states for his book *Black Like Me* (1961). Hung in effigy in his home state of Texas, he moved to Mexico for nine months before returning to Fort Worth.[29] In 2006 Ice Cube, the rapper, actor, and filmmaker, together with Hollywood makeup artist Keith Vanderlaan and R. J. Cutler, produced a six-week documentary television series, *Black, White*, in which a black family is transformed with wigs and makeup into a white family, and a white family is transformed into a black family.[30] First time as tragedy, second time, not as farce, but as "reality television" entertainment.[31]

The skin's tattoos (marks of community and initiation) and brandings (of slaves and prisoners), its circumcisions and scars (marks of initiation, dueling scars, disease and injury marks that are like *qipu*, the Incan color-coded knotted strings), its pigments (racializing categories) and painting (eye kohl and mascara, evil-eye prophylaxis, and sexual "war paint" makeup) have continued to mutate and proliferate since World War I as cosmic-cosmetic, as entertainment, fashion, political statement,[32] and self-marking (on the surface of the body and within).

The camp survivor's request (in *Imaginary Women*) to Molly, whose profession is "supplementing tattoos rather than trying to remove them,"[33] echoes, if in a different key, Flannery O'Connor's meditation on the agon of iconic versus aniconic hermeneutics in her story "Parker's Back."[34] Parker's wife, Sarah Ruth, disapproves of his tattoos. Parker crashes his tractor and, in reaction to his near death, has the Byzantine icon of Christ the Pantocrator (in Greek, "ruler of all," holding the book) tattooed on his back (in hopes of harmonizing his jumble of secular tattoos and finding an emblem his wife cannot resist). She beats him with a broom, raising more welts and causing his raw tattoo to bleed, reinforcing (as does his being put out under a tree in the yard) the connection with the suffering Christ. For O'Connor, Sarah Ruth is the heretic because of her "notion that you can worship in pure spirit."[35] "Thin and drawn" (without flesh), seeing sin everywhere, she is a puritan whose loss of sensuality in worship loses both the world and the divine, making the consubstantial insubstantial.[36] For Parker and O'Connor, in contrast, there is a signifying chain from the book held by Jesus in the icon, and from the book of images of God in which Parker found the icon, to his body as a book on which is inscribed the mysteries of incarnation and which is a rite of initiation and passage attempting to configure or assemble meaning in his life among the tattoos of earlier experiences. Parker, as Dennis Patrick Slattery notes, doesn't use his Old Testament names—Obadiah ("servant of the Lord") Elihue—preferring a New Testament image. Sarah Ruth, in contrast, goes by her (inapt) Old Testament names. Her preference for aniconic worship is treated by O'Connor as abstraction that destroys, disenchants, and disincarnates the world rather than as the Jewish reliance on the plenitude of the word, the signifying chain, leaving endless traces and loose ends everywhere for further lively embodied engagements. (Victor Turner insists that the Old Testament forbids tattoos and bodily marks [Lv 19.28, Dt 14:1] but fails to notice in this context that circumcision becomes a defining rite for Jews [and for Muslims], something that recently Jacques Derrida has turned into a philosopheme.[37] In the contemporary world, tattooing has become a site of negotiation for Muslims and for Jews.[38])

The agon between body marks, the bestial, and the divine plays out in several registers. For Molly and the concentration-camp survivor, the message is like that of Justice Brandeis: The remedy for bad speech is not censorship but more speech. For O'Connor, suffering from an autoimmune disease, the lupus/wolf within, but also a gifted wielder of what Slattery calls a Catholic "theological poetics of corporality," bodily wounds and illnesses are terrains in which one experiences the deeper mysteries of the divine.

Circumcision (male in Islamic worlds or also female in East and West African cultures), anthropologist Vincent Crapanzano points out, is not just a purification and affiliation-initiation ritual, as "native model" explanations often assert, nor a simple change-of-status rite, as Arnold van Gennep would have it. Instead, it is a paradoxical structure of contradictory messages that set up tensions to which the initiate returns again and again in efforts at resolution: at first sexual experiences, at hazing in the army, at marriage, in curing trance dances, and in colonial or authoritarian relations. The period of liminality of the ritual process does not end with the ritual itself but continues through life, as a constant "rite of return."[39] In Morocco, circumcision should not be done until the boy is "old enough to remember" both the pain and the paradoxical messages of "being a man" (e.g., not crying) and yet of submission (like Isaac or Ismail), of becoming a man by being unmanned. Muslim circumcisions are often explicitly referred to as weddings: The blood shed is like that of the wedding night, the boy's hands are hennaed like a bride's, and he wears an anklet. He is called a groom and is led on a horse but obviously is still too immature to be one; he is then handed to his mother for her to hold him during the circumcision and is swaddled like a baby and put on the mother's back afterward for a dance (where there is more pain). It is a rite of vulnerability as much as of making a man.[40] So, too, for women who undergo various forms of female "circumcision" (clitoridectomy, infibulation) in order, it is said, to become excised of ambiguous male features, conform to the beauty of an adult woman, be desirable—even, in some places, to increase rather than decrease sexual desire, but at the same time, also to decrease desire, thereby pacifying, controlling, and repressing young women.

In Toni Morrison's *Beloved*, the relation between bestiality, divinity, and human body marks plays out on yet a different harmonic. Like the tattooed number from the concentration camp, a tree design has been whipped onto Sethe's back, and freedom is achieved through transformative narrating, remembering, imagining. Like the icon on Parker's back, Sethe can feel, but not see the marks on her back, which Amy Denver describes to her as a tree design. Amy's words and her laying on of hands, easing Sethe's pain with spiderwebs (stringing the body like a Christmas tree, making of her body a narrative of the life of Christ), are like O'Connor's theological poetics of corporality. But there is another thematic as well, as Slattery notes: In the novel, bodies and words carry price tags and can be marketed to the highest bidder. Sethe must pay the tombstone engraver for "Beloved" with her body and wonders if for another ten minutes she could have also gotten "Dearly." All these themes reemerge in the stories of 1990s body modifiers, and Slattery observes that Morrison marks many of her female characters' bodies: "In Song of Solon, Pilate has no navel, in Sula, Sula has a birthmark over her eye, in Beloved, Sethe's mother is branded with the cross in the circle and Sethe is marked by the whipping."

Each mark is a shorthand narrative inscribed in a fleshy book of maturing, aging, and regenerative biological life.[41]

Peter Greenaway's 1996 film *The Pillow Book of Nagiko* brings many of these themes together with a foray into Japanese traditions of tattooing (*irezumi*), Japanese-Chinese ideograms, the dialectic between temporary and permanent bodily marks, and the more structural question of oedipal scripting, private/familial scripts, and publishing/marketing, life/immortality versus death/impermanence, and the exchange rates among the currencies of sexuality, desire, money, writing, and relations of power, as well as (in Hélène Cixous's reading) a call to a "feminine" writing from the body.[42] The establishing ritual occurs annually on Nagiko Kiohara's birthday until she is seventeen: Her father paints a greeting on several parts of her head, her aunt reads to her from Sei Shonagon's *Pillow Book*, the gramophone plays music from when her parents married, and her grandmother, mother, and aunt witness. The father whispers a spell about how God calligraphed on his clay model of a human being, and when he was satisfied with his creation, breathed life into it and signed his name. It is a creation rite, with the father miming the original creation. The child smiles when she sees her painted self in the mirror. It is a mirror scene, with even the characters across her mouth cracking open in her smile. There is mirroring throughout the film between the tenth-century Sei Shonagon and Nagiko. Greenaway describes the ritual as affectionate yet disturbing, setting up a ritualized tension, an emotional force that will enroll Nagiko in an economy of desire and productivity that she will pass on to her own child at the end of the film. The oedipal dynamics are, as in Crapanzano's sensibility about the colonial and other hierarchical contexts of Moroccan circumcision, not merely patriarchal but also capitalist: The father's publisher is present at the ritual, and afterward the father is ritually sodomized by the publisher, linking his source of income and power in the external market to the family economy. Nagiko becomes a fashion model in Hong Kong ("for this fashion model this matter of surface is key"). She seeks out a series of lover-calligraphers, all of whom write on her body in nonpermanent ink, until she meets Jerome, a Western translator ("he knows about transfers").[43] She then herself takes up the pen; and in the end, she has a permanent tattoo inscribed on her chest. Each of the experimental calligrapher-lovers has a different relationship to inscription, one even writing in invisible ink, invoking the Islamic prohibition on images and pointing to a television showing a documentary on Islamic calligraphy (which, unlike Chinese and Japanese ideographic calligraphy, is alphabetic, without obvious pictographic legacies). In the end, she is fully scripted into the oedipal frame, with a permanent tattoo and a child to whom to pass on the script. There is unexplored ambiguity, however, both in the substitution of milk for semen as the bodily symbol of cultural transmission (as Victor Turner might have pointed out), at least raising a possibility of countering patriarchal tropes, and in the publisher being killed by a sumo wrestler inscribed with the slogan, "I am old said the book; I am older said the body," suggesting the ongoing dialectic between "life itself," and the "code of life." Greenaway, moreover, as Bruno points out, uses three screen ratios to mirror the diegetic play with writing the geometry of passion impermanently on the skin, and more permanently in the haptic memory traces from her earliest imprinting, which Nagiko seeks to reexperience with her lovers.[44]

The ritual process is often central to contemporary queer, lesbian, gay, and "modern primitive" body-modification "body art" subcultures. As Victor Turner writes and Peter Greenaway screens, body modifiers deploy the bodily markers of the ritual process: shock, adrenalin, endorphins, percussion, heightened breathing, pain, blood, cutting, cicatrization, skin stretching, implants, hook hanging. And these visceral groundings are fused to cognitive-moral imperatives (self-formation, community bonding, identity marking, the productive tension of conflicting messages). Although claiming countercultural status as tribal, personal, political, or erotic, it is remarkable how often women who seek to "reclaim the body" by tattoos, piercing of intimate body parts, branding, or keloid raising scarification (or cicatrization) construct these acquisitions as powerful and communal rites of affliction in response to childhood or marital abuse, feelings of unattractiveness, and reactions to religious and other disciplinary regimes.[45]

It takes but a small grammatical shift to change the point of view of agency without changing the psychosocial ritual process: "Instead of an object of social control by patriarchy, medicine, or religion, the body should be seen as a space for exploring identity, experiencing pleasure, and establishing bonds to others."[46] "Karen," raised in a working-class Italian family, subject to childhood sexual abuse, a single welfare mother in her twenties, a graduate of night school and law school, a lesbian, and a breast cancer survivor, has, among other tattoos and scarifications, a dragon tattooed on her breast as a talisman of overcoming fear, separating from her family, reclaiming her body, and being a person in her own right. The tattoo image came from a girlfriend's suggestion that she visualize a dragon guarding a cave: "The purpose of that visualization was to give one an awareness of what they do with fear . . . and where their courage comes from. . . . Some people actually pick up swords and swipe the dragon mightily . . . My way of dealing with it was to make myself one with the dragon and make the dragon become me." She also had a scarification in the shape of an orchid, done by a professional body-modification artist in a ritualized spiritual process among her sadomasochism (SM) group, which was modeled on a Maui form of tattooing (a sharpened shell is used to cut, then ash is rubbed in). She compares this to Buddhist visualization practices: "I think it is the same or similar state that Buddhists, when they spend hours in a state of prayer—that place of acceptance, floating." As with other ritual processes, paradox and tension are maintained between what is allowable in the ritual versus secular space, intimate surfaces of the skin versus public visibility, as one negotiates employment and other social arenas. For some in queer culture, bodily marks are so visible as to enforce separation from employment outside tattooing and body-piercing shops. Mathew, who runs a profitable body-piercing studio, has stretched earlobes, scarring, and piercings and had himself branded onstage in an SM club by a well-known lesbian professional body modifier. The crowd, he says, was "awed by the power coming off the stage," and he himself felt lighter with each strike of the hot iron, until he got so high he had to separate himself from people and "really go into myself," in what he calls a spiritual experience. For modern primitives, the romantic appropriation of Native American and other "indigenous" rituals is explicit, even if authenticity is devalued in favor of experimentalism.[47]

Body modification is semiotic, somatic, fetishized, commoditized, and flipable (as investments are flipable from desired holdings to quick marketable gains) as pop culture. The ambiguity of what is customized choice, truly individualistic, and what is a conventionalized, limited palette for selection is always at play. As Girl Punk Melissa Klein notes, "Punk female fashion trends have paired 1950s dresses with combat boots, shaved hair with lipstick, studded belts with platform heels . . . We are interested in creating . . . modes of contradiction."[48] Cyberpunk style, drawing on science fiction such as William Gibson's *Neuromancer*, says Pitts, "is distinct in its futuristic, high-tech body projects' use of biomedical, information, and virtual reality technologies" and views the body as a "limitless frontier."[49] This "iconic futurology," usually imbued with an individualistic ideology, is also physically experimental, as with Body Modification Ezine (BME) founder Shanon Larratt's experimentation with small magnetic implants under the skin of his fingers to create an ability to feel the electromagnetic fields around him.

While the ethnic body may be marked in ways discriminatory, it can also, in satirist and performance artist Guillermo Gómez-Peña's hands, cross wires, borders, and categories into techno-ethnographic "undiscovered Amerindians," Mexterminators, Gringostroika warriors, geisha apocalipticas, and other forms of ethno-cyborgs.

The surface of the body, likewise (or Other-wise), is also the register in which the marks of biological crossings between species and technologies make their presence visible: from neurodermatological signs; to retroviral, viral, and bacterial host-parasite relations; to the fashioning of new organs for both the interior and the surface of the body (skin grafts for burn victims, tissue-engineered ears for infants born without them, as well as various forms of prostheses). Bioart, as well as literature and film, follows along behind the advances of science, as well as projecting ahead.

FIGURE 8.3: Guillermo Gómez-Peña, *La Geisha Apocaliptica*: the cross-cultural masked body. Photo: Ric Malone. Reproduced with permission of the artist.

FIGURE 8.4: Guillermo Gómez-Peña, *El Indio Amazonico*: the cross-cultural, hyperreal decorated body. Performance by Guillermo Gómez-Peña. Photo: Ric Malone. Reproduced with permission of the artist.

BESTIAL DEEP BODY MARKS: BIOSURGICAL ARTS, BESTIAL DISORDERS (WOLVES, CRABS), AND THE DERANGED NERVOUS SYSTEM

Kantorowicz had two bodies, with Hawking we have three: the fleshly body, the distributed body, and the sacred body.

Hélène Mialet, "Do Angels Have Bodies?"[50]

What would happen if, instead of eyes, scientists had microscopes in their eye-sockets? . . . They would . . . become angels for then they would ". . . be in a quite different world from other people . . . : the visible ideas of everything would be different."

John Locke, quoted by Hélène Mialet, "Do Angels Have Bodies?"[51]

To restore the human being to a free condition, we must put the body on the autopsy table for a complete anatomical reconstruction.

Antonin Artaud, *To Have Done by the Judgement of God*[52]

La grande pellicule [film, skin] *éphémère*. Opening the Libidinal Surface. Open the so-called body, and spread out all its surfaces: not only the *pellicule* [skin/film] with each of its folds, wrinkles, scars . . . the immense membrane of the libidinal "body" . . . made from the most heterogeneous textures, bone, epithelium . . . [nerves, vocal apparatus] . . . All these zones are joined end to end in a band which has no back to it, a Moebius band. . . . Terror in the labyrinth is such that it precludes the observation and notation of identities; this is why the labyrinth is not a permanent architectural construction, but is immediately formed in the place and at the moment . . . Each encounter gives rise to a frantic voyage towards an outside of suffering . . . there is no intensity without a cry and without a labyrinth.

Jean-François Lyotard, *Libidinal Economy*[53]

Not only the outer skin or hide, nor only the face and eyes, are bodily media for marks and signifiers. So, too, is the material interiority. The more we tinker and experiment within the body, the more the nature of the bestial/divine and libidinal/fantasmic body are redefined, the more these bodies take on new connectivities, significations, intensities, and transductions.[54] Acoustic signatures, neural cell signals, genetic markers, and chemical signatures increasingly are new signifiers of the bestial/divine, ill/well, disconnected/connected body, available for performance and bioartists, physicians and bioengineers, retuned philosophical-anthropological understandings of our place in the world, and bioavailability in biopolitics.[55] Lyotard once claimed that we might lay out the body as surfaces, film, or membranes all the way through, following the lead of the surgeon but with an eye to all the libidinal surfaces, sensors, and intensities.

Gunther von Hagens's Body Worlds (displays of plastinated cadavers) has literalized this flaying of the body, in a revival of public anatomical demonstrations, making available to the general public the muscles, sinews, nerves, vascular system down to the fine capillaries, and organs in a rare immediacy, showing prominently also the black marks on the lungs of smokers and occasionally a cancerous tumor.[56] Surgeries also are being broadcast on television as public education. Dermatology, moreover, is gradually being transformed into photomedicine, as researchers explore the ways in which our epidermal membranes can be penetrated by "noninvasive," "minimally invasive," and less and less invasive technologies including lasers, endoscopes, and swallowable and probe-tipped cameras. Neural signals are experimentally being used to help the disabled control a prosthetic hand. Cells are listened to by nanotechnologists like Jim Gimzewski (see note 15) and by bioartists such as Joe Davis, who claims to be able to identify different bacteria and cells by their acoustic signatures, and Adam Zaretski, who plays music to *E. coli* bacteria to see whether they can be stimulated to produce more antibiotics (or otherwise change their modes of being). And, of course, forensics finds ever more ways of reading signs and marks in the body's ruins.

Explorations with surgery by performance artist ORLAN, nerve and muscle connectivity by Stelarc, and tissue engineering by Ionatt Zurr and Oran Catts juxtapose themselves to more direct expansions of the fragile body, such as the machines that keep the celebrated Cambridge University physicist Steven Hawking functioning, prostheses for war injuries, face transplants for accident and burn victims, laser and regenerative medical techniques for life-threatening hemangiomas (vascular anomaly birthmarks), and tissue engineering to replace a congenitally missing ear.[57] Bioartistic experimentation beyond the current state of the craft of biology is in dialogue both with the (secularized) medical imaginary and with the religious and familial imaginary of people suddenly on dialysis machines, with organs from others, with prostheses, immunosuppressants, and other interventions and pharmakons.

Quite apart from the wheelchair, oxygen, voice synthesizer, and other devices that allow Hawking to write, teach, travel, and appear in public, he has, Hélène Mialet suggests, three bodies.[58] There is the frail body disabled by a form of Lou Gehrig's (amyotrophic lateral sclerosis, or ALS) or motor neuron disease, which destroys the nerve cells in the brain and spinal cord that link to muscles, causing the latter to twitch and atrophy, but does not impair the mind, personality, memory, or senses of taste, smell, touch, hearing, or sight, nor the eye muscles, nor bladder and bowel muscles. This frail body has become cyborg in the sense of being now accessorized with computerized facilitation. (On frontiers in the field of biocybernetics, see the following.) There is, further, a soft "distributed body" (and intelligence) composed not only of all the personnel that attend to his cyborg and biological body but also of his students, who do the calculations and research he can no longer do. And there is what Mialet calls the "sacred body," the scientific corpus of the cosmologist/physicist. She compares this divine body to Kantorowicz's famous 1957 study of the two bodies of the French king ("the king is dead, long live the king"), the physical body that is buried at death and the mystical corpus that embodies the sovereignty of the monarchical state and that is passed on to his successor. In Hawking's case, Mialet glimpsed the making of a small piece of the sacred body by interviewing him. It is a long and very laborious process for him to type out on the computer answers to questions, yet he is an expert at using the Equalizer and Easy Keys systems, as well as saving what he wants and deleting the incidentals. No point, he indicated, for her to tape-record or take notes during her interview with him: He would just give her a printout when the interview was over. However, to her surprise, given her project of thinking about the constitution of all three bodies and their roles in his scientific practices, the printout was already sanitized of the questions he had asked in passing (as to "why" she had been selected to write an article about him) or requests ("legs") he had made to his nurse to adjust his position. So, too, of course, is the relation between scientific articles and the practices that go into their production. How important is his now-canonic anecdote about how his illness, in slowing him down getting ready for bed at an early stage of the disease, provided the time for him to think about "event horizons" and have his insights about black holes? Or how important is his topological or geometric intuition as a shortcut to arduous mathematical equations? In what ways do the three bodies "mark" one another? Certainly the distributed body (now made natural?) takes care to highlight the sacred body and veil the (raging, eliminating) bestial (and yet still natural) body.

Or better: Hawking's three bodies are material-semiotic shifters redistributing the categories and their references.[59]

If Hawking's three bodies do mark one another, what of people on dialysis machines or undergoing organ transplantation with the required lifelong regime of immunosuppression? In interviews with end-stage renal-failure patients in Turkey, Aslihan Sanal elicits a series of exchanges between old and new religious and familial bodies.[60] The psychology of intimate familial relations is put under stress, and in one case, an Alawi young woman, upset by a number of hallucinatory dreams and psychological states, found herself questioning her paternity and identity as an Alawi and identifying instead with significant others in her dialysis cohort. This is more than the "tyranny of the gift," described by Renée Fox and Judith Swazey in their pathbreaking studies of early heart-transplant patients, who had to negotiate feelings about being unable to reciprocate the "gift of life" or being oppressed by unwelcome claims on intimacy by relatives who wanted to visit the heart of a loved one, now in the transplanted person's chest (hence donations in such experimental procedures are now anonymous).[61] At issue in routine organ donations (as with the construction of subcultural communities by body modifiers) are larger social solidarities negotiated through publicly regulated as well as gray- or black-market transplantations, through egalitarian public-hospital social ideologies versus privatized market-hospital systems, and through social boundaries and norms shifted by altruism or refused by transgressive acts.[62] The bodies of the poor (who donate for the money) are often weakened and stigmatized, marked by gender and class inequalities, and actively recruited into bioavailable commodification markets for body parts.[63]

As with the discovery of "phantom limbs" during the American Civil War, the relations between the imaginary body and the physical body can be complex, a matter not only of maps in the brain (researched in the neurosciences) but also of psychodynamic-interpretive and symbolic registers (traditionally explored in ritual and art). Here, some of the experiments of performance artists and bioartists are of value in exploring the shifting boundaries that surgery, neural and muscle connectivity, and tissue engineering make possible. This is, Stelarc claims, more interesting when concrete consequences are at issue, not just science fiction or conceptual speculation, and when "images are cyber skins that transduce the physical body into the phantom entity" no longer just in the mind but in photo and sonic physicality that feeds back into the material body and its nervous system.[64]

Indeed, the idea of reverse phantom limbs—the marking in the brain or the brain's writing of "software" and "wetware" for new bodily functionalities—are technologies already in development for ALS and spinal cord injury patients, quadriplegics, stroke victims, surgeons, and pilots.[65] In these so-called biocybernetic technologies, users train themselves to produce specific brain-wave patterns on demand.[66] These are read by electrodes that can move a cursor up and down, right and left on a computer screen or move a prosthetic limb, becoming after a while part of the brain's body map. Niels Birbaumer of the University of Tübingen has designed a thought-translation device that allows an ALS patient to select and type letters of the alphabet on a computer, which can then produce email messages using common word- and phrase-prediction software, somewhat analogous to the software used by Hawking but here run using brain waves. The "cognitive cockpit" is being developed at Britain's Defense Evaluation and Research Agency to allow pilots with electrodes in their helmets and biosensors in their clothes to fly by directing attention (brain waves) to icon controls. The biosensors monitor vital signs, so if a pilot blacks out, an AI copilot can take over. Somewhat more simply, BIONs (bionic neurons) are being developed by Gerald Loeb at the University of Southern California for therapeutic electrical stimulation to prevent/reverse muscle atrophy in stroke and arthritis patients. These are sealed stimulators the size of a grain of rice that are injected into the muscles and send regular pulses of electricity, powered by a transmission coil and controlled by a handheld computer (called a Personal Trainer) with different programs to vary strength, timing, and duration of pulses. Cyberkinetics Neurotechnology Systems of Massachusetts is testing a BrainGate two-millimeter-square silicon chip with an array of 100 electrodes implanted into the motor cortex that can sample patterns of neuron firings and feed them into a computer. The BrainGate allows Mathew Nagle, who was stabbed in the neck, severing his spine and leaving him paralyzed, to open email, turn on lights or a television, play video games, and move a robotic arm. Jonathan Wolpaw, with help from the Altran Foundation's engineers, is testing a cap with sixty-four electrodes that does not require implantation into the brain to help an ALS sufferer use his eyes to send email.[67]

Stelarc's art performances gesture in the direction of these technologies; they seem crude in comparison yet provide a kind of advance publicity in a realm protected from the ethical frontiers of medicine. One of the most interesting elements in these new technologies is the exploration of the idea that all sensing modalities are *transductions*, mediated transformations between one kind of signal and sensory system and another.[68] Abe Caplan, the technician helping Nagle learn to use the BrainGate by encouraging him and adjusting the computer's settings for averaging the motion-prediction algorithm, asks, " 'Want to hear it?' He flicks a switch, and a loud burst of static fills the room—the music of Nagle's cranial sphere. This is raw analog signal, Nagle's neurons chattering. We are listening to a human being's thoughts."[69]

In this sense the brain also "writes" software for mind-body disorders that are marked as disabling pain, such as, John Sarno argues, tension myositis syndrome (TMS; or really muscle-nerve-tendon syndrome), carpal tunnel syndrome, chronic pain syndromes, back pain, spinal disc problems, allergies, asthma, eczema, bulimia, anorexia nervosa, and neurasthenia/chronic fatigue.[70] Many of these have physical marks that can be seen on X-rays or MRIs or can be documented by electrical tests (carpal tunnel syndrome), but these are not caused by physical disorders and mostly are not resolved by surgery or other such physical interventions. The pain of TMS, he argues, is caused by the reduction of blood flow to the part of the body feeling pain, depriving it of oxygen and thus causing pain. Similarly, carpal tunnel syndrome, he argues, is not, as often thought, due to nerve compression but rather local ischemia. The root cause or source code for these problems is found in unspeakable rage and emotional pain that are displaced via the autonomic, immune, and endocrine peptide systems.

Thus, Sarno argues, Freud's notions of the unconscious and displacement were correct, if as yet underspecified. A similar argument is made by those studying the "second brain," the enteric nervous system of the gut—the neurons, transmitters, and proteins that line the esophagus, stomach, small intestine, and colon—which produces benzodiazepines, the antianxiety and antipain chemicals, now made into Valium, Prozac, and other popular antidepressants, which can thus also affect bowel movements.[71]

If Freud is a founding figure from the early twentieth century for the explanatory tacking back and forth between the neurological and psychological as marvelously complex and worthy of exploration—bodily marks, signals, and transductions leading to new vistas and understandings both of the psychophysical body and of the relative autonomy of the ethical self and limitations thereof—Antonin Artaud's notion of the obsolete body, whose parts react involuntarily and which "needs to be put on the autopsy table for anatomical reconstruction," thereby freeing the human to be a BwO (body without organs), provides a quite different but equally strong "sur-realist" set of frames of reference for the scientific, artistic, and philosophical imaginaries of the second half of the twentieth century.[72] Artaud's (1896–1948) suffering body (meningitis, neuralgia, stammering, depression, opium and heroin addiction, peyote experimentation, electroshocks, colon cancer) provides material settings out of which the human emerges (as Artaud seems to have articulated theatrically and poetically, including in nonverbal sounds).

As with Freud, Artaud can lead in multiple directions. The performance artists ORLAN and Stelarc both invoke Artaud's notion of the obsolete body. Gilles Deleuze invokes the notion of the body without organs in a somewhat different way. In both cases a biological sensibility is emergent and perhaps comes together in the work of bioartists Oran Catts and Ionatt Zurr. Although Catts and Zurr have begun collaborations with their generational seniors, ORLAN and Stelarc, they work more directly with tissue engineering to grow and cultivate the biological rather than to replace or reconstruct it. Tissue engineering is used to grow steaks ("victimless meat") from cells taken from frogs who happily continue to live on, pig wings (a more conceptual play on how the impossible is no longer such), and a third ear for Stelarc, and to colonize someone else's skin for ORLAN. The idea of meat that can be grown rather than killed draws on regenerative medicine, using techniques developed in some of the world's leading tissue-engineering laboratories, such as J. Vacanti's at the Massachusetts General Hospital, where Catts and Zurr were artists in residence for a year.

One might read the curious careers of ORLAN and Stelarc as mid-twentieth-century probings of surgical and cyborgian imaginaries, part of the preparatory but brute technologies for the contemporary biotechnological revolutions (analogous to the "slash, poison, and burn" technologies of surgical, pharmacological, and radiation oncology that are slowly being superseded by noninvasive and regenerative technologies). Born a year apart, just after

World War II, ORLAN and Stelarc each now have a career corpus produced partly in conjunction with developing surgical and robotics technologies. The corpus of each is generationally marked also by highly literate vernaculars. For ORLAN, one key reference, in addition to Artaud, is the Lacanian psychoanalyst Eugenie Lemoine Luccioni (in her conferences, a performance format in which she is seated at a desk, or occasionally standing, reading from her texts, while slides and videos of her work are projected, creating a distance between herself and her body in the images, "despite her black lipstick and half black, half yellow hair, [she] comes across as a middle aged, French academic with self-depreciating humor").[73] For Stelarc, it is Marshall McLuhan ("Well I certainly have read Deleuze and Guattari, Baudrillard, Virilio, Lyotard . . . [but still McLuhan's] notion of externalizing our nervous central system is for me a central tenet"). Stelarc (born Stelios Arcadiou in Australia in 1946, named honorary professor of art and robotics at Carnegie Mellon University in 1997, and artist in residence for the city of Hamburg in 1999) has worked with roboticists at Waseda and Tokyo universities. He has moved from prostheses (a third hand moved with leg and abdominal muscles, a left arm attached to electrodes that act as muscle stimulators), to "suspended animation" performances to make the body move and fly within electromechanical exoskeletons, to hybrid-human machines (including a brief interaction with the U.S. National Aeronatics and Space Administration's project on cyborgs), to more recent explorations of parasitism (both the body within a networked and remotely controlled system and the body as host to agentive objects within). ORLAN (born in 1947, now professor of beaux arts in Dijon) has had a long career in painting, sculpture, dance, and public or street theater, from small disruptions to rush-hour traffic as a youngster, to the 1977 scandal performance (she lost her teaching job) *Kiss of the Artist* (sitting in front of a life-sized photograph of her nude body and one of her as the Madonna, she sold kisses for five francs), to her 1990s "carnal art" series of nine facial surgical *Interventions* (the seventh broadcast live by satellite to galleries and museums in Paris, New York, Toronto, and Banff), and her computerized *Hybridizations* of her own face with images of very different ideas of beauty from many other cultures (such as pre-Colombian Olmecs with "flattened skulls, cross-eyed vision, and false noses"). She invokes as well the Mayan priest of the god Xipe Totec, who dons the flayed skin of his human sacrificial victims, an image that von Hagens also evokes in one of his dramatic plastinations, in which a real male body holds its own skin. Like Stelarc, ORLAN requires collaboration with surgeons and information technologists.

Much of what counts as experimental knowledge production emerges in these careers like much biological knowledge (and like biological evolution) incrementally out of a play and feedback from the exhaustion of sets of structural possibilities (successive projects unfolding as sets of variations) more than a precise working out of a long-term intentional plan. As art performances, such knowledge production requires engineering (careful planning, design, and coordination), but it partakes centrally as well of the inventive spirit of hacking and experimentation (science), often like the latter in small steps. This incrementalism (and ambiguity between knowledge and play, the trivial and the profound, the monstrous/bestial and the redemptive/divine) is what provides the time for ethical reflection, as does the slowness of the hard work of the biomedical sciences and biomedical engineering (as opposed to urgent temporality of the hype and fund-raising that is one of the necessary but also disorienting conditions of technoscience and its constant production of novel tools and platforms).

Pain and redemption are not signifiers in the performances of either ORLAN or Stelarc: This is not ritual that uses pain to inscribe the body with significance (neither truth through torture nor identity and healing as among body-art folks or traditional religion). ORLAN says she wants to put the "naked body in the spaces opened up through scientific discovery," "realized through the technology of its time," and that her art, "lying between disfiguration and figuration . . . is an inscription in flesh, as our age now makes possible. No longer seen as the ideal it once represented, the body has become a 'modified ready-made.' "[74] Commenting on the surgeries, and some audience members' inability to look, she advises, "When watching these images, I suggest that you do what you probably do when you watch the news on television. It is a question of not letting yourself be taken in by the images, and of continuing to reflect about what is behind these images" ("Autobiography"). In her operating theater "I can observe my own body cut open, without suffering. . . . I see myself all the way down to my entrails; a new mirror stage."[75] A new mirror stage, of course, is a reference to Lacan and psychoanalytic ideas about the misrecognition of the self. A text by Lemoine Luccioni inspired her: "The skin is deceptive . . . There is an error in human relations because one never is what one has. . . . I have the skin of an angel but I am a jackal . . . the skin of a crocodile but I am a poodle, skin of a black person, but I am white, the skin of a woman but I am a man; I never have the skin of what I am. There is no exception to the rule, because I am never what I have." She revels in the idea of being able to carry on a discussion with an audience by video feed while her body is open to the surgeon: "I read the texts as long as possible during the operation, even when they are operating on my face, which gave during the last operations the impression of an autopsied corpse that continues to speak." Her Web site tells us that "famous designers, such as Paco Rabanne and Issey Miyake, have designed costumes for ORLAN to wear during the surgeries. Poetry is read and music is played while she lies on the operating table fully conscious of the events taking place (only local anesthetic is used)."[76] No pain: "In my work, the first deal with my surgeon is 'no pain.' I want to remain serene and happy and distant," and she wants to send "images of my body opened up with me at the same time having a completely relaxed and serene expression able to answer any questions."[77]

ORLAN's trajectory from blaspheming to "technologies of our time" and the "naked body in the spaces opened up through scientific discovery" is recapitulated within her surgical series.[78] In 1993, her seventh surgical performance was called *Omnipresence*, punning on the attribute of the Christian God and the telepresence technology. Silicon implants were put into her temples, and the operation was broadcast to four interactive sites in Europe and North America. Her "carnal art," she says, works through augmentation, not self-mutilation. She draws on Greek (pagan) mythology: the implants in her temples as Dionysian horns, her bloodied lips as Bacchus' grapes, Dionysian rites of *sparagmos*, or tearing the body apart; the anarchism of her father (anticlerical, resistance fighter, and Esperanto enthusiast); and on Artaud ("I also use a lot of Artaud, because I am interested in the concept that the body is obsolete").[79] Her 1997 show, *Ceci est mon corp . . . ceci est mon logiciel* (this is my body . . . this is my software), using images from the 1993 *Omnipresence* as well as the 1998 *Hybridization* (done collaboratively and transatlantically with programmers in Canada), points both to an increasing concern with connectivity, telepresence, and the public sphere and to the body without organs, releasing the human from its obsolete primate body. She also, along the way, gestures to writing, marking, with the blood that is part of the surgical release.[80]

A more conflicted psychoanalytic reading of ORLAN's work than ORLAN's own accounts is given by Parveen Adams. Adams points to a performative contradiction, the "work of anamorphosis":[81] ORLAN, Adams writes, "directs under local anesthetic . . . [and] while ORLAN experiences little pain, she makes sure that we experience a more substantial pain . . . of *jouissance*," of the *gap* between signifier and signified.[82] Adams suggests we pay attention to three splits in the speech act: first, between the performative effects experienced by spectators (like Freud's horror looking down Irma's throat in his famous dream) as opposed to those experienced by ORLAN ("while lying on the surgical table, what she experiences is the overwhelming desire to communicate"); second, "her subjective account is that with each operation she becomes more distant from her body, it is as though she sloughs off her body to enter the pure subjectivity of speech";[83] and, third, while ORLAN claims to morph her face with art-historical images of women's faces ("flesh become image"), Adams thinks that instead she is showing that the image is empty, that the mask has nothing behind it, unhinging the grounds (even "all relations") of representation. While ORLAN "sets off on the highway of information," the spectator "is forced to tarry in this circus to witness something else, something which insists."[84] Like Freud, Adams experiences horror as she watches the surgeon separate the ear from the face, shocked by the realization and recognition that the face is detachable. The "gap" or "space" that is opened as the skin is lifted "unhinges"[85] the stabilities of inside/outside, mask/face, and any notion that the face represents something deep or interior. Adams quotes Lacan's description of Freud's dream of Irma:

> There is a horrendous discovery here, that of the flesh one never sees, the foundation of things, the other side of the head, of the face, the secretory glands par excellence, the flesh from which everything exudes, at the very heart of the mystery, the flesh in as much as it is suffering, is formless, in as much as its form in itself is something which provokes anxiety.[86]

In contrast to normal cosmetic surgery or that of transsexuals, where the goal is completion ("until my surgery I am unfinished"), Adams concludes, ORLAN undoes any sense of such completion: "As the face becomes detached, it no longer projects the illusion of depth, it becomes a mask without any relation to representation."[87] Of course, one might reply somewhat pragmatically, first that this dissociation is something medical students pass through in their training (ORLAN's "technologies of our time . . . spaces opened up through scientific discovery")[88] and, second, that facial transplants can be life restoring, even if they require lifelong immunosuppression (see note 57).

Still, Adams crucially draws attention to the differences between the surface marks of the illustrated body and the deep marks of the traumatic body. The latter is increasingly visible (as epidemiological demographics as well as individual bodies) both through the cancers and other diseases that may be environmentally caused by our own productive activities and through the spreading, pervasive effects of torture, war, and rape as a tactic of war, visible no longer only in the men and women on crutches (World War I, *Kandahar*) but also in the silent ones (World Wars I and II, rape victims), the symptomatic ones (Vietnam War, Gulf War), and their stressed families.[89]

If ORLAN has set out on the information highway, Stelarc has long been centrally concerned with these connectivities, with placing the "obsolete body" within the wider nervous and muscle systems in which the human body is being enveloped, intensifying Walter Benjamin's observations of the immediate aftershocks of World War I, concretizing Marshall McLuhan's anticipations of telepresence and "externalizing our nervous central system." William Gibson gives an account of him that is not unlike Ayer's description of ORLAN—"an utterly conventional looking man," who radiates "calm and amiability"—and adds, "But what I recall experiencing" from having watched Stelarc in performance with his third robotic arm, "was a vision of some absolute chimera . . . I sensed that the important thing wasn't the entity that Stelarc evoked but the labyrinth that the creature's manifestation suggested."[90] Stelarc had swallowed a camera with great difficulty and medical help, and while it had unfurled properly, it did not refurl; Stelarc, says Gibson, took the likelihood of needing surgery in stride.

Today, there are swallowable cameras in pill-sized capsules for gastrointestinal exploration.[91] Stelarc's *Third Hand* (Maki Galleries, Tokyo, 1982) was triggered by electromyographic (EMG) signals from abdominal and leg muscles, and he was able to write in a coordinated way the word *evolution* with all three hands, while the interior sounds of his body were transduced and amplified through a speaker system. He describes this connectivity not as a simple insertion but as an implosive effect of technological miniaturization: "With the desire to measure time more and more accurately and minutely, the necessity to process vast amounts of information, and the impulse to catapult creatures off the planet, technology becomes more and more complex and compact. This increasing miniaturization creates an implosive force that hurtles technology back to the body, where it is attached and even implanted."[92] (Today, inversely, the noisy signals of Matthew Nagle's brain are parsed for functional feedback so he can do things without his limbs.)

As with ORLAN, pain is separated from the performance. In *Fractal Flesh* (1995), Stelarc was in Luxembourg, wired to a muscle stimulator that controlled the left side of his body, thereby splitting the nervous system. Parts of his body were under the control of external triggers; other parts were under the control of his brain and central nervous system. Participating viewers in Paris and Amsterdam could enter signals via a touch screen that would affect the left side of the body. In *Extended Arm* (2000), one hand on a keypad controlled long aluminum fingers, while the other hand was controlled by muscle stimulators that forced it through a choreographed set of motions. Pain, Goodall remarks, "can be seen to pass in waves through the face, but is not part of the overall aesthetic." These waves of pain, she suggests, "are a reminder of how much is going on."[93] Even more generously, Timothy Murray suggests that "the trauma of the body in regimes of high-technology is at the basis of Stelarc's practice."[94] But Amelia Jones demurs, insisting Stelarc's claims are ideological masculine denials and desires to escape the body.[95] Brian Massumi draws attention to the transduction and extension of the body, arguing that even the early body suspensions on hooks were carefully calibrated to minimize pain and tearing of the skin (with eighteen hooks to distribute the body's weight), the skin stretched to the limit, the body in "suspended animation," and the body sounds transduced and propagated to fill the surrounding space. The "amplified body processes include brainwaves (EEG), muscles (EMG), pulse (plethysmogram), blood flow (Doppler flow meter). Other transducers and sensors monitored limb motion and indicate body posture. The body performs in a structured and interactive lighting installation which flickers and flares in response to the electrical discharges of the body."[96]

Over time, Massumi notes, Stelarc has moved from exploring prostheses (*Exoskelton*, *Goggled Eyes*, *Third Hand*, *Virtual Arm*, *Extra Ear*), all of which are extensions rather than substitutions for the body, to cybernetic networks rewiring the body's motion (*Split Body*, *Fractal Flesh*, *Stimbod*, *Ping Body*, *Virtual Arm*, *Virtual Body*, *Parasite*, *Movatar*). Always, Massumi suggests, Stelarc has recognized the idea of a single body evolving to be an absurdity, and conceptually all the experiments were stand-ins for extension into a collectivity, which "truly begins to unfold when the audience is let back in,"[97] but also pragmatically as a function of the fact that most of his performances require medical and technical collaborators.[98] Indeed, it is often most interesting when these collaborations and transductions fail, thereby showing the otherwise-hidden work and fragility of the connections. In 1995, for the *Fractal Flesh* performance, in which participants were located in Paris and Amsterdam and Stelarc in Luxembourg, the early Web was too slow, and a dedicated network of modem-linked computers had to be set up. But soon, with faster and broader-band Internet connectivity, for *Ping Body* (1996), *Parasite: Event for Invaded and Involuntary Body* (1997), and *Parasite* and *Movatar* (2000), Stelarc could allow pinging over forty or more sites to feedback into the performance, varying the modalities of four kinds of body movement: voluntary, involuntary, controlled, programmed.[99]

In using brain waves and EMG signals, Stelarc operates on the margins of neurological and biocybernetics research to re-enable sufferers of Lou Gehrig's disease (ALS) and similar conditions of paralysis to move prostheses and screen cursors via brain implants or via merely reading electrical waves on the skull like an EEG.

With *Third Ear*, Stelarc is beginning to move toward tissue engineering, and with *Prosthetic Head* toward "a conversational system." The latter is an extension of old computer science experiments with expert systems technologies, beginning with the wildly successful first artificial intelligence psychotherapy program, Eliza (1966), whose creator, Joseph Weizenbaum, left the field for fear such technology would be misused. The *Prosthetic Head* uses phrase recognition and prediction, selecting phrases from its human interrogator and mirroring conversational fragments, so that, as Stelarc quips, the artificial intelligence is only as smart as its interrogator. But "Baldi," a conversational agent that, in conjunction with a cochlear implant, teaches deaf children at the Tucker Maxon Oral School in Portland, Oregon, is already much more productive. "Baldi" (designed at the University of California at Santa Cruz) listens to a student, runs what it hears against a speech-recognition system, notes mispronunciations, and shows the student how to pronounce the "sh" sound by puckering its lips, the "th" by sticking its tongue between its teeth, and so on.[100]

Suppose, Stelarc says, you add biorhythms so that the *Prosthetic Head* is grumpy in the morning and tired in the afternoon and has information about you through a vision system (à la Steve Mann) and can comment on your clothes or expression.[101] Suppose you build in affective computing devices to read emotion, such as IBM's *Blue Eyes* software (to detect excitement in the eyes of consumers), Rosalind Picard's affective computing devices at the Massachusetts Institute of Technology (MIT) Media Lab, or the software developed by Paul Eckman (University of California at San Francisco) and researchers at the Salk Institute for detecting deception signaled by facial muscle cues.[102]

Mimicking and interactive interfaces that read body marks move in one direction. But, alternatively, suppose, Stelarc continues, one developed *e-motion* (not emotion) through sensors, such that an avatar could move your facial muscles, turning you into a surrogate body for its *e-motions*, causing your facial expressions.[103] While this is a bit bizarre as a goal, it is an inversion not unlike his ideas about synthetic skin, which again extends what medically exists today to help burn victims. Suppose, he says, such a synthetic membrane not only were permeable to oxygen but also possessed some photosynthetic capabilities. It could then produce nutrients and dispense with the need for the gastrointestinal tract, circulatory system, and lungs, allowing the body to be hollowed out, making space to host technological components: a body without organs,[104] something perhaps he already gestured toward by having a camera introduced into his stomach to not only see but also "play" the music of the body (the opening and closing of heart valves, the slosh of blood) through Doppler ultrasonic transducers. Perhaps this is a retro-gesture toward Manfred Clynes and Nathan Kline's original ideas about cyborgs as preparation for space flight, and the idea that long-distance space flight would require the human body to undergo changes and machine-body hybridization.[105]

In an interview with Ross Farnell, Stelarc says he is not trying to live as a posthuman body but rather that

> it's important that it is not purely a fanciful idea or science fiction speculation, but rather . . . you try to cope with the precision and the complexity and the speed of this technological terrain, and you . . . live with the consequences. [mutual laughter]. . . . What constantly pleases me is that in creating these unstable situations with the body you generate unexpected outcomes with new interactive possibilities.[106]

In this interview, rather than the idea of a hardened skin with a hollowed-out body, he suggests that with nanotechnology, we will "have the possibility that the body becomes the host for colonies of micro-miniaturized machines,"[107] particularly (or initially) nanobots that circulate in the blood, sensing problems, doing repair, delivering adjustable dosages of medication—an idea that some medical scientists also hold.

The *Third Ear* is iconic for the way in which art lags behind reality while suggestive of future possibilities. Ignoring current functional technologies such as the cochlear implant, the *Third Ear* is nonfunctional or reverse directional (it could come with a chip that emits rather than receives sound). The *Third Ear*, however, is visually and materially a reference to the famous tissue-engineered ear successfully implanted on the back of a mouse, proof of the concept that one can chemically engineer polymers as matrices for three-dimensional tissue growth, an interdisciplinary collaboration between chemical engineers Robert Langer and Linda Griffiths (both of MIT) and surgeons Joseph and Charles Vacanti (of Boston's Children's Hospital and the Massachusetts General Hospital). The tissue-engineered ear was designed for infants born without an ear and is connected to efforts by Anthony Atala's tissue-engineering lab to grow bowels, as well as the attempts by his and a half dozen other such labs, including Vacanti's, to grow or self-repair a variety of organs. With complex organs like kidneys or livers, the layering of tissues with polymer matrixes and lithographic-like techniques to guide capillary growth are among the key challenges.

The possibilities of repair and regeneration are at the heart of projects by the Perth, Australia, based bioartists Oran Catts and Ionatt Zurr, as noted above, as is their pedagogy for learning about ecological relations. Photos of their bioart now hang on the walls of J. Vacanti's lab where they spent a year expanding their tissue-engineering skills and working on their victimless meat project. Catts and Zurr intend their projects as public education, not just about new biotechnologies, but more importantly about ways in which we can learn to live with, rather than consume or destroy, the ecological and reproductive forces of biology. In their early work growing tissue around small dolls as a "green" form of Guatemalan "worry dolls" (to which one tells, and off-loads, one's anxieties), they began the effort to install a biology lab bench at art shows to demonstrate how the tissue engineering is done, to demystify and counter the anxieties publics have about this technology. In their more recent work growing meat from harvested cells rather than slaughtering animals, the effort is to think about how we can harness biological processes in symbiotic ways.

GATTACA: BACK TO THE FUTURE BODY

> Imagine the universe in expansion: does it flee from terror or explode with joy? Undecidable. So it is for the emotions, these polyvalent labyrinths to which only after the event, the semiologists and psychologists will try to attribute some sense. . . . We have nothing to do any more with the heraldry of the tragic. . . . It is not the tragedy of a destiny, nor

> the comedy of a character (it can be presented in this way, of course) . . . rather the strangeness of fictive spaces, Escher's waterfalls whose point of impact is higher than their source.
>
> Jean-François Lyotard, *Libidinal Economy*[108]

> Maybe the problem is not biogenetics as such but, rather, the social context of power relations within which it functions . . . Enlightenment remains an unfinished project that has to be brought to its end, and this end is not the total scientific self-objectivization but—this wager has to be taken—a new figure of freedom that will emerge when we follow the logic of science to the end.
>
> Slavoj Žižek, *Organs without Bodies*[109]

> Why not walk on your head, sing with your sinuses, see through your skin . . . Find your body without organs. Find out how to make it.
>
> Gilles Deleuze and Felix Guattari, *A Thousand Plateaus*[110]

The haptic and proprioceptic body is natural (indeed the ground of perception), the body out of control is bestial, the transported body is divine. Like the old chain of being, or the merit-reincarnation sequence, of bestial, natural, and divine, the contemporary topology of natural, bestial, and divine is like a Möbius strip twisting back on its own implications. It is of anthropological-philosophical interest that while analytic philosophy and Habermasian civic responsibility have remained "modernist" and suspicious of this regenerative twisting topology, so-called continental philosophy (Nietzsche, Freud, Merleau-Ponty, Bergson, Lacan, Lyotard, Deleuze) has become enlivened with a biological sensibility. It is this biological sensibility more than the irrationalist tonality (of, for instance, Heidegger), against which Habermas is still (correctly) vigilant, that informs the productivity of much contemporary thinking and poesis (*Dichtung*). And it is that productivity, the *Dichtung* or weaving back and forth between contemporary biotechnology and ethical/anthropological stakes, that I have aimed to capture in the eye, surface marks, and deep marks, which are also the face, the communicating body, and the traumatic body. It is anthropologically and ethnographically (historically and socially) critical that they be understood in their social dynamics (class, gender, postcoloniality, asymmetrical power, etc.), not just as individualized bodies or codes.[111]

Reading Bodily Marks

My hair stands on end, I blush or maybe only redden (in embarrassment, love, anger), I wink or maybe only blink (voluntarily, involuntarily), I blink more rapidly (am I lying?), my pupils widen, nostrils flare, my cells scream, the cancer tumor heats up, the two genomes in my chimeric cells fight[112]—do you read me? When you feel me, do you also feel for me? When you catch my scent, can you tell if I'm afraid? To whom do my body marks speak: to you, to me (in my fantasies and misrecognitions of self-knowledge), to the instrumentation that listens through my skin to deep rhythms, vibes, heat, and other codes? Kismet—the robotic cartoon with camera eyes that follow my motion, that moves its head and eyebrows and has a voice synthesizer that responds to my comments (but no nose or other facial features)—responds to my body's voice and motion, but where in this "sociality" does projection start and communication end? Is it part of *my* libidinal body?[113] The space suit, pacemaker, and nanotech uniform that read (and adjust) my vital functions, even repairing some of my wounds, the hospital tubes and electrodes that keep me alive, are clearly exoskeletons. But what of the fingerprints, exfoliating skin cells, hair, and other marks that my body leaves behind so that DNA typing can ensure GATTACA-style surveillance?[114] Do you read me, will you still need me when I'm sixty-four?[115]

Ethical Spacing of Emergent Bodily Marks: Bestial, Natural, Divine

What will the near and far future body be? How much of the past is projected forward to make sense of the new, like catacoustis[116] or like the artist's glimpses of past source images in current confusions and emergences?[117] How much of the future is reconfigured from the past, like genetics reaching back into evolutionary history to track alternative branching points, alternative biological solutions? Is this the Nietzschean eternal return out of which liberation can emerge (experimentally) or be reconstructed (reverse-engineering style)? And is this the temporal spacing of ethics, the interactive "face of the other,"[118] the "given time"[119] to think about what we are doing, the feedback, and warning signals, signs of wisdom? In Ray Bradbury's *Illustrated Man* (1951) the narrator meets a man with tattoos all over his body, each with a story that predicts the future. One bare spot fills in with the future of anyone who spends time talking to him. When the stories stop, the bare spot fills in with an image of the Illustrated Man choking the narrator to death, and he runs toward a town he knows he can reach before morning. Bradbury's cautionary tales about our technoscientific world do not contain inevitable predictions: We can escape the nightmare possibilities. Bradbury says that to survive we must build "empathy machines" or "compensating machines" and that the artist has a key role to play.[120] Such machines are perhaps like the robot grandmother in "I Sing the Body Electric" (1960), who gives the impression of imperfection but anticipates human desires, or Isaac Asimov's famed ethical programming commands for robots to ensure that they do not hurt human beings.

Marked Bodies of Biopolitics

Tortured bodies, the wounded walking, and the living dead (zombies) are significant social and epidemiological categories today, the way the poor, the serf, the untouchable were in the past. So, too, are statistical groupings of the ill, often now networked or organized by disease for political leverage to get research done, drugs to market, insurance to pay. So, too, are the marketing categories of the "patients in waiting," the bodies whose genetic markings register probabilities of predispositions for future illness and who thus in the projected world of "individualized medicine," of pharmocogenomics, could be prophylactically on "drugs for life" (Lipitor being a current example).[121] All these are marked bodies.

Markings of, and Markings by, the Body

The bodily marks can be parsed in many ways. Maurice Merleau-Ponty reminded us that the body is always a third term in the figure-background Gestalt of perception, Henri Bergson that perception comes with memories that the active body accumulates, anthropologist Mick Taussig that synesthesia bleeds the senses into one another's zones of expectation, musicologist Steven Feld that sound can overwrite sight in cultural cosmologies, anthropologist Constance Classen that cultural cosmologies focus bodily attention on different senses, neurologist Oliver Sacks that case materials show the senses operate in different ways in different people.[122]

Marks of Our Time

We end this anthropological pointillism of the bestial, natural, and divine marks of the bodies of our times—the transcribed body (in various codes like GATTACA), the psychotropic body (like Sacks's testimony to his own enhanced and precise control of visual imagery and memory under the influence of large doses of amphetamines[123]), the testimonial body (of aging, scars, torture, etc.), the libidinal body (of transferable intensities), the automatic body (nervous, wired, raging with hormones), the topological body (of misrecognition), the collective technobody, the signing, signaling, enigmatic, communicating body, the multiple body—with a return to the ethics of our puzzlements.

Against Habermas's warnings, deploying modernist common sense, that to manipulate the human body is to threaten the "ethical self understanding and can disturb the necessary conditions for an autonomous way of life"[124] (compare the opening epigraph from Barker), both contemporary bioscientists and philosophers who attempt to think through these biotechnologies posit the need to follow further the Enlightenment demands of inquiry: "there is no return to the preceding naïve immediacy."[125] Following the American chronicler of twentieth-century popular-culture speculations, Ray Bradbury, the artist, anthropologist, and philosopher can all play a role making sure that the consequences of the speculations are not mere abstract nightmares or utopian fantasies but are lived in their micro-multiplicities, feedback, reroutings, and reincorporations, in a density of experience and trial. The world into which we enter is emergent precisely in this mundane/profound sense that we run quickly out of reasons but still must act, and those actions have serious consequences that create forms of sociality that incorporate the face of the other, the call to be responsive to the other. This remains true as we create new sensing biotechnologies, even as the overloads of sensations cause us to inscribe ourselves ever deeper into the ruins of the past, out of which we must emerge.

CHAPTER NINE

Dissolution, Reconstruction, and Reaction in Visual Art, 1920 to the Present

ANA CARDEN-COYNE

The 1920s signified a period of social, political, and economic reconstruction in Europe, following the devastation of World War I. Reconstruction impacted on the human body not just in terms of physical and vocational rehabilitation but also in cultural practices. Artists and writers, many of whom had been combatants, ambulance drivers, or volunteers in hospitals or paramilitary organizations, had witnessed the impact of modern war on the flesh of men. In Britain, official war art and mass-media photography mediated concerns for images to reveal the "truth" about war.[1] In a plethora of images, the human body was visualized as the target of masterful technologies designed for the purpose of mass obliteration. Sensitive to the ironies and abuses accompanying modernity, artists responded to the impact of war. Bodies were culturally and philosophically reassessed in a range of ways. For surrealists and Dadaists, the body was made even more absurd and grotesque, redefining beauty beyond recognition. There was an "irrational modernism" with its own body language, evident in Charlie Chaplin's staccato posturing as much as the sensual dystopias of Marcel Duchamp and Man Ray.[2] Jazz bodies symbolized libidinal sexuality; for some it meant corporeal liberation, for others moral decline. Contorted bodies, black bodies, and sexual bodies threatened bourgeois morality and taste. Luis Buñuel's sliced eye of a woman was an exemplary "vile body" (to appropriate Evelyn Waugh). Abjection and degradation edged the borders of beauty, as Julia Kristeva has suggested.[3] In a period when the embodied perfection of the *rappel à l'ordre* (return to order) dominated postwar Europe, artists embraced and reacted against the conventions of classical beauty. Still, others engaged with reflexive irony on the capacity for visual spectacle to find meaning in mass death.

Politically, the ordered body represented the cornerstone of postwar healing. Culturally, it was evident in the new classical turn across Europe, as the body became a site of physical repair and aesthetic rebirth.[4] New objectivists and constructivists and a host of others abstracted the body to its core geometry, so that shapes elicited profound hopes for restored humanity projected onto essentialized forms of embodiment. Architectural ideals from the past were jettisoned into the future. The human body cried out for a new materialism. Objectivity could be found in machine aesthetics, inspired by urban industrialism, Taylorism, and Fordism. Valorization of industrial efficiency and human perfection emerged from out of the chaos of war. Even the very weapons that weakened mankind could not exhaust the possibilities of rational progress. The idea of a New Man permeated the European imagination.[5] Automatons (man-machines) were modern warriors, drawn from engineering models, as well as prosthetic technologies, reconstructive surgery, puppets, and mannequins. We see this in Britain, France, Italy, and Germany, among other combatant nations. Winners and losers found that machine bodies had stories to tell, from Weimar to New York.

These are just some of the ways in which modern artists represented the body in the 1920s. Objectifying, classicizing, and abstracting the body formed three main strands in visual representations of the body in the first half of the twentieth century, and these constitute the focus of the first part of this essay. Throughout, I follow the themes of the body's dissolution, reconstruction, and reaction across the twentieth century, considering representations of the body in the last fifty years, through changes in society, visual media, and technology. Gender, race, and sexuality are key components in defining and inscribing bodies, as are physical integrity and disability. I draw attention to those factors when they are of prime significance to the intention and delivery of the work and the cultural context in which works were conceptualized.

OBJECTIFYING, CLASSICIZING, AND ABSTRACTING

World War I forced artists and intellectuals to confront their theories with practices; speculation about the object, and experimentation with the material world, confronted extreme experiences surrounding death and survival.[6] Technology's impact on real people broke much of its romantic associations. Many artists and writers were mobilized during the war, and many suffered terribly. Umberto Boccioni, Henri Gaudier-Brzeska, and Wilfred Owen were killed, while Georges Braque and Fernand Léger were wounded. Blaise Cendrars had a limb amputated, Roger de la Fresnaye suffered complications for seven years until his death, and Guillaume Apollinaire died on Armistice Day, after a battle with influenza.[7] At the other extreme, Pablo Picasso and Juan Gris were outcasts, labeled non-Allied aliens. While living in France as noncombatants, they endured abuse accorded to foreigners, intensifying their abhorrence of war. Picasso and Gris's close friend, the art dealer Daniel Kahnweiler, was exiled from Paris during the war.[8] After 1918, many artists and their various circles now wished to promote peace, social stability, and human progress: to make productive use of the global catastrophe. As the Belgian writer Michel Seuphor declared, "in the place of the romanticism of speed . . . we put the slow pace of human awareness. In the place of revolution, we put order and the will to perfection."[9] At that time, such meditations were not as reactionary as phrases such as will and perfection seem to us now, with the benefits of hindsight.

Modern European artists struggled to redefine the body in this period. War had forced a peace with their uneasy relationship to the past. Classicism, for instance, had been emblem of the conservative, the reactionary, and the stultified conventions of the humanist tradition, which emphasized mimicry and privileged the human form.[10] Reworking classicism was not simply a "return to man," then, but also a reorganization of the body. As abstraction and object, flesh was reconfigured into an "architectural" aesthetic.[11] Underscored by the experience of war, there was an unassailable connection between the human body and these artistic developments. Objectivism, abstraction, and classicism appealed more than ever for a common belief in certainty over sensation and logic over emotion, qualities reflected in the treatment of forms. In Russia as in Europe, what had been seen and witnessed were horrors testing the capacity for truth, justice, and resolution. Many artists remained conscientious revolutionaries, bonded to the hope of Socialism, despite the war. In 1920, a group of artists led by the Russian Vladimir Tatlin proclaimed: "We declare our distrust of the eye and place our sensual impressions under control."[12] This was the revolutionary end of the wedge.

The human body was important in a number of artistic developments in this period that were less radical and yet shared some of the same values. In Britain, for instance, Clive Bell and Roger Fry's "Significant Form" had originated from a French cubist rationale. Key to this thinking was that "from the fat to the muscle it was cubism that gave us the anatomy of the picture itself." In 1923, however, Bell critically differentiated from cubism, linking the human body and the evolution of modern art. Artists were trying, in his view, to "clothe the bones with flesh" without forgetting that "the bones must be there."[13] Similarly, the critic Herbert Read described the cubist as "naked before the world" by his "stripping of the object," relying solely on form.[14] Returning to the human form, artists applied new theories to formalism and abstraction. Architectonics and volumetric forms borrowed from the classical precedence of Praxiteles, Poussin, and Ingres. Against the horror of war, bodies were beautified through structured forms; given solid geometric arms, heads, and legs. Flesh became sensual but austere, sometimes linear, and often monumental.

Depictions of the human body in the new classical style often possessed a lofty quality. Dame Laura Knight's portrait of a woman *Blue and Gold*, published in a British art magazine in 1930, appeared more goddess than woman; an aloof facial expression evokes spiritual qualities, accentuated by the backdrop of heavenly skies. Abundant breasts pose her as a mother. Notably, the classicism in this work foregrounds the idealization of motherhood across most Western countries in the postwar period. With French fears over *denatalité* (subreplacement birth rate) and the influence of pronatalism, we should recall other classical images of women depicted as monumental bearers of the future race. Evocative images of women with one breast revealed or suckling children were common in this period. Picasso's *Maternity* paintings (1920, 1921) and *Mother and Child* (1921), André Lhote's *Nude* (1920), Amédeé Ozenfant's highly abstract painting *The Source: Woman at the Spring* (1927), Albert Gleizes's cubist abstraction *Mother and Child* (1920), Roger de la Fresnaye's *Mother and Child* (1923), and Léger's *Mother and Child* (1922) are some examples of the female body as goddess of the past and mother of the future.[15] Repopulation was literally within women's bodies. Visual art extended this powerful symbolism to one of Western social and cultural rebirth.

Artists may well have been influenced by the 1921 Exposition Nationale de la Maternité et de l'Enfance (National Maternity and Childhood Exhibition). As Romy Golan has shown, this perennial French fear, exacerbated by the loss of 1.3 million men, reinforced the identification of peace and restabilization with women's maternal role. Procreation and the home were definitive messages of the French return to order.[16] Alongside pronatalist concerns with female biology were important social and global contexts in which the idealization of *la grand mere* occurred.[17] Similarly, classical images of female bodies were produced in Britain, the United States, and Australia. Monumentalism in female portraits was taken up by London-based painters such as Savely Sorine, William Roberts, and Dod Proctor, whose works were published in popular magazines such as *British Painting* and *The Studio*.[18] Fecund depictions of "woman" with colossal breasts and enlarged womb appeared in the work *Reclining Nude* (1925) by the Bloomsbury sculptor Frank Dobson, and Eric Gill's stone sculpture of a female figure, with its evocative title *Mankind*

(1927). Modern artists reflected cultural preoccupations with the female body as an agent of recuperation. At a time when sex reformers such as Marie Stopes and Margaret Sanger, alongside physical culturists such as Ettie Rout and Fred Hornibrook, were positing the benefits of exercise, female orgasms, and motherhood, artists visualized reproductive sexuality, affirming the sentiments of life after death and peace after war.[19]

At the same time, monumentalizing the female body refracted the process of humanizing found even in hard machinic forms. Léger's work became more organic, with curvilinear rather than rectilinear bodies. The stacked, interlocking geometries of earlier works gave way to simplified planes. We can compare two paintings of French poilus, *La Partie de cartes* (1917) and *Le Soldat à la pipe* (1916), with the later work *L'Homme à la pipe* (1920). Significantly, the 1917 work was painted while he was convalescing from war wounds. Soldiers playing cards were depicted with a fragmented machine aesthetic.[20] Configured as robots, limbs appear as burnished steel confronting each other in a space where the table replicates a war zone.[21] In the later picture, the man with the pipe is no longer a soldier. His body is less segmented and more curvilinear, and his face is recognizably human. Tubular volumes give way to flat space and elemental mass. Léger returned humanity to machine ideology, at the same time applying a superhuman quality to bodies in art. Automatons proposed man and machine in symbiotic relationship. Intended as modern Olympian Gods, such images were nevertheless fragile. In this period, the prosthetics industry and government rhetoric attempted to pitch technology as empowering the body of the disabled man. At the same time, Sigmund Freud sensed impotence and awkwardness in cultural obsessions with man becoming a "prosthetic god."

By 1929, Léger, the guru of the mechanical aesthetic, moved away from dehumanizing Taylorist iconography toward a monumental humanity. We find this in the tranquility of his prodigious *Bather* (1931), the grand stature of his circus acrobats, or the three Olympian goddesses in *Le Grand Déjeuner* (1921). Golan has noted this transformation in the work of Le Corbusier, where linear Purist geometry gave way to softer, billowing circles.[22] Although Léger attributed his machine consciousness to the war, inspiring admiration of the "beauty of the fragment," the objectification of his subjects was combined with classicism, especially in the pursuit of absolute form and utilitarian beauty.[23] These two aesthetic demands came under the catch-cry of "purity," meaning pure and essential form, an idea he later shared with his great friends Le Corbusier and Ozenfant. In 1918, they associated the end of war with a social shift across the world, where everything would become ordered, lucid, and pure.[24] In his 1923 publication *Vers une architecture*, Le Corbusier extolled principles of classical architecture for modernism, a doctrine of geometric order, austere simplicity in design, and harmonic beauty.[25] Transition from the glorified fragment to absolute and timeless forms occurred within the cultural context of postwar reconstruction.

Utopia appeared on the horizon. Visual artists hoped for a new cultural order, one created by their ability to perceive universal qualities of humanity and communicate them in essential structures. War and revolution gave rise to an optimistic need to form bonds with other peoples. Belief in human unity led artists to identify with people of different classes and nations and to universalize human experience. In France, Léger claimed that his war experience gave him a chance to know his fellow poilu and to be renewed through them. Drafted into the engineering corps with workers, laborers, and miners, Léger thought that war enabled him to be close to ordinary people. Although Léger identified with the plight of his men, he also valorized them with a gaze commonly seen in bourgeois art of the late nineteenth century. In *Three Comrades* (1920), he envisioned the grand humanity of his men. Larger than life, they are modern Olympian deities transposed onto the idea of democratized warriors. War seemed particularly authentic as an equalizer. Léger found that war machines (cannons and airplanes) had become his last hope, inspiring him in the face of dread and death, just as pure bright color lent escape from the "grayness during that war, in so much mud."[26] Color planes not only were a structural element in the unity of the work but also symbolized the optimistic values of reconstruction, defying the bleak experience of war.

Artists from defeated and victorious countries shared this sense of war. Oskar Schlemmer saw the opportunity for art's healing powers to invoke a spirit of equilibrium within the chaos of contemporary life. Schlemmer had been wounded on the Eastern front in 1915. At that time he wrote, "I'm not the fellow who volunteered in August anymore. Not physically and especially not in terms of attitude."[27] War had changed him, and this continued to be evident in his work, affecting his desire to create peace in the universe. Universal order would synthesize the opposing forces within art and society alike. Elsewhere, neoplastic artist Theo van Doesburg argued that the destructive reality of war would eventually give way to a new spirituality, based on perfection and absolute harmony. He had served at the Belgian Frontier and, after the war, under the influence of his friend Bart de Ligt, Holland's leading pacifist, published a series of articles denouncing war's brutality.[28] Unity of the world would arise not only from socioeconomic development but also from the values of this spiritual harmony, which demonstrated this new relation between art and society. In a similar vein, the Russian pioneer of abstract minimalism, Kasimir Malevich, proffered "the birth of a new universal step" calling "humanity to unity." Such all-encompassing notions proposed an integrated plan for aesthetics and society based on restorative assumptions and humanitarian progress:

> Our being may advance to the single unity and wholeness on the path of universal movement. . . . We wish to form ourselves according to a new *pattern, plan and system*; we wish to build in such a way that all the elements of nature will unite with man and create a single all-powerful image.[29]

Desire for unity was an expression of one of the most important tenets of this traumatic period. Unity of man with nature (Van Doesburg called it the "synthesis of life") created a balanced existence paralleled in abstract geometry and visual art.

Modernists might disagree bitterly about the aesthetic exactitudes of the "pattern, plan and system," however many continued to accord with principles of unity and universality. For all Malevich's idealizing, such as by harmonizing classes and aesthetics following the October Revolution, he concurred with classical notions of rebirth, human unity, and totality. Human progress was an investment in humanity itself. This meant that artists were mortal and that art was the essence or sum of real life. Hence in 1921 neoplastic leader Piet Mondrian labeled this the "Abstract-Real," and in 1926 Léger named the "New Realism" a connection between human beings and machine objects.[30] Color and geometric form created unity within the artwork. This resembled the unity within the cosmos, the very thing that gave art its social purpose.

Despite dissension between Malevich and constructivism, the spirit of reconstruction was evoked in the universalist ideal for art and humanity by Russian emigrant Naum Gabo and his brother, Antoine Pevsner. Although constructivism had originated in prerevolutionary Russia, Gabo was mindful of war-wrecked Europe in his writings on the subject: "It is sufficient when Art prepares a state of mind which will be able only to construct, coordinate and perfect instead of destroy, disintegrate and deteriorate."[31] Gabo had completed figures during the war that were committed to the universal values of constructive art, such as *Bust* (1916) and *Head of a Woman* (1917). In this period, other artists perfected the body with unsentimental formalism. Corporeal abstraction and unity were the goals Pevsner's constructions *Head of a Woman* (1925), *Portrait of Marcel Duchamp* (1926), and *Torso (Construction)* (1924–1926).[32] Bodies were pared down from weak flesh. Instead, inner structure served as pure essence to convey strength. Constructivist corporeality was created out of the "materials of modern classicism," such as plexiglass and steel, reinforcing the impression of heroic armature. While artists had their own aesthetic agendas and political differences, various social pressures and cultural environments also impacted on ideals of embodiment and representations of the human body.

By the 1930s and 1940s, the radical dissolution of the body via abstraction on the one hand, and the optimistic reformulation of the body through constructive and monumental forms, on the other, was becoming marginal to the overarching political demands of militarism and totalitarian aesthetics. In Germany, the National Socialists staged an exhibition of art that just a few years earlier epitomized German modernism. Abstraction, cubism, expressionism, Die Brücke, surrealism and Dadaism, Bauhaus, and Weimar art were all banned and labeled "degenerate art" (*entartete Kunst*). Artists were forced to flee, alter their style, or cease practice.[33] Some artists became anti-Fascist activists, such as Fritz Cremer, Kurt Schumacher, and Oda Schottmuller, eventually executed in 1942.[34] Never before had culture and representation become a matter of life or death.

Degenerate art was bound to racial, mental, sexual, and moral stereotypes of decadence and primitivism.[35] Biologically determined in character, "degeneracy" meant that the body would be crucial for the National Socialist representation of German culture and civilization. To the Nazi "way of seeing," abstraction was a crude dissolution of formalism. Radical expressionistic lines appeared as contortions, while cubist fragmentation symbolized not just the decrepit, disabled, and racially inferior but also the corruption of impure bloodlines. As Klaus Theweleit has shown, masculinity among German veterans was directed into a brutal and unforgiving culture of violence and hardened self-representation.[36] Germans had long located their cultural heritage in classical civilization. While in the Weimar Republic emphasis was on the cultural achievements of fifth-century Athens, under the Nazis there was stronger preference for Spartan militarized fitness. Hardening the body in brutal exercise regimes was intended to create fearless warriors with high pain thresholds. In addition, they built a hero cult based on enforced unity and self-sacrifice.[37]

Reality and representation merged under totalitarian thinking about bodily integrity. Against modernism's deformities, then, Nazi art proposed the hyperbole of a whole and yet clinical classical body, bulging with hard muscles but also stripped of sexuality. The work of Arno Breker, Joseph Thorak, Georg Kobe, and others depicted German youth in the "hygienic" classical style. Muscularity was also used in depictions of females, such as Fritz Klimsch's *The Beholder* (1932) and Georg Kolbe's *The Chosen One* (1942), where naked women are deeroticized into a fantasy of moral virtue, racial purity, and maternalism.[38] As George Mosse argues, Nazi art presented beauty without sensuality—the nude became a naked ideal of purity, respectability, and *völkisch* (ethnic) corporeality.[39] In Fascist aesthetics, too, representations of bodies, especially that of the New Man, delivered a "chill beauty," whether in heroic images of Italian youth, in sterile depictions of Mussolini, or in graphic design and advertising iconography.[40] Fascism exploited "the reciprocal exchange that pertains in any state formation between the body politic and the formation of the individual body."[41] In the Soviet Union, classical formalism was revolutionized by the Socialist aesthetic, stylizing and valorizing the bodies of workers, sportsmen and -women, peasants, and fighters. In totalitarian art generally, "the people" appeared as the new gods, symbols of their brutal regimes and unyielding national politics.

During World War II, avant-garde art went underground not only as a result of totalitarian politics but also from the overwhelming authority of military culture.[42] Democracies at war, however, similarly activated the shift toward hypermuscular bodies, especially in depictions of men as war machines. Armored, phallic, and muscular, the male body became public art since it was the focus of propagandist aesthetics in posters and films, and through the projection of "real men's" bodies in Hollywood films, sponsored by the U.S. Department of Defense.

Cultural forms, ideas, and aesthetic styles in Europe were translated outside its borders. In the United States, Lewis Hine had already made a reputation by photographing women laborers and child workers in factories. In 1920, he completed a series of images of workers with their machines that spoke of the machine aesthetic fashionable in Europe at the time—muscles, masculine brawn, and technology—merging into one dynamic pulsating rhythm of sexual potency; and yet it is a body without real social or labor power. By 1942, the concept of production signified all things masculine. Men were valuable units of industry; their bodies could measure their productivity. The U.S. government both defined and deployed the symbolic male body. Male bodies were the "embodied symbols of the nation," influenced by "prevailing gender, racial and other cultural norms."[43] Modernist graphics used styles asserting the power of American industrialism. Jean Carlu's poster "America's Answer—Production," commissioned by the Division of Information, Office of Emergency Management, in Washington D.C., delivered the rhetoric of productivity bound to the valorized worker. Capitalist mythology was nevertheless reminiscent of totalitarian art in the same period. The grip of the arm and spanner around the letters of the word *production* presented masculine brawn and capitalist modernity as a mighty force.[44]

TECHNOLOGY, SOCIETY, AND PUBLIC BODIES

In the 1950s, the single most influential technology with respect to the representation of the body was not the machinery of war but rather television. The impact of mass industrial killings at Auschwitz and elsewhere, and the impact of the atomic bombs, relegated the body in pain to the silent memories of the traumatized and disempowered. Corporeality, too, seemed to dissolve in many cultural forms, especially painting. From Jean Dubuffet's scratched figures to

Jackson Pollock's drips, the body was in pieces. After World War II, did television symbolize yet another postwar rehabilitation? Or was it an anaesthetic of entertainment and consumerism? Certainly after World War I, production and consumption promoted a new evocation of bodily integrity. Now, however, the "modern electronic world" beamed real and ideal bodies into ordinary households, especially in the United States, where the focus of the art world had shifted.[45] Art in this period was dominated by American minimalism and purist abstraction; however, artists such as Mark Rothko suggested that not even abstraction offered a way to remove the body from visual representation. Rothko said that "the blood flowed out of his body" onto his large monochromatic canvases.[46] The 1950s also produced massive social and economic changes to the structures of capitalist modernity and the position of the United States in Western cultural practices. Some artists responded to these shifts as new technologies developed and the Cold War set in.

Mass-media images became the fodder for a new visual repertoire of the body. Against the heritage of abstraction, pop art emerged, with antecedents in comics, cartoons, pin-ups, and caricatures mass produced in magazines and billboards. In 1947, Eduardo Paolozzi's *I Was a Rich Man's Plaything* included postcards, advertisements, and the cover of a sensational magazine, *Intimate Confessions*. A long-legged "exmistress" is the target, as the words "pop" smoke from the gun aimed at her head. Although pop had inherited absurdist strategies from surrealism and Dadaism, inspired by the collages of Picasso, Kurt Schwitters, and Marcel Duchamp, the human body was not a specific object of attention. In contrast, pop art's female figures were glamorized and sexualized. Dragged from unreal catalogs of mass-media femininity, they appeared as ironic, even hysterical, stereotypes from 1950s American society and popular culture. Women were also staged in theatrical displays of modern life in visions of domestic banality. Pop culture appropriated ideal and superreal images of women from advertising and mainstream commercial culture. Richard Hamilton, Roy Lichtenstein, James Rosenquist, Andy Warhol, and Tom Wesselman found great humor in such representations of the female body located in the consumer paradise of 1960s America.

Within the decade, the women's liberation movement began to radically critique representations of women and female bodies in the mainstream media. Combating historical idealization of the female body, and indeed women's exclusion from the elite world of high art, including their lack of representation in museums, feminists passionately challenged social and aesthetic misogyny. Artists became activists. Engaging with feminist debates about beauty, sexuality, gendered codes of behavior, and proscriptive attitudes toward women's biological destiny and their intellectual capabilities, they saw political change as the aim of their practice. One major consequence was that the female body became a site of resistance. Feminist artists used their own bodies to create provocative works, whether in public happenings or street actions. Valie Export's *Touch Cinema* (1968) invited the public to fondle her breasts through a box strapped to her chest: "My body is the screen. But this cinema hall is not for looking—it is for touching."[47] This public performance occurred well before the publication of Laura Mulvey's seminal essay on "Visual Pleasure and Narrative Cinema" (1975), which theorized the compelling exactitudes of the necessarily masculinist and objectifying gaze.[48]

Across artistic and counterculture circles, feminist and Marxist critiques of bourgeois aesthetics were performed. While some exposed "the male gaze," others pressed capitalist and masculinist voyeurism to its limits. Performance and body artists attacked the power structures invested into "looking" in the 1960s and 1970s. Audiences were crucial in playing out the logic of the gaze, and artists implicated viewers in their tortures. Viennese actionism was a particular movement at the radical edge of this thinking. Given the violation and suppression of the body in recent German history, this reaction was not entirely surprising. German artistic responses to the burdens of history can be seen in the cultural use of the body, especially the artist's own body. Across the Atlantic, artists were critiquing racist and gender stereotypes, especially in representations of black men. Dressed in an Afro wig and bell-bottom pants, Adrian Piper cross-dressed as a black man and documented reactions on the streets of Manhattan. The "extreme otherness" of the race-gender matrix was revealed in the fear and animosity such a figure could generate.[49] In the 1990s, hip-hop culture became the definitive signifier of black heterosexuality. Some artists inverted the masculinist rhetoric of Black Power to challenge the cultural association of black youth with violence, drugs, and machismo.[50] Other artists have evoked the beating of Rodney King, such as the video art of Portia Cobb's *No Justice, No Peace* (1993), returning agency to young black people discussing the case.[51]

Performance art was the major arena where the body was enacted as text, subject, and object. Maurice Merleau-Ponty wrote an influential essay theorizing the body "as expression and speech."[52] Writing its own obscure language through objectifications and history, the body was described as gesturing beyond fleshed existence to the fundamentals of meaning.[53] Historians associate this practice with the "performative" aspect of (inter)subjectivity and theatricality, especially given the antecedent works of Antonin Artaud's *Theatre of Cruelty* (1930s). Artaud replaced linguistic and narrative conventions with a spectacle of gesture, space, and sound, collapsing the distinction between artist and audience. The significance of the body's relation to sculpture and to questions of dimension, figure, and ground might also be incorporated into our understanding of performance and body art.[54] Physical endurance and self-mutilation were ways of confronting the dialectic between body and society, self and other, text and flesh, subject and object. Visceral, sexual, and angry bodies staged a radical display in performance art, often captured on film. In Valie Export's *Genital Panic* (1969), she slouched in a chair brandishing a machine gun and exposing her pubis to the camera. Threatening the viewer with the reverse gaze, the work confronts so as to induce panic in the voyeur.

Feminist artists (many of whom took up performance and body art) supplanted masculinist modes of representation with a female-specific practice, reclaiming the female body for women and asserting the "aesthetic pleasure" of not just representing women's bodily experiences but "seeing the body through women's eyes."[55] In the public performance *Interior Scroll* (1975–77), Carolee Schneemann read a text on "Vulvic Space" that she pulled from her vagina. It became an iconic feminist piece, relating "the womb and the vagina to 'primary knowledge.' "[56] Women's bodies could represent their own Western history and traditions—their own *logos* and self-determined eroticism. During the Vietnam War, feminists and body artists worked their own form of protest through body art and performance. Schneemann's video *Viet-Flakes* (1965) drew on atrocity images as the ultimate metaphor for the corrupt politics of American imperialism.

Feminist art was antithetical to the Euro-American modernist tradition. Instead, it proposed a practice that was personal, political, autobiographical, and nonformalist. Although male artists continued to work within their own frameworks of exploration (both formalist and nonformalist), the legacy of feminist radicalism was that it entered mainstream artistic practice, was taught in art schools all over the West, and ultimately repackaged artists such as Barbara Kruger, Jenny Holzer, Sherry Levine, and Cindy Sherman within "postmodernity."[57] The French artist ORLAN is a good example of this hybridity and transformation. She was the first artist to incorporate surgical alteration of her own body into artistic practice. Undergoing a series of plastic facial reconstructions, her work explored deformity, monstrosity, and selfhood. Initially concerned with the violence of beauty and rejuvenation, her work was increasingly concerned with the radical mutation of identity in a series such as *Refiguration—Self Hybridisation* (1998). Indeed, historical questions of identity reached across feminism and postcolonialism throughout the 1990s. Significant changes in medicine and science were widely discussed in the media, to which artists responded. As the project to map the human genome was debated, and popular films such as *Gattaca* proposed the erasure of difference and disability in genetic engineering, ORLAN claimed her work was "opposed to DNA" and the programming of human evolution.[58] Her work also spoke of the body as a literal sculpture—to be molded and transformed—which may have been influenced by developments in sculpture itself.

Certainly by the 1970s, the dominant sculptural mode of Henry Moore, with its organic and monumental corporeality, had been challenged by the industrial urbanism of Anthony Caro and the pop-hyperrealism of Duane Hanson's

FIGURE 9.1: Carolee Schneemann, *Interior Scroll*, 1975. Performance image. Photo: Anthony McCall. Reproduced with permission of the artist.

mannequins of iconic American bodies: tourists, policemen, and waitresses. In Britain, the St. Martin's art school produced Gilbert and George, who used their own bodies in sculptural installations such as *Singing Sculpture* (1970), naming their self-installations "Living Sculptures," just as Grecian dancers and posers or tableaux artists had done at the turn of the twentieth century. Their bodies remained the critical feature of their work over years concerned with religious intolerance of sexuality and the decay of civil society under Thatcher. Delivering a lecture on "Sculpture Now: Dissolution or Redefinition," Richard Cork echoed the concerns of artists in this period, and yet to some extent there were shared sentiments with the generation of World War I artists. Cork concluded that "true dissolution could never be a conceivable reality," since there was so much to gain from "a redefining process."[59] Indeed, the redefinition of the sculptural object in the 1980s saw a new way of considering the body. Material culture available in backyards and city streets redefined the body, such as plastic detritus in Tony Cragg's *Red Skin* (1982), a stencil of a stereotypical warring chieftain attached to the gallery wall. Although the human figure was not an overarching preoccupation, some artists saw it as a significant reference point. Antony Gormley cast his own body in plaster, turning the individual into a universal figure. Anatomical realism had long been transgressed; however, Gormley took the figure and transformed it into a fiberglass vessel, a "container of feeling" rather than real flesh, blood, and organs.[60]

From the 1980s, postmodern artists and photographers broke into new areas of critique and taboo, especially concerning adolescent sexuality (Sally Mann) and homosexuality (Robert Mapplethorpe), as well as erotic danger (Nan Goldin) and sexual intimacy (Tina Barney).[61] Indeed, the gay rights movement, coupled with the burgeoning international AIDS crisis, once again turned many artists into activists. Gay male bodies became public property, just as female bodies have been. Gay sex became the topic for discussion on the evening news. Every gay man was a sacrificial body at the altar of public health. Religious reactions against homosexuality became intellectual fuel for artists. Ron Athey used his own infected blood in performances, and Andres Serrano attacked the sacred taboos of religious aesthetics. Serrano's *Piss Christ* (1989) was at the center of a public-funding crisis within the National Endowment for the Arts, a factor that anthropologist Carol Vance described as part of the right-wing "war on culture."[62]

Activism demanded an aggressive return to the body as visual cultures, media, and television seized the view of the gay male body as socially aberrant. As a site of medical scrutiny, disease, and discrimination, the gay male body became a spectacle of horror, even in government education programs supposedly designed to inform. Artists such as Felix Gonzales Torres, John Lindell, and Donald Moffat used personal narratives as gay men living through the AIDS "crisis" in a society that loathed them with the historical stereotyping that linked plague with threat.[63] In 1987, an exhibition titled *Morality Tales: History Painting in the 1980s* marked a new beginning, as did the National Day with Art, which was a national festival of action and mourning. Exhibitions became education campaigns not just about the "facts" of HIV, initially reported as a "gay cancer," but about resisting media representations, demanding government funding for medical research and social services, countering prejudice, and offering advice and assistance. In 1992, *From Media to Metaphor: Art about AIDS*, an exhibition of artists both living and dying with AIDS traveled to cities throughout the United States and Canada, aiming "to stem the spread of HIV infection, and foster more humane treatment for people living with AIDS."[64] In the reactionary era of Reagan and Thatcher, followed by Bush's neoconservative revolution, artist-activists courted controversy to have their message heard, such as the SILENCE = DEATH project.[65] The art collective known as Gran Fury, initially formed from ACT UP (the New York–based AIDS Coalition to Unleash Power), created a series of photographs of kissing couples intended for a public art project on the sides of buses. With political messages such as *Kissing Doesn't Kill. Greed and Indifference Do* (1990), the work signified a merging of billboard advertising (Benetton-style) and grassroots politics. The success of this work led to an invitation to exhibit at the Venice Biennale; however, Gran Fury was threatened with prosecution by the Italian authorities for their savage critique of the Catholic Church.[66] Bodies were the battleground on which new medical and moral wars were fought, especially as AIDS was characterized with military language as a "social enemy."[67]

By the 1990s, the body had become a diffuse and diverse centerpiece of contemporary art practice, recognized in critical exhibitions such as Documenta IX (1992) and the Whitney Biennial (1993). Bodies exploded onto the Internet, in new digital media formats, and were a major public and social issue. In 1993, an article in *Flash Art* asked "What's All This Body Art?," claiming that a new generation of body artists had arrived without postmodern reference to the past. Happenings and performance formats were reinvigorated, although less concerned with the "historic" body. Bodies were no longer unmediated sites of identification for artists. Nor were they authentically lived material representations. Instead, electronic and digital media, video, and interactive formats transformed the signifying practices of bodies from reliance on the material fact of fleshed existence to what was labeled "posthumanism."[68] Flesh was seen as obsolete and malfunctioning, especially given developments in scientific and medical technology. Medical ethics regarding the mapping of the human genome, in vitro fertilization, cloning, and other genetic technologies infused cultural and representative practice with a new desire to comprehend the boundaries of flesh and self. Influenced by developments in nanotechnologies and microminiaturization, body artist Stelarc reinvigorated the idea of the cyborg as a new species.[69] Stelarc's "internal sculpture" inserted into the stomach transformed the body from a cavity into a sculpture. The term *cyborg* was coined in the 1960s to describe NASA's exploration of modifying the bodies of astronauts. Celebration of the cyborg, however, follows long-standing artistic, philosophical, and social fantasizing about futuristic bodies, automatons, and robots, which we saw at the beginning of this essay in the aftermath of World War I.[70]

Representations of bionic bodies may have been concerned with the fusion of man and machine; however, nerves, feelings, emotions, and the sensory faculties bound to corporeality were to come under new scrutiny. Aptly titled *Sensation*, an exhibition of young British art held in London in 1994 presented a radical shift away from the boundaries of the modernist and postmodernist body. Sensational for its confronting topics, materials, and handling of subjects and themes, the exhibition also hoped to bring into play the five senses—to feel, taste, see, hear, and smell society's obsessions, many of which were corporeal. Jake and Dinos Chapman's *Tragic Anatomies* (1996) created a horrendous plastic world of child's play and incest in the Garden of Eden. Evoking Rothko's "embodied" practice, Norman Rosenthal wrote in the Royal Academy's catalog for the exhibition that "the blood must continue to flow." *Sensation* was not the endpoint of the modern/postmodern project but an ambitious new turn.[71] Ironically, while the exhibitors aimed to engender empathy, reactions from the mainstream press poured scandal, heightening the effect of *Sensation*. It could have been postmodern "degenerate art." Indeed, ten years later when Charles Saatchi's warehouse burned down, destroying much of the collection, the press gloated obscenely in the demise of this sensational body of what became known as Brit Art.

Influencing artists around the world, *Sensation* delivered radical hope for the power of the body as interlocutor of the social, political, and subjective. Nowhere was this seen so vitally as in China. In the 1990s, Chinese contemporary artists transformed photography and video media into a form of resistance to the conservative traditions still reverberating from the Cultural Revolution (1966–1976). Prohibitions on pornography continued to be arbitrarily exercised against artists who depicted the nude body, including raids on studios and the confiscation of works. Now, artists used their bodies to explore the legacy of cultural repression through images of childhood entwined with historical and popular memory. Demonstrations in Tiananmen Square in 1989 generated dissident artists in Europe and Australia, giving rise to new markets in what was seen as Chinese political art. Artists, however, were interested in practices of selfhood and conceptions of personal identity in the context of the overwhelming burden of collective consensus and political repression. Self-expression became a medium of social critique and a deeply personal voyaging around the changing political and social landscape, as China's modernity hastened. Performance artists also explored mutilation and wounding, masochism, and gender play. Qiu Zhije (*Tattoo 1* and *Tattoo 2*, 1994) said, "our bodies have become merely vehicles," collapsing the illusive boundary between body and space. Initially practiced in Beijing's East Village, masochism, pain endurance, humiliation, torture, and gender-role inversion were explored by artists such as Cang Xin, Ma Liuming, Zhang Huan, and Zhu Ming. Performances were photographed. Rong Rong, for instance, photographed Can Xin's *Trampling on the Face*, where guests were invited to destroy numerous masks of his face.[72] Experimental photography has engaged with explicit fantasizing about the body or treating it as illusion. The body appears as an ephemeral space where the boundaries between subjectivity and the cultural and political environment (historical and heritage sites such as the Great Wall, the Forbidden City, and Tiananmen Square) are tested and redrawn.

Representations of the body have continued to engage with the potential and limitation of human flesh. Whether through science and biomedicine, or digital media and the Internet, development and proscription of the human body is real, biotechnological, and cyberspatial. Fringe and emerging artists have been afforded a voice, accessing wider audiences and participating in global artistic dialogues. Some institutions have engaged with this digitized cultural turn. The Wellcome Trust, the Gulbenkian Foundation, the Royal Society, the Science Museum, Imperial College, and the Medical Research Council, for instance, have worked alongside artists to critically engage their disciplines in such intellectual and artistic inquiry, especially given the outreach potential of such ventures.[73]

Body art has been sustained in contemporary visual culture as bodies are exposed and explored in new and different ways. Historian Amelia Jones has traced the powerful historicity and yet political potential for body art. The demise of the individual represented an embodying and yet decentering of subjectivity—incoherence and diffusion rather than powerful, overarching strategies of knowledge and power—the grand narratives of social, political, and medical bodies.[74] Donna Haraway explained in the "Cyborg Manifesto" (1991) that technology represented the destabilization of the individual and a "radical politics of non-normative subjectivity," which has been influential in new media practices.[75] In another direction, the issue of human rights abuses and global warfare has placed the body as the crucial site of suffering and critical enquiry. Printmaker Sue Coe's *Bomb Shelter?* (1991) depicted an Iraqi mother shielding her child from an incoming missile. Representations of non-normative bodies are being grounded in desire for political and community change. As postcolonial, queer, disabled, indigenous, immigrant, and refugee bodies enter the social and cultural sphere, demanding presence and engagement, the complex lives of subjects and objects appear to be locked together as never before. This is not appropriation and subordination, or the valorization of the "subaltern body." Nor is it fetishizing the suffering of others or

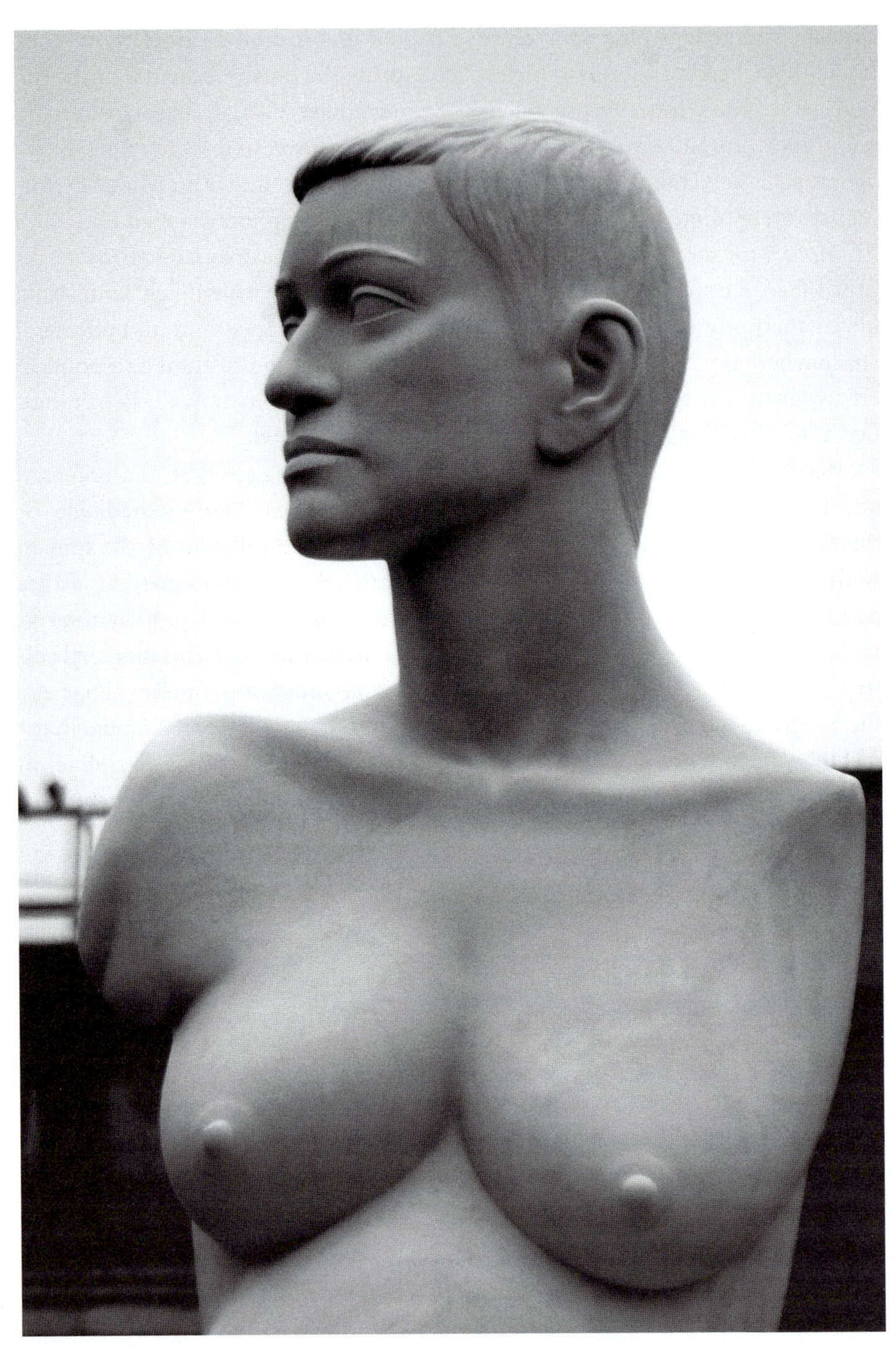

FIGURE 9.2: Guy Feldman, photo of Marc Quinn's statue of Alison Lapper pregnant, Trafalgar Square, London, 2005. Reproduced with permission of the artist.

overidentifying with history's victims. Arguably, this is visual culture adapting what Dominick La Capra has called "empathetic unsettlement," described in relation to secondary witnessing.[76] Marc Quinn's public sculpture in Trafalgar Square, *Alison Lapper Pregnant* (2005), appropriated traditional forms of classicism and modern formalism to show disability as beautiful, iconic, and heroic. Inverting the traditions of that public space as military, masculinist, and physically complete, Quinn's sculpture signaled the new public presence of disability at a time when disability legislation struggled to be enacted.

In a climate of political and social conflict, as Western wars are waged in Afghanistan and Iraq and the "war on terror" is seemingly endless, and while neoconservatives renew attacks on homosexuality, immigration, welfare, and the liberal imagination, many cultural practitioners have responded by engaging the body. The human body remains the major zone where sociopolitical questions of both liberty and legitimacy are contested. As a result, representation now faces burgeoning social questions of embodiment in relation to intersubjectivity, identity, and belonging. This critical mass inflects social and academic enquiry and yet is also defined by cultural tensions arising from shifts in economies and social relations, as well as demands for political change at local and global levels. This fallout (or cultural capital depending on your view) often collapses distinctions between fleshed experience and the world of signs. History and reality continue to mark experiences, representations, and meanings of the body.[77]

CHAPTER TEN

The History of the Body

Self and Society, 1920–2000

D. M. VYLETA

There is hardly a song in modernity more tired than that of man's alienation from the social and natural world. The theme traverses the entirety of the nineteenth century. The Romantics—especially in Germany—looked back to the Greeks for a life in which religion and art, person and polis all stood integrated in an organic whole; in which the gods lived on Mount Olympus, a few days' journey yonder, and Aristotle's *Ethics* could plausibly be conceived as the prequel to his *Politics*. Young Marx dreamed of a society that was as genuinely postscarcity as the biblical Eden and free of the state that he saw as an unworthy intermediary between one human being and the next. Dostoevsky's Elder Zosima in *The Brothers Karamazov* preaches a vision of pure Christian love against a world in which humanity stands alienated from itself, its environs, and its creator, a world that brings forth fretting, diseased Underground Men who can locate a sense of worth and freedom only in quasipathological spite. The neoromantic *Wandervögel* movement in pre–World War I Germany united a generation of youths who yearned for a sense of communal harmony within a setting of natural splendor, far removed from the self-interested conflict and partisan political squabbles that they identified as the ills of the modern age.

Remarkably, these various formulations of the longing for harmonious interaction and an organic connection between individual selves and society as a whole coincided with the rise of what Marxists used to call "bourgeois individualism"—depicted in the realist fiction of Balzac, Elliot, and Tolstoy with their fine psychological portraits of self-contained agents—and the attendant triumph, in many corners of the modern world, of parliamentary democracy, a political system that implicitly endorsed the assumption of such individualism. After all, parliamentary democracy, despite occasional rhetoric to the contrary, conceives of politics as a mode of compromise made on the basis of a head count of the incommensurable individual wills that make up the electorate: that is, it has abandoned hope for a political consensus that would be the "natural" outcome of shared communal goals.

In the twentieth century the tensions between a sense of self shaped by the socioeconomic forces of industrial capitalism and its ideological rejection both by the far right and the far left came to a dramatic head. Dystopian visionaries of the 1920s saw the future as populated by atomized individuals who had been converted to soulless man-machines by the needs of productive efficiency. In the 1930s both Nazi Germany and Stalinist Russia put their considerable resources to use in order to shape New Men and Women according to their respective ideological specifications, both promising the destruction of any boundaries that estranged self from society. Decades later, Margaret Thatcher, distilling centuries' worth of liberal political discourse into the nutshell of a now-famous sound bite, would unmask the very concept of society as a left-wing hoax, an empty abstraction that did little other than distance policy making from the individual selfhood of the end consumer: "There is no such thing as society. There are only individual men and women, and there are families."[1]

Conflicting ideas about the status of the body were integral to this clash of conceptions of selfhood. Was the body—its racial characteristics, its state of health, its mode of dress—an appropriate concern for politics? Was it a private possession that its owner was free to shape and manipulate in any way suitable for communicating his or her individuality to others? Could the body serve as a metaphor of how individual and society interrelated? This chapter explores and elucidates some of the answers given to questions such as these in the years following World War I. Before it does so, however, it will be useful to review, albeit schematically, the notions of selfhood bequeathed to the twentieth century and the status of the body within these.

MODELS OF THE SELF

The twentieth-century West inherited a wealth of contradictory conceptions of selfhood and society. On the one hand, Enlightenment models continued to have a strong hold on both the popular and scientific imagination. In one of its most influential formulations—that of Thomas Hobbes's *Leviathan*—the individual became the primary unit in which society could be thought and analyzed. The self that was invoked here stood in epistemic isolation from its fellow human beings. Motivated by the wish for survival and a wealth of private yearnings, it was ill-suited to live in harmony with the competing desires of its neighbors and could be compelled to do so only by the artificial order imposed by the state on the strength of its coercive powers. The body was here conceived in mechanical terms: Hobbes used a hydraulic model, and later theorists envisioned a system of levers and pulleys. This meant not only that the radical independence of the individual was physiologically anchored but also that something like a "science of man" could be dreamed in which human beings' desires and aversions—their "passions"—could be described by the Newtonian laws of mechanics. Much Enlightenment thought was poured into the question of how human beings' reason interacted with this physiological machine it inhabited, and it was in the *cogito* that true selfhood was anchored. Eighteenth-century thinkers as diverse as Cesare Beccaria and Immanuel Kant described ways in which reason could learn to control the passions and thus free men and women of an existence tied to the inexorable laws of cause-and-effect.[2]

In time this discourse began to focus on the role of socialization in the process of "taming" the body, that is, on educational practices that allowed the self to achieve true independence from stimuli acting on its physical shell. This idea is evident, for instance, in the debate surrounding prison reform from the late eighteenth century on: Lack of rational self-governance here became the hallmark of the criminal in thrall of his or her beastly passions; self-governance was to be impressed on his/her soul through a regime of isolation, silence, food deprivation, and work. Habituation and bodily discipline could thus serve to free man's self and foster independent rational actors who were fully functional in a liberal society, and who—qua *reason*—left behind the perspectival limitations engendered by their race and sex, as well as their social and historical position.[3]

A related logic can be located in the Freudian narrative of the self as formulated in his early studies at the cusp of the twentieth century. Here, the developing individual has to negotiate a sexual obstacle course whose successful mastery expresses itself not only in correct adult sexual practice but also in the high degree of control exerted by the conscious, rational self over its coarser rivals within an individual's psyche. Failure to master the experiences of childhood and adolescence, in contrast, may turn one's body into an enemy as it transforms from a tool wielded by the rational self into the site of symptoms, that is, into the inarticulate messenger boy of the smutty-minded unconscious.[4]

In the course of the nineteenth century, the Enlightenment narrative of the self had, however, come under sustained attack from a variety of thinkers who described a self deeply penetrated by social realities and thus were able to rethink the individual as an emanation of society in its specific historical configuration rather than treating society merely as an abstraction designating the sum total of individuals bound together artificially by shared contractual obligations vis-à-vis the state.[5] Among these propagators of a theory of self as penetrated by society, Marx towers over the history of modernity, both because of the depth of his thought and because of the various attempts to turn his and his successors' theories into political realities. Marx's famous credo that an individual's material realities—his or her "being" (*Sein*)—shaped his or her understanding and experience of self and world—"consciousness" (*Bewußtsein*)—suggested that any attempt at a description of human psychology and the relationship between body and mind had to be rooted in historically specific socioeconomic analysis. "Socialization" here no longer was conceived as a process that might free the rational self from external constraints and thus point to the possibility of a metaindividual and transhistorical perspective but instead was regarded as shaping the individual's mental world and alienating different social groups from one another. Furthermore, this process of socialization was, at least initially, always a physical event: *Sein* implied a bodily practice of work, not just the abstract experience of an economic order.

While Marx's focus on the "productive relations" as the most significant aspect of an individual's social existence hardened into a central truth of the ideologico-political movement he helped to found, its central conception of a self inscribed into social experience continued to be adopted and adapted by a slew of thinkers throughout the latter half of the nineteenth century, many of whom worked within the emerging field of the social sciences, from sociology through anthropology to cultural criticism. Georg Simmel, for instance, gave a spatial twist to this narrative in his 1903 essay "The Metropolis and Mental Life."[6] Denying any deep divisions within the experience of modern life by different social classes, Simmel described how the constant stimulation of the nerves and the flowering of the money economy that took place in the modern metropolis shaped individuals who translated all differences in value into purely quantitative terms, while also breeding in them an attitude of latent hostility toward their fellow citizens. Once again, the individual self could be understood only within the parameters of its social experience, and again social experience was physical in nature—partaking in the contemporary narrative of neurasthenia, Simmel invoked men and women whose nervous systems were literally "overloaded" as they walked down crowded city streets, dodged traffic, and rubbed shoulders with fellow metropolitans in trams and omnibuses.

Émile Durkheim's 1912 study *The Elementary Forms of the Religious Life* may serve as a third example of this shift in perceptions of selfhood. Here, Durkheim formulated a theory of religion based on ethnographic reports of Aboriginal rituals that displaced the origin of Kantian conceptual categories such as "space," "time," "number," and "cause" into the social realm: Rather than being "hardwired" in the human brain, these categories were described

FIGURE 10.1: Soldier and peasant woman. Relief carving on one of sixteen sarcophagi, Soviet War Memorial, Treptower Park, Berlin. Completed 1949. Photo: Dan Vyleta.

as being culturally contingent and as learned and internalized through ritual practice. More overtly than in Simmel or Marx, the social was here directly seen as shaping cognitive acts; the self literally apprehended its world through socially constructed categories. Once again, socialization was implicitly physical: Spatial categories, for instance, were incarnated in architecture but also enacted in communal dances and other kinds of ritual.[7]

Marx's writings, Simmel's essay on the city, and Durkheim's study of religion also share a certain teleological dimension that conjures up a point of history in which the social factors shaping the self and the self's "authentic" needs reach a point of harmony.[8] This is particularly obvious in Marx's "humanist" writings pre-1848, where Communism—the endpoint in history—is described as facilitating a self that is finally able to develop fully its human potential and individuality. In other words, models that saw the self as deeply penetrated by society did not simply affirm a vision of human diversity in the manner of Herder, Ranke, and other historicists; rather, they tended to posit (or at least hint at) a human ideal and hence a program of social change. The ideal itself was often not of an entirely different nature than the one upheld by a liberal humanist, but here individual self-development had become an insufficient means of achieving it. Many of the utopian projects of the twentieth century thus implicitly drew on philosophical anthropologies that made the self a function of social life rather than a mere agent negotiating its constraints and promised not simply an improved standard of living but the genuine recasting of human souls.

Finally, the nineteenth century also witnessed the medicalization of the languages of self and society, a development that was to prove very influential—and often fatal—in the century that followed. This discourse was particularly focused on deviant selves, that is, selves whose socialization had failed. In contrast to the reformist models discussed in the preceding, medical discourses of deviance left little conceptual space for education as a device for achieving integration. Diseased selves could only be contained or destroyed. Criminology and psychiatry formed the vanguard of disciplines that propagated this logic, and it came to have an increasingly prominent status in popular as well as political culture across Europe and the United States by the early decades of the twentieth century. Often the disorders described were attributed with epidemic potential, that is, were cast as social threats rather than individual problems. For instance, degeneration theory, first articulated in the mid-nineteenth century and omnipresent in the 1890s, posited that lifestyle choices associated with modern city life (sexual promiscuity, masturbation, alcoholism) precipitated physiological decay, which would then be passed on to one's progeny. Pathologies could thus be both acquired and inherited, and latent pathologies could be activated by unsavory social circumstances. Much of this discourse carried overtones of class prejudice and allowed articulation of anxiety about the urban masses who were increasingly making their presence felt in the political arena. In effect, it facilitated a bifurcated narrative of two distinct types of self—one healthy and socially integrated, the other diseased and dangerous—in which both social engineering and eugenic measures could be employed to eliminate the threat of the second.[9]

At the same time, this medicalized language of the self facilitated easy metaphorical leaps from patient body to social body and thus revived an antique simile in a newly literalized form and for distinctly modern, nationalist purposes. As Paul Lerner has shown in a recent study on Germany's psychiatric response to World War I male hysteria, the medical professions themselves were fond of this transposing of physiological language onto the political sphere. The point of such rhapsodizing about the people's body ("*Volkskörper*") was to describe (or else prescribe) the dissolution of the individual's identity and will in the national project:

> the entire Volk has been transformed into a single, unified, closed organism of a higher order, not only in a political-military sense, but also for the consciousness of each individual. The telegraph lines are the nerve endings of this great new body, through which identical feelings, identical strivings of the will oscillate at the same moment without regard for time and space.[10]

Unlike ancient and medieval usages of the "body politic" metaphor, this medical vision of the national body did not depend on the mutual respect between different, specialized agents within an organic hierarchy (head, hands, stomach, and others). Rather, the individual had become an interchangeable (and hence expendable) avatar of the national body. The collective will was to dominate individuals and help them transcend their limitations; hierarchies became insignificant since the same national thought animated subordinate and superior. Indeed, it was argued that this destruction of the individual soldier's egocentricity before his love for the Fatherland (one notes the gendered dimension of this often deliberately mystical narrative in which the love for the father forged a band of brothers capable of transcending their subjectivity) would prove psychologically healthy for a people troubled by the selfishness of modernity. In this manner a harmonious relationship between self and society could be invoked in biological terms, and war could be greeted as a great national healer.

At the beginning of the second decade of the twentieth century, the individual's degree of independence vis-à-vis his or her environment thus remained a highly contentious issue, as did the precise role the body played within this relationship. If a segment of liberal bourgeois intellectual opinion tended to make a stark division between their own achievement of aloofness from both bodily need and historical contingency and the masses' dangerous—and potentially congenital—inability to follow suit, the Marxist critique of an age that created a working class shattered in both body and soul found echoes in other prophets of social doom who saw modernity above all as a period of disintegration in which the unity of self and society achieved in a traditional society stood disrupted. It was along these axes that theoretical descriptions (and political prescriptions) for twentieth-century selves developed.

THE BODY AND THE MACHINE: THE 1920S

In the 1920s the discourse about self and society ran headlong into modernism's fascination with technology.[11] The technology in question was no longer primarily the train tracks and smokestacks of the initial industrial revolution but the wonders of electricity and X-ray technology, of skyscrapers, machine guns, mustard gas, food tins, and the airplane. The old simile that likened men and women's physical body to a machine was revived. But now this simile no longer simply served the purpose of describing the hold of the physical universe of cause and effect over the lives of human beings; rather, the transformation of man into machine became the utopian goal of any number of visionaries, starting as early as Marinetti's 1909 Futurist Manifesto.

The underlying goal of such a transformation was typically expressed in terms of efficiency: Humanity was to imitate the tireless, efficient rhythm of machine production. It gained practical application in the writings of Frederik Winslow Taylor (*The Principles of Scientific Management*, 1911) and the production methods introduced by Henry Ford from 1913 on. Taylor's time-and-motion studies amounted to far more than the simple extension of the system of division of labor, which, after all, had attracted admiration ever since Adam's Smith's 1776 *Inquiry into the Nature and Causes of the Wealth of Nations*. In Taylor's system, any superfluous motion was to be eliminated: Even the most elementary of movements was subdivided into units of one-hundredth of a minute, analyzed, and reconfigured. Efficiency thus took on an aesthetic value all its own whose superiority could be defended without any explicit reference to productivity. Taylor's ideas were taken up and applied not only in Ford's factories but by dietary reformers such as John Harvey Kellogg and Horace Fletcher, who preached digestive efficiency rooted in the time-tabled intake and processing of food by the body-machine. In various guises, Taylor's ideas quickly spread around the globe: In 1919 Alexei Gastaev set forth a vision of a Taylorized community of Communist workers whose very thoughts had been mechanized and whose emotions could be measured by "the pressure gauge and the tachometer." Gastaev was to turn his ideals into educative practice in the 1920s when he established the Soviet Union's Central Institute of Labour.[12]

The vision of the self as efficient machine could thus be grafted onto capitalist as well as Socialist ideas; in its most radical versions, it gleefully advocated the deindividuation precipitated by mechanization; that is, it envisioned selves whose inner lives were as much governed by an external timetable and a set of societal rules as their outer movements. Naturally, such a vision precipitated a strong reaction in the humanist camp. The 1920s thus also became the decade of dystopian writing against this fetishization of the machine as the model for human life. One of the earliest examples of this trend was Karel Čapek's 1920 play *R.U.R.* (*Rossum's Universal Robots*), which premiered in Prague's National Theatre in early 1921. In fact, Čapek—whose brother Josef was responsible for the invention of the word *robot*—had already rehearsed *R.U.R.*'s key theme as early as 1908: Anyone wishing to turn workers into machines would have to not only educate their bodies to perfect efficiency but also remove "all feelings of altruism and camaraderie, all familial, poetic and *transcendental* feelings," that is, all those qualities the humanist identifies as most worthwhile and human.[13]

R.U.R.'s plot relates the invention of androids through the efforts of two generations of inventors: the father, old Rossum, a scientific materialist out to disprove God, and the son, an engineer with an interest in productive efficiency. The result of this unholy cooperation is the creation of a superior worker who promises to abolish scarcity and usher in an era of universal economic well-being. Blind to this promise, human workers across the globe rebel against

FIGURE 10.2: "In Praise of Technology"—scientists working for the benefit of the family. East German mural on the side of the Haus des Lehrers (House of the Teacher) on Alexanderplatz, Berlin. Completed 1964. Photo: Dan Vyleta.

their mechanical competitors. In response, the Robots are armed by their owners and squash the rebellion. Unfortunately, once victorious, they continue to wage war against all of humankind; simultaneously, it is discovered that human women have become infertile. The play's plot sees the Robots attack the factory that produced them: Their goal is to learn the secret of reproducing themselves. Meanwhile, it is revealed that the company director's wife, Helena, has prevailed on the chief researcher and Robot "designer," Dr Gall, to alter the constitution of the Robots the factory produces—they are to be endowed with "souls" to facilitate understanding between Robots and humankind. The play ends with all but one of the humans killed. The last survivor, the chief engineer Alquist—a "Tolstoian" humanist—is to teach the Robots how to reproduce their kind but finds he is unable to reconstruct the formula that Helena destroyed in an attempt to stop the production of Robots once and for all. Intelligent life seems doomed to total annihilation, when Alquist discovers that two of the Robots have fallen in love, that is, have acquired authentic human emotions and—implicitly—the key to future life.

The play's most salient theme, therefore, is that of the body overcoming the modern mantra of efficiency. Sexual love, rooted in the body, here becomes a bastion of authentic humanity against all attempts to reengineer the self according to the needs of productive processes. This is evident not only in the play's final resolution but also in the role of the sole female character in the play, Helena: Her instinctive, maternal-feminine disgust with human-made processes of reproduction is a constant theme. She also becomes the emotional focus of the entire male cast, whose work life is marked as sterile and no less inhuman than that of their machines. The aesthetic ideals of Taylorism are thus attacked as hollow, while work is marked as an activity that has human value beyond its productive results. In this manner, *R.U.R.* formulates its humanist answer to the age of machines and progress. Significantly, the body—explicitly gendered—is no longer a stumbling block to rational self-realization as in so many Enlightenment narratives but instead the inalienable repository of human emotion. Even "transcendence"—the orientation of human beings beyond the physical—is thus given a physiological basis. Society, in this vision, can only be thought starting from the emotional needs of the individual; it cannot be constructed according to abstract ideals.

Virtually contemporary with *R.U.R.* is Yevgeny Zamyatin's dystopian novel *We*, which was finished in Russian in 1921 and first saw publication in a 1924 English translation. *We* connects the ideals of Taylorism to political totalitarianism, though its emphasis is on the former rather than the latter. An inspiration for Orwell's later *1984*, Zamyatin describes life in OneState, a hyperrational society that is like a "single mighty organism" run by an all-powerful "Benefactor" who is aided by a small army of watchdog-spies.[14] In

OneState, men and women are reduced to number codes, their lives governed by the "Table of Hours," a Taylorian timetable system that regulates every minute of the day and that has been deeply internalized by the Numbers. Sexual acts are as regulated as the intake of food—down to the number of "statutory chews for each mouthful"—and transparent chamber pots are the witty symbol for a society in which the private self is thought to be eliminated and its total penetration by society achieved.[15] In the course of the novel, its narrator—D-503—undergoes a transformation from loyal subject to reluctant coconspirator in a rebellion staged by a coalition of dissenting Numbers and the hairy forest people who live nature-bound lives beyond the well-guarded city limits.

As in *R.U.R.* the key to rebellion against the standardized rhythms of OneState lies in sexuality and the recognition of a physiologically rooted individuality implicit to it. The narrator's total absorption in the communal machine comes to an end when he is sexually seduced by I-330, a manipulative femme fatale. Their unregulated and passionate sex act precipitates emotions of jealous devotion to a specific individual, and thus marks a break with a system in which Numbers are to be treated interchangeably. Through consciousness of the beloved's individuality, consciousness of his own self arises in the narrator.[16] Similarly, the love experienced by the female Number O-90 for the narrator, and the desire to illegally bear his child, turns this placid, apolitical, maternal figure into a revolutionary. Indeed, women—both as seductresses and as mothers—are continuously marked as ill-suited to the sterile rationalism of the age: Their physiologies put them at odds with a system that tries to convert them into machines.

In the final analysis, however, this physiological basis for individuality—and this rebellion against any attempts at dissolving self in society—is present in both men and women alike. From the very beginning of his narration, D-503 has to grudgingly admit that physical differences between individuals cannot be obliterated, and that these form the basis for authentic—that is, nonrational—attachments. Similarly, privacy of thought and desire is marked as a physiological given: Despite all attempts to eradicate it, the Numbers enjoy an inner life that cannot be accessed by the State. Even the narrator's attempts to write out his experiences and doubts—to make public his private self—are again and again frustrated by the inadequacies of language; the more confessional his entries attempt to be, the more disjointed they are, the implication being that interiority and individuality are existential mysteries, which by definition resist rational analysis or even expression. In his despair, the narrator blames a somatic disease from keeping him from fully dissolving into OneState's community; it is indicated to the reader that the nature of this "disease" is nothing other than human nature. Like Čapek's dystopian play, *We* thus argues for the truth of the humanist vision of self in which individuality and the self's ultimate independence from the state are physiologically rooted, and sex represents the first line of defense against any attempts to machinize men and women.

A similar narrative is taken up by yet a third contemporary work of fiction, albeit in a less disciplined and consequently more ambivalent manner. Fritz Lang's 1927 film *Metropolis* was based on the kitschy and melodramatic novel by his wife and—collaborator, Thea von Harbou, that was serialized during production in the *Illustriertes Blatt* and published as a whole only concurrently with the film's debut. From its opening shots, *Metropolis* systematically connects the rhythm of machines to the ticking of a great clock to the robotic goosestep of an army of uniformed workers who arrive for their shift to replace an identical group of colleagues. Later on the film dwells on the ballet of worker bodies going through their repetitive tasks in perfect synchronization. The synchrony is destroyed when one of the workers is physically unable to keep up this machine rhythm: His physical limitations thus precipitate an explosion in the factory. The workers' only way out of this dehumanizing existence in which they have become interchangeable, deindividualized cogs comes through the preaching of Maria, the symbolic mother of the workers' children. However, Maria's role is usurped by an identical-looking machine-Maria who entices the workers to partake in a self-destructive rebellion. This false Maria also performs erotic dances in the city's pleasure district, literally entrancing the men who watch her and baiting them into acts of violence. The film was made for the then-staggering sum of 5.1 million *Reichsmark* and received the endorsement of being *Volksbildend* ("educative for the people") by the Weimar censors, although adolescents were barred from seeing it, presumably because of its steamy, seminude dance scenes.[17]

With its blend of themes that include class warfare, oedipal struggle, the horror of machinization, occultism, and so on, *Metropolis* is perhaps best understood as a pastiche of late-1920s cultural bugbears than an original contribution to them. In fact, its main value may be to illustrate how much some of the themes discussed in this chapter had frozen into recognizable tropes as early as 1926–1927: They could be clad into bold new images but no longer excited with their content, as even the most enthusiastic of the film's reviewers (and there weren't too many of those) admitted. What is striking is how much the juxtaposition of sexuality and the gendered body on the one hand with the machine-like working body on the other had become conventionalized. At the same time, the film's most enduring specter—that of the sexualized machine—is worth noting: It warns not only against the dangers of dehumanization implicit in treating human beings like machines but also against the hypnotic fascination elicited by the aestheticized machine. It is the latter danger that precipitates the film's final catastrophe. In both her dancing and her rhetoric to the workers, the false Maria manages to turn men into an atavistic mob: that is, to dissolve whatever remaining sense of self they might have had in self-destructive mass action. Against this danger, *Metropolis* pins an old-fashioned romantic (which is to say, individualistic) hero, who can differentiate between lust for a sexualized machine and love for the human priestess/mother. The fact that this tale of the dangers of falling in love with machines is told in the medium of film, and produced by one of the medium's most technocratic directors, adds an element of irony to its heavy-handed didacticism—for does not cinema turn the viewer into one of the crowd, seduced by the wonders of technology?[18]

Of course, the utopian and dystopian dreams about machine bodies were not the only discourse about self and society that marked the 1920s. As already indicated, medical languages about society were going strong, borrowing their terminology from bacteriology as well as the science of heredity. Psychoanalytic language was also in the ascent and beginning to penetrate popular culture. Moreover, psychoanalysis was beginning to move away from being confined to the diagnosis of individual patients and was fast developing into a mode of cultural criticism in which modern societal norms were both marked as the outcome of primordial sexual struggles and criticized as ill-suited for the psychological well-being of man and woman.[19] One pioneer of this development was the anarchist Otto Gross, the son of the famed Austrian criminologist Hans Gross. Drawing on both Freud and Nietzsche, Otto Gross began to argue as early as 1913 that what society called "education" and "socialization" were, in fact, overbearing "suggestive" forces that alienated us from our authentic natures, limiting our innate mental powers and perverting our notions of sexuality until we understood all heterosexual acts as forms of rape. Only a chosen few managed to hold out against society's program of suggestion and hold onto some semblance of their authentic selves. Consequently, this maladjusted minority became outcasts who were often stigmatized as insane. The only way forward was to instigate a revolution that would return society to its natural state of Communist matriarchy. Such a revolution would not be brought about by democratic forces, since democracy by definition could only represent the opinions of the well-adjusted (that is, psychologically crippled) majority. Gross thus produced his own blend of Communist vanguardism, psychoanalysis, and the tale of humans' alienation in modernity. He died in February 1920 from a combination of pneumonia, malnutrition, and morphine addiction and was buried (despite being a baptized Catholic) in the Jewish cemetery in Berlin-Pankow. Though he was, in life as in death, an eternal outsider, his taste for antidemocratic utopian revolutionism would prove a potent force throughout the 1920s and may thus serve us as a bridge to the 1930s and 1940s, the decades in which both Germany and the Soviet Union were attempting to forge real-world utopias through policy, baton, and gun barrel.[20]

ENGINEERING THE SOUL: SELF AND SOCIETY IN NATIONAL SOCIALISM AND STALINISM

Nazi Germany

There are worse places to start delving into the Nazi vision of self and society than Leni Riefenstahl's propagandistic documentary of the 1934 sixth annual party congress in Nuremberg, *Triumph of the Will*, with its pictures of shirtless eighteen-year-old lads shaving under clear Bavarian skies and rubbing each other down with sponges and its long camera pans over a faceless, seemingly unending crowd chanting "Heil Hitler" in unison. It was a vision of a racially homogeneous state in which the self had been dissolved into mystical union with the nation; in which community implied the organic camaraderie of adolescent play; and in which the physical—young, virile, masculine, *heroic*—body of the individual could stand as an incarnation of a national body whose will, in turn, was fully embodied by its Führer ("Hitler is Germany, just as Germany is Hitler").[21]

Intellectually, there was little to this vision that was innovative. Pronouncements about the mystical, organic union of self and nation date back, at the very least, to the fiery mid-nineteenth-century rhetoric of the Italian Pasquale Mancini, even if the xenophobic contempt for other nations was a later addition. Race science, and the attendant division of the world into a hierarchy of racial groups, also had its roots in the 1850s, though it was not until the dawn of the twentieth century that the word *race* took on an unambiguously biological meaning and that university research began to be conducted in this area. Medicobiological language that allowed the Nazis to label the targets of their exterminatory policies as "bacilli" and "bacteria" whose destruction was vital for the survival of the Germanic body politic had also been in vogue for several decades by the time the Nazis ascended to power, and its use was by no means the purview of the right. Eugenic and social Darwinist discourses had been going strong since the war years and were equally open to both the political right and left.[22]

What was new to the years of Nazi rule was the successful silencing of all rival discourses by their total domination of the public sphere, and the ruthless implementation of policies informed by this vision of society by a regime that in its ambitions, if not always in practice, was totalitarian. The result was legalized social engineering on an unprecedented scale.[23] Concretely, this articulated itself in laws that allowed the state to keep criminals incarcerated beyond the end of their prison sentences and to intern "professional" criminals in concentration camps without a court order (Law against Dangerous Habitual Criminals, November 1933; Ordinance on Preventative Police Custody, November 1933);[24] in pronatalist policies that rewarded genetically untainted, "Aryan" mothers for their fecundity;[25] in the establishment of Hereditary Health Courts (*Erbgesund-*

heitsgerichte), which ordered a total of 400,000 sterilizations of those suffering from a variety of often loosely defined "hereditary diseases";[26] in the establishment of the Research Institute for Eugenics and Population Biology in 1936 that busied itself—among other things—with establishing a comprehensive database of Germany's Sinti and Roma, complete with anthropometric measurements, blood samples, and genealogical tables;[27] in the categorization, isolation, and eventual incarceration of thousands of "antisocial" elements (mass arrests and deportation to concentration camps started in June 1938);[28] in the murder of approximately 70,000 people between 1939 and 1941 in "T-4" euthanasia programs targeting asylum inmates;[29] in the identification, discrimination, and mass murder of "racial enemies"—above all Jews—whose total extermination throughout the territory controlled by the Reich was ordered in the winter 1941–1942, though systematic killings in Poland, Russia, and elsewhere had started much earlier than that.[30]

An enumeration such as this runs into the danger of reducing Nazi politics to the implementation of ideology and suggesting an ideological core that was monolithic, unchangeable, and internally coherent. In fact, the story of the gradual creation of the desired *Volkskörper* via elimination of so-called alien elements was much complicated by factitious in-fighting among various Nazi organizations and luminaries; by political compromise aimed at appeasing the party's various constituents; by practical, logistical considerations; and by local factors that could often decisively influence the speed and diligence of policy implementation. It is worth remembering that thousands of political enemies were murdered despite their flawless racial pedigrees. Far-reaching innovations such as the introduction of the Nuremberg Laws of 1935 were the result of pressures by party conservatives to replace spontaneous, violent, and often-unpopular attacks against Jews with a system of legal discrimination that would appeal to the majority of Germans. The very definition of Jews adopted in a supplementary decree in November 1935 considered both ancestry and religious denomination, in obvious contradiction to racial science.[31] After the invasion of Poland, the reclassification of two million Poles as ethnically German married ideological directives to the pragmatic need to create a loyal segment of the population and a new source of *Wehrmacht* conscripts. The National-Socialist program for national regeneration was thus shaped by a wealth of contingencies, without, however, ever abandoning its core ideological tenets.

Even more important, all of the Nazis' most aggressive policies were carefully hidden from the public eye, including the killing of asylum inmates, whose death certificates claimed the patients had succumbed to pneumonia, blood poisoning, or strokes, and the mass murder taking place in death camps. In other words, for all their rhetoric concerning the necessary removal of weak and alien elements from the body social, its implementation was a guarded affair. This, in turn, points to the assumption of a lack of public consensus regarding both the Nazi social ideal and their means of achieving it. Only in front of a closed audience of senior SS officers could a fanatic like Himmler openly describe the murder of women and children as the "extermination of a bacterium" and "an unwritten and never-to-be written page of glory in our history."[32] The SS, under Himmler, also oversaw the establishment of *Lebensborn* ("well of life") homes and maternity hospitals in which Aryan mothers raised their purebred stock, or they invented neopagan, "Nordic" rituals that served to underline their self-perception as a racial elite that had overcome Christian emasculation.[33] The general population lived lives far removed from such shenanigans, and much historical research has gone into the question of the degree to which the Nazi social vision was internalized "on the ground." Did the regime successfully shape Nazi selves?

Part of the problem in answering this question is that much of the Nazi propaganda envisioning society as an organic *Volkskörper* was purely negative, that is, assumed that the "surgical" removal of foreign and degenerate parts would reveal a society in which the self was naturally subservient to the primordial nation: This bond was literally inscribed in the blood. Such a vision demanded acquiescence from those who were not directly affected by the Nazi's program of excisions and rewarded complicity, above all in the form of denunciations that formed a backbone of the regime's machinery of terror.[34] No major reorganization of society took place that would have asked the population for active repudiation of all formally held social values—no social revolution that would have obliterated class divisions, no sweeping exchange of elites. As for the remolding of the individual self, here the Nazi demand boiled down to a fuzzy vision of a nation of soldiers—disciplined, classless, physically fit to fight, with little notion of private life or a private sphere, and, above all, obedient. Through the creation of youth organizations like the *Deutsches Jungvolk*, the *Deutsche Jungmädel*, the *Hitler Jugend*, and the *Bund Deutscher Mädel*, this creation of citizen-soldiers could be pursued on a broad scale; before long, of course, the better part of the male population were to be turned into soldiers proper, with all the socializing effects this entailed.

The effect, both short term and long term, of Nazi ideology on the population is hard to judge. As indicated previously, cooperation, both opportunistic and value-driven, was common, as were everyday acts of noncooperation and resistance. Party rallies and comparable public displays of power, unity, and near-religious fervor no doubt found their converts—diaries testify to the sense of "magical splendor" they inspired in participants—but it is hard to imagine that the day-to-day realities of life were sufficiently obliterated by the limited number of public rituals and spectacles; nor was the participation in these anything close to universal. Attempts to organize adults' leisure time in ideologically meaningful ways, such as "Work through Joy" holidays, could just as easily backfire and gave rise to tour operators who offered trips to locations where one was guaranteed to remain untroubled by such party-organized pleasure seeking. Those Germans with strong affiliations—say, the class loyalty among skilled industrial workers or confessional identification in traditionally religious (typically Catholic) areas—tended to retain these across the years of Nazi reign. Among the party members, rational opportunism was probably more common than internalization of ideology, and, in direct contradiction to the regime's rhetoric, it is often assumed that for many the Nazi years signaled a retreat into the private, apolitical sphere of family life. Ideological penetration was likely to be deepest among the young, who were pressured into joining party organizations early on (membership in one of the youth organizations from the age of ten became compulsory in 1939). School education was, of course, also inflected by Nazi ideology, and attempts were made to actively turn children into spies on their parents, threatening familial privacy and no doubt speeding along the process of atomization. Even so, there is significant evidence of rejection and resistance among certain strata of adolescents. In the end, the relative brevity of the Thousand-Year Reich as well as the level of destruction and despair it precipitated for its own people may have precluded its notions of self and society from becoming all that deeply rooted.[35]

The Soviet Union under Stalin

Given that the creation of the Soviet Union was underpinned by Marxist theory, albeit a modified version of it, it will not come as a surprise that it was the site of one of the twentieth century's most protracted efforts to forge new selves along with a new mode of social and economic existence. While the long-term transformation of the self would, according to Marxist materialism, be the automatic outcome of changes within the economic base, Bolshevik ideologues also argued that, during the initial period of transition, it could be fostered and accelerated through active educational measures. In the words of Lenin's wife, Nadezdha Krupskaia, "Socialism will be possible only when the psychology of people is radically changed. To change it is the task standing before us."[36] In the immediate aftermath of the revolution, and above all during the years of the New Economic Policy (1921–1928), which saw the reintroduction of limited free-market structures in the service of economic reconstruction after the ravages of the Civil War, this program of "social education" showed a remarkable willingness to experiment and innovate, as well as a considerable amount of pluralism. This is particularly obvious in the cultural sphere that was overseen by the "Commissariat of Enlightenment." While the regime was well aware of the propagandistic potential of literature, film, and other arts, it nevertheless embraced a policy in which no single movement or aesthetic conception was granted a monopoly, in part as a gesture of reconciliation with the bourgeois "specialists" whose technical and scientific expertise it depended on.

Formal experimentation, anathema to the Soviet Union's later artistic development, thrived and was often propagated by the most orthodox of Marxists. In the schools, some radically innovative pedagogical strategies were embraced: The punishment of pupils was to be replaced by an emphasis on each child's individual creativity, a vision that owed much to young Marx's promise that under Communism unknown creative reserves would be freed in the great mass of humanity. Like the state, institutional schooling should "wither away" in the long term; as a starting point the party abolished grades eight through ten. Evaluation began to be based on participation rather than tests; religious education classes were scrapped along with Greek and Latin. If the Communist self was to be modeled, therefore, the methods employed were not entirely of a different order than those that would at one point become popular in liberal-capitalist countries; nor were struggles, contradictions, and a limited amount of sociopolitical criticism banished from the cultural sphere.[37]

Indeed, the "liberalism" of the regime was even more obvious in policies that directly remodeled traditional social institutions, such as the family. From the Revolution's earliest days, a policy of equality between the sexes was pursued (even if implementation and popular acceptance among both men and women often lagged far behind legal change). Women were given the vote; divorces were made a matter of mutual consent in 1918. Abortion was legalized in 1920, and there was talk of a dedicated system of day care that would free women from lives tied exclusively to their maternal functions. The "bourgeois" concept of the patriarchal nuclear family was denounced as a historical phase with no greater permanence or justification than capitalism. Sexuality was to be freed from the distortion it suffered under a system of marriage that was little more than a financial arrangement.[38]

All this changed with the ascendancy of Josef Stalin to the leadership of the party. From 1928 to the start of World War II—a decade marked by the program of forced collectivization, "dekulakization" (that is, the persecution of wealthy, entrepreneurial peasants), and the single-minded sponsorship of heavy industry as well as by mass "purges" within the party and without—Stalin made it his explicit goal to build Communism at the level of the individual and to enlist the help of culture producers as engineers of human souls. The aim was to bring about a society of New Soviet Men (and Women) who valued collectivism over "egoistic individualism" and who lived public, transparent lives, actively involved in building Socialism, rather than introspective, private existences. In the early 1930s much propagandistic effort was poured into convincing the population to embrace an attitude of open-hearted optimism and exuberance, emulating positive Socialist heroes presented to them in novels and films. In built-from-scratch cities such as the "steel town" Magnitogorsk in the Ural Mountains, urban planning reflected the regime's collectivist ideal: The functionalist architecture stressed egalitarianism, while communal kitchens

FIGURE 10.3: Idealized portrait of the "Peasant and Worker State." Worker, peasant, and businessman are working together in organic harmony, for the benefit of the state. East German mural, Ministry of Finance (formally the Reich Air Ministry under the Nazis). The mural dates from 1952. In June 1953 the site witnessed massive demonstrations by East German workers against the raising of production quotas; Soviet occupation troops helped to subdue the unrest. Photo: Dan Vyleta.

and other services helped abolish the private sphere. (In old cities apartment space had long been "rationally" divided, allocating a certain number of square feet of floor space to each tenant, a system that often led to highly irrational results). The period was also marked by a new "prudery," favoring comradeship over sexual devotion and instigating harsh punishments against pornography. At the same time, more traditional family structures were strengthened by making divorces more difficult again and outlawing abortion.[39]

On the surface of things the Soviet project of social engineering was less focused on the body than its Nazi equivalent. For one thing, its official rhetoric was marked far less by organicism: Nature was not the final point of reference, and technological metaphors proved influential in a country much taken by the visions of scientific production elaborated by Taylor and put into practice by Ford. Indeed, the productive worker's body played a central part in the narrative of the Soviet self, celebrated in such novels as Valentin Kataev's 1932 *Time Forward!*, set in Magnitogorsk and narrating the protagonist's attempt to break the "world record" for pouring concrete. Indeed, heroic "shock workers"—that is, workers who exceeded the production estimates specified by their managers—became a central part of Soviet mythmaking. After Aleksei Stakhanov, in answer to a challenge put down by local party leaders, set a new record in hewing coal in a Ukrainian mine in August 1935, the Stakhanovite movement swept the country. Sponsored by a state longing for an ever more productive industry, it was locally often transformed into an articulation of worker criticism of plant management and ironically often proved disruptive for production order. The worker-hero found pictorial representation in the burly, plain-featured figures of socialist realism and revolutionary romanticism: These depicted wholesome, clean working bodies cheerfully unaffected by fatigue. The goal here was to invoke an anticipated, idealized future and create role models whose physical aspect intimated their Soviet qualities. This representation was deliberately didactic and literal in an era in which escapism was branded as a symptom of bourgeois decadence and modernist formalism was characterized as both obscurantist and depraved. It was a bluff era, celebrating bluffness.[40]

If Nazi terror was directed above all against political and "racial" enemies and explained itself in terms of purifying the *Volkskörper*, some of the worst excesses of Soviet terror found victims primarily among its own party members—a "tradition" initiated in the late 1920s, when purge commissions began to be set up that required all party members to justify themselves before them. Thus, 98 out of the 139 members of the 1934 Central Committee fell prey to witch hunts, and something like half the party membership experienced arrest during the years of the Great Terror (1936–1938). At the same time, looking at the Terror in terms of regional target numbers of "anti-Soviet elements" that local authorities were to arrest, it does emerge as a form of social engineering on a grand scale. The accusations against Terror victims were those of dissimulation—that is, of a discrepancy between the enacted and the real self—and of not having dissolved in society but of harboring self-interested ambitions and placing these before the community. More specific accusations included being identified as a "wrecker," that is, a saboteur of the great productive effort of the worker-state, or having obscured one's class origins. Thus, while class characteristics were not regarded as hereditary, one's sociological genealogy—whose class designations were controlled and ascribed by the Soviet state—could nevertheless precipitate one's downfall.[41]

Even more so than in the case of National Socialism, there is no historiographic consensus about the success of the Soviet project of "engineering souls." On the one hand, there exists evidence of real internalization of the Soviet vision of self, like the diary of Stephan Podlubnyi. Podlubnyi was a Moscow worker/student who was tortured by being unable to overcome "[his] daily secretiveness, the secret of [his] inside" and judged himself and others according to Soviet ideals such as the total transparency of the self and the conflation of personal and regime goals.[42] Similarly, there is evidence of Magnitogorsk workers evaluating and classifying themselves according to Soviet categories and of Magnitogorsk wives judging and comparing their husbands' "worth" according to the criterion of productivity.[43] On the other hand, there is a long historiographic tradition—upheld by dissident novelists such as Mikhail Bulgakov and Yuri Dombrowski—that maintains that the discrepancy between the regime's rhetoric and the actual demands of life in 1930s Russia created a society of atomized, "liberal" subjects whose self-interested entrepreneurialism was actually well served by an era of unprecedented social mobility. Even the mass denunciations that, as in the Nazi case, were a key motor of Soviet terror in this light appear as the opportunistic weapon of a Hobbesian battle of all against all within a world of strained resources. Rather than breaking down the wall between private and public, the regime thus produced a nation of cynics who had learned to "speak Bolshevik" and achieve their goals within the rules laid down by the system. Theorists such as Slavoj Žižek have even suggested that it was precisely this cynicism that gave the Communist regimes of the Eastern Bloc their long-term stability, since it engendered conformity in speech and action while isolating people from each other as they pursued lifestyle improvements rather than politics.[44]

In fact, 1930s Soviet society was marked by the emergence of a new middle-class elite whose tastes and educational values were often not miles away from those of the old bourgeoisie. Women were once again celebrated primarily as mothers, though they retained their place in the workforce. In schools, the educational experimentation of the 1920s was replaced by a return to a more traditional curriculum implemented through traditional pedagogical maxims. The classical Russian literary canon was coopted, with the exception of a few archconservative and gloomy figures like Dostoevsky. For all the political radicalism of the 1930s, there were thus also signs of a society settling down. It was not until the postwar era that a new cultural offensive would be started.

The example of the Soviet Union shows a society negotiating a theoretical vision of socially penetrated selves with ambiguous results. Contemporary representations of the body stressed the Soviet virtues of productivity and artless simplicity, but the regime's ambitions of social engineering were not focused on physiological markers or metaphors in the manner of the Nazis. At the end of the 1940s, the tale of body, self, and society received a further twist by the official approval of the behaviorist theories of the anti-Bolshevist Ivan Pavlov (1849–1936). The regime thus updated its Marxist materialist vision of the self by embracing a psychological theory that explained the self purely as a result of physical conditioning. More than ever, control over the experiences the individual body underwent was marked as a crucial part of any project of society building.

FROM LIBERAL DEMOCRACY INTO THE FUTURE: THE SELF AT THE END OF HISTORY

Looking Back from the End of History

In contrast to both Fascism and Soviet-style Communism, postwar liberal democracies were and are invested in a notion of self as to some important degree autonomous and impervious to social conditioning. This is particularly true in those societies, principally the United States and increasingly Great Britain, that have embraced free-market capitalism most openly. Individualism—the notion of one's personal uniqueness and the consequent need for each to find a personalized course of action in his or her negotiation of social conditions—here becomes a central, all-pervasive value, articulated both by the mainstream of society and (in a different key) by its self-proclaimed rebels. Even the political "left" nowadays frequently has recourse to such liberal ideals, arguing that the pursuit of individual happiness is better and more fairly facilitated by an interventionist state than by its absence. Libertarians, in contrast, see the state as hamstringing economic growth potential. Given that—pace Ludwig von Mises and other luminaries of libertarian thought—the desire for material well-being is the only common denominator in a world of radically self-determined individuals harboring a myriad of different dreams and ambitions, society should be organized in such a way that prosperity can take its course. For those for whom the triumph of liberal democracies and capitalism represents the endpoint of history, this triumph is in good part facilitated by the fact that, finally, after millennia of struggle, the sociopolitical framework for life corresponds to the needs of our indelible human constitution.

Any such hindsight look at the victory march of liberal-democratic ideology threatens, of course, to gloss over the fact that postwar Western Europe was built out of an amalgam of liberal and Socialist-inflected thought, with all sorts of conceptual compromises in terms of assumptions about the interrelation of self and society this implies. Particularly in the 1970s, sociological narratives of selfhood stressing the importance of society and community were going strong in any number of countries. Nor should one forget that liberal rhetoric about meritocracy and the right to individual self-fulfillment was (and is) fully compatible with systematic discrimination against women, ethnic minorities, and members of "lower" social classes. In Western academic circles, liberal discourses of the self have long been questioned: "Postmodern" theories that challenge the unity of the self, insist on the power of language to shape our identities, and imply that the experience of an autonomous self had to be as much learned as any other have been studied by thousands upon thousands of university students since the 1980s, just as Marxist ideas were influential a generation earlier. Still, judging from the views aired in mainstream popular fiction, television, films, and music at the end of the twentieth century, the notion of the self-determined individual retains a very powerful place in our imaginations.

In the liberal idea of the self, there is a tendency to uphold the depoliticization of the body. The liberal state, for the most part, makes no claim over or demand on the body (though consumption pressures and the economic order underlying the state undeniably do); nor is the metaphor that likens the physical body to the body politic much in vogue anymore. The body, one might say, has become private property, a means of self-expression, that is, part of the self-reflexive project of deliberately constructing one's self. As such, the body has become a crucial space for representing both one's individuality—one's difference from other members of society—and one's participation in a specific social subgroup. In the latter function, the body can also mobilized against the state and thus repoliticized: Among members of protest and countercultural movements—mods, punks, "left" and "right" skinheads, segments of the gay rights movement—dressing the body becomes a tool of both self-identification and provocation. In fact, in some protest groups, for example, in the neo-Nazi scene, objectionable ornamentation does not merely reference their political convictions but virtually replaces them—political content is here dissolved in symbology. A different sort of repoliticization of the body takes place when ethnic groups brand state or society as discriminating against them on grounds of physical difference: They argue that the system's alleged ideological indifference toward the body is little more than a smokescreen. Finally, ethically controversial areas relating to the individual's freedom over his or her body—questions of abortion and euthanasia, for instance—have traditionally seen liberal democratic states police the body. More recently, health issues such as smoking or obesity have precipitated interventionist policies on the part of various nation-states, in part due to financial pressures on their national insurance systems. In this sense the triumph of economic neoconservatism is, in some countries, accompanied by "big" rather than "small" government.

The body also took center stage in yet another narrative of the self that grew in popularity in the last two decades of the twentieth century, namely, the vision of human beings as controlled by their genes. This narrative clearly had its roots much earlier in the century, when deterministic, biological models were also in vogue. Scientific progress in gene technology has promised to deliver the final blueprint of the biochemical machine human, hence the tremendous popular interest precipitated by the Human Genome Project and comparable commercial ventures such as the research of the U.S.-based company Celera Genomics. Even though the evidence remains controversial in many individual instances, social phenomena such as homosexuality are increasingly explained with reference to genes rather than socialization. Some scientists—most famously Richard Dawkins—have urged us to view our genes' priorities in evolutionary terms and suggested that the evolution of culture could analogously be viewed in Darwinian terms, coining the term *meme* for a given culture unit. An attendant development was the (re)emergence of evolutionary psychology that similarly tries to explain social phenomena—for example, varieties of courting behavior among the sexes—in terms of evolutionary needs. In effect, evolutionary psychologists claim that cultural differences are incidental and can be reduced to a common biological thread and that our bodies lead us through life according to criteria we are not consciously aware of nor can control. Ideologically, these biologically determinist models are clearly related to Enlightenment visions of the mechanical body subject to physical laws that shape the individual's desires and aversions. As such, they are just as much part of a liberal tradition as narratives of self-determination; their cheerful coexistence is therefore less of a paradox than it may first appear to be.

Looking Forward: Future Selves of the 1970s, 1980s, and 1990s

Let us finish this gap-toothed survey of postwar liberal discourses of the self by a brief and admittedly equally eclectic glimpse of the future, the future, that is, as imagined by a handful of popular authors and directors in the final three decades of the twentieth century.

Stanley Kubrick's 1971 rendering of Anthony Burgess's 1962 dystopian novel *A Clockwork Orange* conjures up a future in which criminal deviants are disciplined (and hence "cured" of their criminality) by turning their physical bodies against their violent fantasies of abuse. In a near-future Britain, ruled by a thinly veiled version of the Conservative party and suffering from late-capitalist social malaise, the malcontent rapist and murderer Alex undergoes conditioning via the "Ludovico" method, which provokes physical nausea as the automatic response to desires of inflicting violence as well as those of a sexual nature. After his release Alex undergoes a series of encounters in a brutalized society in which he becomes a helpless victim: The film lingers on the spectacle of his retching, burping body as truncheons or boots rain down on his prone figure. Throughout the film, the body as an internalized policeman contrasts with the total failure of socialization of the young man, whose violence seems an expression of nihilist rebellion. Society is portrayed in disarray, littered with pornographic imagery that seems to conflate all sexual acts with acts of violence, an equation that Alex's captors/nurses reinforce during the conditioning. Integration of the deviant self into society is pursued purely on the level of enforced conformity—the only alternative to the Ludovico method to which Alex is subjected are the hellfire sermons of the prison chaplain, which substitute the body's rule over criminal desires with a rule of fear. The film's famous use of music—virtually all acts of violence within the film are juxtaposed to an incongruous score, and Alex's conditioning takes place to the sounds of Beethoven's Ninth symphony, making him incapable of listening to this, his most beloved symphony, without experiencing extreme pain and terror—further underlines the failures of socialization. Classical music, symbolic of a cultural education that, according to many a bourgeois's dream, should make the recipient incapable of acts of barbarism, here becomes the soundtrack of murder and violation, as well as the ruling classes' tool of subjugation. It is also telling that the Ninth is introduced into the conditioning program precisely when Alex is forced to witness Nazi atrocities; the failure of culture to inoculate a people against committing the gravest of crimes is thus underlined. At the same time, the film is more than conscious that its own representation of violence is ambivalent; the parallel between Alex's program of conditioning in front of a giant movie screen and the viewer's own position is too obvious to be ignored. The film ends up offering a choice between the violence of rebellious individualism and that of "normalization" through physiological conditioning. Only a total reconstruction of society promises relief.

If we jump a decade toward 1984 and switch mediums from film to novel, we can study the emergence of the "cyberpunk" genre in William Gibson's *Neuromancer* and with it a somewhat different narrative of body, self, and society.[45] Here, too, Burgess's/Kubrick's dystopian theme is in evidence: Once again, society of the not-so-far future is marked by the total triumph of unbridled capitalism. Everything is for sale; multinational corporations have openly replaced national governments as the true sources of authority. All social services, including policing, have become privatized. The inhabitants of this future are either loyal serfs of one of the economic conglomerates, members of a despairing underclass, or belong to an elite group of outsiders who have no allegiance but to themselves and operate outside the law as imposed by the corporations. *Neuromancer* thus re-rehearses a "neoromantic" notion of individuality familiar from late-1920s and 1930s American hard-boiled literature in which the (more or less) lone hero is the solitary incarnation of integrity and self-determined authenticity in a world that is corrupt to the core.[46] In such a world the self has to create its own moral codex that is sustained by no reference to social practice. Interestingly, the body here emerges as a crucial tool in the protagonist's struggles for self-preservation: It undergoes technological enhancements, either turning it into an efficient, self-contained fighting machine or allowing a digitalized version of the self to "plug into" and negotiate "cyberspace," that is, the worldwide "matrix" of computer information. Despite the novel's critical political overtones against liberal capitalism, it thus celebrates the triumph of the ingenious liberal self (aided by an updated, technologized body) against social conformity.

A further decade forward brings us to another dystopian near-future scenario, in Neal Stephenson's 1995 Hugo Award-winning *The Diamond Age*.[47] It yet again describes a world that has succumbed to the total primacy of economics. Society, no longer sharing any values and (in part) actively upholding total relativism and subjectivism as its core ethos, has split up into associations—called *phyles*—based on shared ethnicity, religion, or lifestyle choices. The only law respected by all is the common economic protocol governing all actions that affect a person's ability to fend for themselves economically. The demise of society is so pronounced that it takes place even on the level of information: Most individuals receive news that is personalized to a level where no shared information about the social world might be held with their neighbors. In fact, many *thetes* (an underclass not belonging to any phyle) spend as much time as they can afford in so-called *ractives*—interactive, escapist entertainment in which the self enters a virtual world and experiences virtual adventures. The exception to this trend toward total atomization are phyles whose members have decided to self-regulate and coordinate their social lives, above all the neo-Victorians, whose value system, work ethos, and gender relations are based around a romanticized vision of nineteenth-century British life. Technologically, the future is overshadowed by the proliferation of nanotechnological devices that serve as security systems, can be utilized to shape and enhance one's body, or are programmed to construct complex, "living" structures (buildings, islands, etc.). Indeed, nanotechnological devices form an entire ecosystem: They mutate and evolve independently, according to Darwinian rules of survival. The novel's plot is initiated by a problem faced by one of the most eminent neo-Victorians, Alexander Chung-Sik Finkle-McGraw: How can he provide his granddaughter with an education that will make her competitive and successful in such a world? He faces the paradox that the careful socialization taking place in the gentle middle-class haven that is neo-Victorian society cannot generate the subversive individualists who would be able to truly compete in the technological dog-eat-dog Wild West that is the future. As in *Neuromancer* and *Clockwork Orange*, socialization stands in conflict with free, rebellious individuality. The body—in symbiosis with nanotechnological organisms that enhance its potentialities—again emerges as a tool that helps the individual in his or her self-determined pursuit of happiness.

Finally, let us finish with a look at Iain M. Banks's "Culture" novels, the first of which—*Consider Phlebas*—was published in 1987, with sequels appearing throughout the 1990s.[48] In contrast to the other works discussed here, Banks presents the reader with a distant future scenario, whose most salient feature is that the depicted society—referred to by the inclusive term *Culture*—is genuinely postscarcity. Material resources are available in such abundance that private thoughts and memories are the only form of recognized private property. Genetic manipulation guarantees exceptional physical and mental attributes. Consequently, Culture citizens are free to pursue lives of pure hedonism, virtually without limits. In this liberal utopia, physical difference—including racial, special, and sexual difference—not only is tolerated but forms an important playground for self-exploration and construction. Sex changes, for example, are a ubiquitous practice, initiated by Culture citizens themselves, who have the ability to "gland" hormones and drugs into their bodies (an ability mostly used to heighten experiences of pleasure) and to otherwise control their bodies' "physiological settings." Culture citizens are effectively immortal, if they so desire: Not only are they protected from any mishaps by discreetly watching, hyperpowerful "Minds" (artificial intelligences), they can have their present "mind-state" recorded at virtually any time and have it reinstated in a new body should the old one be destroyed. The self is thus equated with the brain that controls the physiological machine that is the body. Via a "neural lace," the brain is also connected to the world of the artificial intelligences. The "Minds" are just as much self-conscious, individual personalities as their human/alien counterparts: They chat, enjoy art, and make electronic love to each other and are, of course, full citizens of the Culture. Society is depicted as a network of purely voluntary associations, governed only by a common ethos that values the self-fulfillment of all. The novels' plots are typically generated by the encounter of the Culture with other civilizations that know scarcity, struggle, and hierarchy and are hostile toward the Culture's decadent pluralism. There are also hints that the human/alien members of the culture are duped into lives of escapist quietism, while the Minds—above all, the Special Circumstances and Contact sections (i.e., the Culture's secret service)—determine policy and shape the future of the universe. Indeed, by Banks's 1996 Culture novel *Excession*, it is the Minds who are the true protagonists who are playing for existential stakes with all the self-confidence and eccentricities of warlords. In the perfect liberal future, in which the body serves little function but to house the self and facilitate self-exploration, the most perfect intellects threaten to relegate carbon-based life-forms to meaningless existences. Only the exceptional individual joins the Minds' adventures and, typically, enjoys his or her experience of "real" struggle.

From the body as an enemy used by society against the individual, through the body as helper in the battle for independence, to the body as accessory, the futures investigated here thus complement the liberal narratives of self that dominate the close of the twentieth century and help connect it to its Enlightenment roots. Whether the pleasure taken in our self-contained individuality will be overtaken by a renewed yearning for organic connection to the social world is not for a historian to judge.

NOTES

Introduction

I would like to thank the British Academy for their support of my research on Eric Avery, conducted in 2005 with a Small Research Grant. I would also like to thank Irene Rafanell and Pablo Schyfter for their numerous discussions of Butler, Foucault, and the Strong Programme, which have had an important impact on my ideas. I would also like to express my gratitude to Eric Avery for permission to use his works in this book and for the stimulating times spent with him discussing art and medicine.

1. See Judith Butler, "Foucault and the Paradox of Bodily Inscriptions," *Journal of Philosophy* 86 (1989): 601–7.
2. See particularly Norbert Elias, *The History of Manners: The Civilizing Process*, Vol. 1 (New York: Pantheon Books, 1978); and Michel Foucault, *Discipline and Punish: The Birth of the Prison*, trans. Alan Sheridan (New York: Vintage, 1977). The series of lectures Foucault gave at the Collège de France, recently published as *Society Must Be Defended*, *Abnormal*, and *Psychiatric Power*, add another dimension to our understanding of Foucault's thinking. Other important formative works in body history include Marc Bloch's *The Royal Touch: Sacred Monarchy and Scrofula in England and France*, trans. J. E. Anderson (1924; London: Routledge and Kegan Paul, 1973). Anthropological approaches include Marcel Mauss, "Techniques of the Body," *Economy and Society* 2 (1973): 70–88; Mary Douglas, *Purity and Danger: An Analysis of the Concepts of Pollution and Taboo* (London: Ark Paperbacks, 1966); and Pierre Bourdieu, *Outline of a Theory of Practice* (Cambridge: Cambridge University Press, 1977). Mauss's text is particularly important, as it is at the basis of both Elias's and Bourdieu's later works.
3. For example, Barbara Duden, *The Woman beneath the Skin: A Doctor's Patients in Eighteenth-Century Germany*, trans. Thomas Dunlap (Cambridge, MA: Harvard University Press, 1991); Thomas Laqueur, *Making Sex: The Body and Gender from the Greeks to Freud* (Cambridge, MA: Harvard University Press, 1990); and Ludmilla Jordanova, *Sexual Visions: Images of Gender in Science and Medicine between the Eighteenth and Twentieth Centuries* (New York: Harvester Wheatsheaf, 1989).

4. Dorinda Outram, *The Body and the French Revolution: Sex, Class and Political Culture* (New Haven, CT: Yale University Press, 1989); and Joanna Bourke, *Dismembering the Male: Men's Bodies, Britain and the Great War* (London: Reaktion, 1996).
5. The work of Gilles Deleuze is particularly important in this respect. See his collaborations with Felix Guattari, *Capitalism and Schizophrenia: Anti-Oedipus* (1972; Minneapolis: University of Minnesota Press, 1983) and *A Thousand Plateaus: Capitalism and Schizophrenia* (1980; Minneapolis: University of Minnesota Press, 1987).
6. Joanna Bourke, *An Intimate History of Killing: Face-to-Face Killing in Twentieth Century Warfare* (London: Granta, 1999); and Ana Carden-Coyne, "American Guts and Military Manhood," in *Cultures of the Abdomen: Diet, Digestion and Fat in the Modern World*, ed. Ana Carden-Coyne and Christopher E. Forth (New York: Palgrave, 2005), 71–85.
7. For two key early works, see Thomas Laqueur, "Amor Veneris, Vel Dulcedo Appeletur," in *Zone 3: Fragments for a History of the Human Body*, ed. Michel Feher, Ramona Naddaff, and Nadia Tazi (New York: Zone Books, 1989), 90–131; and Londa Schiebinger, *Nature's Body: Gender in the Making of Modern Science* (Boston: Beacon, 1993).
8. Bev Skeggs, *Formations of Class and Gender: Becoming* (London: Sage, 1997); and Mike Featherstone, "The Body in Consumer Culture," in *The Body: Social Process and Cultural Theory*, ed. Mike Featherstone, Mike Hepworth, and Bryan S. Turner (London: Sage, 1991), 170–96.
9. Andy Warwick, "Exercising the Student Body: Mathematics and Athleticism in Victorian Cambridge," in *Science Incarnate*, ed. Christopher Lawrence and Steven Shapin (Chicago: Chicago University Press, 1998), 288–326; Bruce Haley, *The Healthy Body and Victorian Culture* (Cambridge, MA: Harvard University Press, 1978); and J. A. Mangan and James Walvin, *Manliness and Morality: Middle-Class Masculinity in Britain and America, 1800–1940* (New York: St. Martin's, 1987).
10. See Harry Collins, "The TEA Set: Tacit Knowledge and Scientific Networks," *Science Studies* 4 (1974): 165–86.
11. See T. S. Kuhn, "The Function of Measurement in the Modern Physical Sciences," in *The Essential Tension* (Chicago: Chicago University Press, 1977); and, especially, Barry Barnes, *T. S. Kuhn and Social Science* (London: MacMillan, 1982).
12. See Michael Polanyi, *Personal Knowledge: Towards a Post-Critical Philosophy* (London: Routledge, 1958).
13. Michel Foucault, "Nietzsche, Genealogy, History," trans. D. F. Bouchard, in *Language, Counter-Memory, Practice*, ed. S. Simon and D. F. Bouchard (Ithaca, NY: Cornell University Press, 1977), 139–64. As an aside, Butler's essay "Foucault and the Paradox of Bodily Inscriptions" makes an interesting connection between Foucault's conception of inscribing on the body and Franz Kafka's *In the Penal Colony* (1919) that is worthy of further thought.
14. Foucault, *Discipline and Punish*, 138. See also Georges Vigarello, *Concepts of Cleanliness: Changing Attitudes in France since the Middle Ages*, trans. Jean Birrell (Cambridge: Cambridge University Press, 1988), for a specific discussion of the body in relation to such regimes.
15. Butler, "Foucault and the Paradox," 602.
16. Michel Foucault, "Body/Power," in *Power/Knowledge: Selected Interviews and Other Writings, 1972–1977*, ed. Colin Gordon (Brighton, UK: Harvester, 1980), 56

17. Robert Baden-Powell, "Campfire Yarn No. 17. How to Grow Strong," in *Scouting for Boys: A Handbook for Instruction in Good Citizenship* (London: Horace Cox, 1908).
18. See Ana Carden-Coyne, "From Pieces to Whole: The Sexualisation of Muscles in Post War Bodybuilding," in *Body Parts: Critical Explorations in Corporeality*, ed. Christopher E. Forth and Ivan Crozier (Lanham, MD: Lexington Books, 2005).
19. See, for general examples, David Pomfret, "The 'City of Evil' and the 'Great Outdoors': The Modern Health Movement and the Urban Young, 1918–1940," *Urban History* 28 (2001): 405–27; Paul Weindling, *Health, Race and German Politics between National Unification and Nazism, 1870–1945* (Cambridge: Cambridge University Press, 1989); and Alison Bashford and Carolyn Strange, "Public Pedagogy: Sex Education and Mass Communication in the Mid Twentieth Century," *Journal of the History of Sexuality* 13 (2004): 71–99. This list is far from exhaustive.
20. For naturism, see Michael Hau, *The Cult of Health and Beauty in Germany: A Social History, 1890–1930* (Chicago: Chicago University Press, 2003); and Maren Möhring, "Working Out the Body's Boundaries: Physiological, Aesthetic, and Psychic Dimensions of the Skin in German Nudism, 1890–1930," in Forth and Crozier, *Body Parts*, 229–46. For vegetarianism, see Arouna Ouedraogo, "Food and the Purification of Society: Dr. Paul Carton and Vegetarianism in Interwar France," *Social History of Medicine* 14 (2001): 223–45. For yoga, see Elizabeth De Michelis, *A History of Modern Yoga: Patañjali and Western Esotericism* (London: Continuum, 2004); much work remains to be done on yoga and embodiment in the West, but see for starters Benjamin Richard Smith, "Adjusting the Quotidian: Ashtanga Yoga as Everyday Practice," 2004, http://wwwmcc.murdoch.edu.au/cfel/docs/Smith_FV.pdf. Other bodily health techniques (and a panoply of other interesting stuff) can be found in Tim Armstrong, *Modernism, Technology, and the Body: A Cultural Study* (Cambridge: Cambridge University Press, 1998).
21. In thinking about the maintenance of social institutions, I am influenced strongly by Barry Barnes, "Social Life as Bootstrapped Induction," *Sociology* 17 (1983): 524–45. For further exploration of Barnes's themes, see David Bloor, *Wittgenstein: Rules and Institutions* (London: Routledge, 1997); and Martin Kusch, *Psychological Knowledge* (London: Routledge, 1998).
22. Carden-Coyne and Forth, *Cultures of the Abdomen*.
23. See, for example, Susan Bordo, *Unbearable Weight: Feminism, Western Culture and the Body* (Berkeley: University of California Press, 1993).
24. Largesse, the network for size esteem, http://www.eskimo.com/~largesse/.
25. Havelock Ellis was particularly adamant that eugenics needed feminism, and considering the number of his feminist friends who were also eugenicists, it seems they agreed. See Ellis, "Birth Control and Eugenics," in *The Philosophy of Conflict and Other Essays in War-Time* (London: Constable, 1919), 128–41.
26. See Naomi Wolf, *The Beauty Myth* (New York: Doubleday, 1991).
27. See Featherstone, "Body in Consumer Culture."
28. Carolyn Ward Comiskey, "Cosmetic Surgery in Paris in 1926: The Case of the Amputated Leg," *Journal of Women's History* 16 (2004): 30–54.
29. For a discussion of the feminist implications of cosmetic plastic surgery, see Kathy Davis, *Reshaping the Female Body: The Dilemma of Cosmetic Surgery* (London: Routledge, 1995).

30. For further attention to aging bodies, see Mike Featherstone and M. Hepworth, "The Mask of Ageing," in Featherstone, Hepworth, and Turner, *The Body*, 371–89; Mike Featherstone and A. Wernick, introduction to *Images of Ageing: Cultural Representations of Later Life*, ed. Mike Featherstone and A. Wernick (London: Routledge, 1995), 1–19; and Margaret Morganroth Gullette, *Aged by Culture* (Chicago: Chicago University Press, 2004).
31. Although gender performances do exercise power. This sentence is influenced by finitist philosophy, particularly that of David Bloor (*Wittgenstein: Rules and Institutions*) and Barry Barnes (*T. S. Kuhn and Social Science*), in which past performances (of usages of concepts) cannot determine future performances/usages.
32. For an overview of the corporeal turn in feminism, see A. Witz, "Whose Body Matters? Feminist Sociology and the Corporeal Turn in Sociology and Feminism," *Body and Society* 6 (2000): 1–24.
33. Judith Butler, *Bodies That Matter: On the Discursive Limits of "Sex"* (New York: Routledge, 1993), 95.
34. Here, *social institution* is being used in the finitist sense of Barnes and Bloor.
35. See Tony Ballantyne and Antoinette Burton, eds., *Bodies in Contact* (Durham, NC: Duke University Press, 2005).
36. See, for example, Heliodorus's *Ethiopians*, which is full of ideas about racial differences.
37. See, for example, David Arnold, *Colonizing the Body: State Medicine and Epidemic Disease in Nineteenth-Century India* (Berkeley: University of California Press, 1993); Julie Livingstone, *Debility and the Moral Imagination in Botswana* (Bloomington: Indiana University Press, 2005); and David Hoyt, "Fin-de-Siècle Ethnographic Discourse in Western Europe," *History of Science* 39 (2001): 331–54.
38. See David Arnold, *Warm Climates and Western Medicine: The Emergence of Tropical Medicine, 1500–1900* (Amsterdam: Rodopi, 1996); and Warwick Anderson, "The Trespass Speaks: White Masculinity and Colonial Breakdown," *American Historical Review* 102 (1997): 1343–70. On alcoholism and drug use, see Andrew Balfour and Henry H. Scott, *Health Problems of Empire* (London: W. Collins and Sons, 1924), esp. 333–56. The issue of drug use in the colonies is discussed by Jim Mills, *Madness, Cannabis and Colonialism: The "Native Only" Lunatic Asylums of British India, 1857 to 1900* (Basingstoke, UK: Palgrave, 2000).
39. See Dane Kennedy, *Islands of White: Settler Society and Culture in Kenya and Southern Rhodesia, 1890–1939* (Durham, NC: Duke University Press, 1987).
40. See Anna Crozier, "Sensationalising Africa: British Medical Impressions of Sub-Saharan Africa 1890–1939," *Journal of Imperial and Commonwealth History* 35 (2007): 393–415.
41. An extreme example of this was Kenyan psychiatrist H. L. Gordon's obsessive cranial measurements for eugenic purposes. See H. L. Gordon, "The Mental Capacity of the African," *Journal of the African Society* 33 (1934): 226–42. This work is discussed by Jock McCulloch in *Colonial Psychiatry and the African Mind* (Cambridge: Cambridge University Press, 1995).
42. John Hoberman, *Darwin's Athletes: How Sport Has Damaged Black America and Preserved the Myth of Race* (New York: Houghton Mifflin, 1997).
43. Holding can be seen at http://www.youtube.com/watch?v=MhYnYbvF9fo (accessed March 10, 2007).
44. Cricinfo, "Michael Holding," http://content-usa.cricinfo.com/westindies/content/player/52063.html (accessed March 10, 2007).

45. Martin Kusch refers to these situations as artificial kinds; see his *Psychological Knowledge.*
46. See, for an overview of sexuality in the twentieth century, Angus McLaren, *Twentieth-Century Sexuality* (Oxford: Blackwells, 1999).
47. See, for example, the works of Havelock Ellis (*Studies in the Psychology of Sex* [1897–1928; Philadelphia: F. A. Davis, 1935]) and Iwan Bloch (*Beiträge zur Aetiologie der Psychopathia Sexualis*, 2 vols. [Dresden, Germany: H. R. Dohrn, 1902–1903]).
48. An important example here is Luce Irigary, *The Sex Which Is Not One*, trans. Catherine Porter and Carolyn Burke (Ithaca, NY: Cornell University Press, 1985).
49. See Roy Porter and Lesley Hall, *The Facts of Life: The Creation of Sexual Knowledge in Britain, 1650–1950* (New Haven, CT: Yale University Press, 1995).
50. Marius Turda and P. J. Weindling, eds., *Blood and Homeland: Eugenics and Racial Nationalism in Central and Southeast Europe, 1900–1940* (Budapest, Hungary: Central European University Press, 2007); and Daniel Kelves, *In the Name of Eugenics: Genetics and the Uses of Human Heredity* (Cambridge, MA: Harvard University Press, 1995).
51. Weindling, *Health, Race and German Politics.*
52. Charis Thompson, *Making Parents: The Ontological Choreography of Reproductive Technologies* (2005; Cambridge, MA: MIT Press, 2007).
53. For an ideal account of a new reproductive technology that addresses the body in an exemplary way, see Nelly Oudshoorn, *The Male Pill* (Durham, NC: Duke University Press, 2003).
54. See Robert A. Nye, "The Evolution of the Concept of Medicalization in the Late Twentieth Century," *Journal of History of the Behavioral Sciences* 39 (2003): 115–29.
55. In particular, medicine was represented in this way in feminist criticisms of the 1970s and 1980s. See Rita Arditti, "Feminism and Science," in *Science and Liberation*, ed. Rita Arditti, Pat Brennan, and Steve Cavrak (Boston: South End Press, 1979), 350–68.
56. For more about Avery, see Michael M.J. Fischer, "With a Hammer, a Gouge, and a Woodblock: The Work of Art and Medicine in the Age of Social Retraumatization—The Texas Woodcut Art of Dr. Eric Avery," in *Para-Sites: A Casebook against Cynical Reason*, ed. George E. Marcus, Late Editions, 7: Cultural Studies for the End of the Century (Chicago: Chicago University Press, 2000).
57. For a context of AIDS in the United States, see Steven Epstein, *Impure Science: AIDS, Activism, and the Politics of Knowledge* (Berkeley: University of California Press, 1996).
58. Eric Avery, "Blood Test," 2005, http://docart.com/catalog/BloodTest.html (accessed January 10, 2007).
59. See Eric Avery, "Liver Die," 2005, http://docart.com/exhibitions/LiverDie/read.html (accessed January 10, 2007).
60. See Eric Avery, "Liver Die Book," 2005, http://www.docart.com/printed_matter/LiverDie/book.html.
61. Julia Kristeva, *Powers of Horror: An Essay on Abjection*, trans. Leon S. Roudiez (New York: Columbia University Press, 1982).
62. See Thomas Laqueur, "Diary," *London Review of Books*, September 7, 2006, http://www.lrb.co.uk/v28/n17/laqu01_.html.
63. Douglas, *Purity and Danger*, 35.

64. See Pablo Schyfter, "Tackling the 'Body Inescapable' in Sport: Body-Artifact Kinesthetics, Embodied Skill, and the Community of Practice in Lacrosse Masculinity," *Body and Society* 14, no. 3 (2008): 81–103.
65. See Nick Crossley, "Mapping Reflexive Body Techniques: On Body Modification and Maintenance," *Body and Society* 11 (2005): 1–35.
66. Francesca Alfano Miglietti, *Extreme Bodies: The Use and Abuse of the Body in Art* (Milan, Italy: Skira, 2003).
67. See James Miller, *The Passion of Michel Foucault* (New York: Simon and Schuster, 1993), for ways of thinking about sadomasochism in Nietzschean terms that add a dimension to our understanding of Foucault.
68. The two preceding paragraphs were adapted from Houdini Connections, "The Spanner Trial," 2004, http://www.houdini-connections.co.uk/4-info/Topics/spanner-info.htm (accessed March 14, 2007).
69. For an argument against the prohibition of pleasure, see Michel Foucault discussing suicide in "The Simplest of Pleasures," in *Foucault Live: Collected Interviews, 1961–1984*, ed. S. Lotringer (New York, SEMIOTEXT(E), 1996), 295–97, available at http://www.thefoucauldian.co.uk/simple.pdf (accessed March 14, 2007).
70. See Mike Jay, *Blue Tide: Search of Soma* (London: Autonomedia, 2002).
71. On apotemnophilia, see Sander L. Gilman, *Making the Body Beautiful: A Cultural History of Aesthetic Surgery* (Princeton, NJ: Princeton University Press, 1999); and John W. Jordan, "The Rhetorical Limits of the 'Plastic Body,'" *Quarterly Journal of Speech* 90 (2004): 327–58.
72. See Andrew Cunningham, *The Anatomical Tradition: The Resurrection of the Anatomical Projects of the Ancients* (Aldershot, UK: Scolar Press, 1997).
73. I use the term *natural kinds* in Barnes's sense. See Barry Barnes, "Social Life as Bootstrapped Induction."
74. Elsewhere, Chris Forth and I have argued that the body can be reduced to a single part. See the introduction to Forth and Crozier, *Body Parts*.

Chapter 1

1. Anthony Trollope, *Barchester Towers* (1857; Harmondsworth, UK: Penguin, 1987), 205.
2. For medical aspects, see Aldred Scott Warthin, *The Physician of the Dance of Death: A Historical Study of the Evolution of the Dance of Death Mythus in Art* (New York: Hoeber, 1931).
3. John M. Eyler, *Victorian Social Medicine: The Ideas and Methods of William Farr* (Baltimore: Johns Hopkins University Press, 1979); and Edward Higgs, *Life, Death and Statistics: Civil Registration, Censuses and the Work of the General Register Office, 1836–1952* (Hatfield, UK: Local Population Studies, 2004).
4. Karl Pearson, "The Chances of Death," in *Chances of Death and Other Studies of Evolution*, 2 vols. (London: Edward Arnold, 1897).
5. Raymond Pearl, *The Biology of Death* (Philadelphia: Lippincott, 1922).
6. August Weismann, *On Heredity* (Oxford: Clarendon, 1891).
7. B. Charlesworth, "Fisher, Medawar, Hamilton and the Evolution of Aging," *Genetics* 156 (2000): 927–31.
8. John Donne, "Divine Poems X," in *The Complete English Poems*, ed. C. A. Patrides (London: Everyman, 1991), 440–41.

9. Thomas McKeown, *The Modern Rise of Population* (London: Edward Arnold, 1976); and S. H. Preston, *Mortality Patterns in National Populations* (New York: Academic Press, 1976).
10. Robert S. Katz, "Influenza 1918–1919: A Study in Mortality," *Bulletin of the History of Medicine* 48 (1974): 416–22; and John M. Barry, *The Great Influenza: The Epic Story of the Deadliest Plague in History* (New York: Viking, 2004).
11. Irvine Loudon, "On Maternal and Infant Mortality," *Social History of Medicine* 4 (1991): 29–73.
12. From *Annual Reports of the Registrar General for Scotland*, as calculated by M. McCrae, "The Scottish Roots of the National Health Service" (PhD diss., University of Edinburgh, 2001), 18.
13. Richard Wilkinson, *Unhealthy Societies: The Afflictions of Inequality* (London: Routledge, 1996).
14. Daniel Dorling, *Death in Britain: How Local Mortality Rates Have Changed: 1950s to 1990s* (York, UK: Joseph Rowntree Foundation, 1997).
15. Charlotte Perkins Gilman, *The Living of Charlotte Perkins Gilman: An Autobiography* (1935; New York: Arno, 1972), 40.
16. Wolf Hass, *Komm, Süsser Tod* (Vienna: Rowohlt, 1998), better known to English audiences as a film of the same name (2002). See also Barbara Whitehead, *Sweet Death, Come Softly* (London: Headline, 1993).
17. Gilman, *Living*, 334–35.
18. Dignity in Dying, http://www.dignityindying.org.uk/information/index.asp?id=54 (accessed May 2006).
19. Linda Beecham, "BMA Opposes Any Change to Law on Assisted Suicide," *British Medical Journal* 325 (2002): 66.
20. *BMA News Review*, July 6, 2002.
21. Thomas B. Macaulay, "Horatius," in *Lord Macaulay's Essays and Lays of Ancient Rome* (London: Longmans, 1905), 853.
22. Wilfred Owen, "Dulce et decorum est," in *The War Poems of Wilfred Owen*, ed. Jon Stallworthy (London: Chatto and Windus, 1994), 29.
23. Gary Giddins, *A Pocketful of Dreams: The Early Years, 1903–1940* (New York: Little, Brown, 2001), 3.
24. Stanley J. Reiser, "The Science of Diagnosis: Diagnostic Technology," in *Companion Encyclopedia of the History of Medicine*, ed. William F. Bynum and Roy Porter (London: Routledge, 1993), 826–51.
25. "Conference of Medical Royal Colleges, 'Diagnosis of Brain Death'," *British Medical Journal* 273 (1976): 1187–88.
26. Rosemary Rhodes, "Death and Dying," in *Encyclopedia of Life Sciences* (Chichester, UK: Wiley, 2002), 5:343–50.
27. Bryan Jennett, "Brain Death and the Vegetative State," in *Encyclopedia of Life Sciences*, 3:410–13.
28. The Ad Hoc Committee of Harvard Medical School, "A Definition of Irreversible Coma," *Journal of the American Medical Association* 210 (1968): 337–40.
29. A. Bagheri and S. Shoji, "The Model and Moral Justification for Organ Procurement in Japan," *Journal Internationale de Bioéthique* 16 (2005): 79–90, 194–95.
30. W. R. Albury, "Ideas of Life and Death," in Bynum and Porter, *Companion Encyclopedia*, 272.
31. Robert Veatch, "The Impending Collapse of the Whole-Brain Definition of Death," *Hastings Center Report* 23 (1993): 18–24.

32. Carson Strong, "Consent to Sperm Retrieval and Insemination after Death or Persistent Vegetative State," *Journal of Law and Health* 14 (1999): 243.
33. J. Roberts, "US Rape Case Leaves Ethical Uproar," *British Medical Journal* 312 (1996): 329; *New York Times*, "Coma Victim Gives Birth," March 19, 1996, http://query.nytimes.com/gst/fullpage.html?res=9D05E2DC1739F93AA25750C0A960958260 (accessed May 2006).
34. David Adam, "What Is a Green Burial?" *The Guardian*, July 8, 2004.
35. Peter C. Jupp, *From Dust to Ashes: Cremation and the British Way of Death* (Basingstoke, UK: Palgrave Macmillan, 2006).
36. L. Kellaher and D. Prendergast, "Resistance, Renewal or Reinvention: The Removal of Ashes from Crematoria," *Pharos International* 70 (2004): 10–13.
37. Jean-David Jumeau-Lafond, *Carlos Schwabe: Symboliste et Visionnaire* (Paris: ACR Édition, 1994), 65.
38. For instance, Terry Pratchett, *The Colour of Magic* (London: Smythe, 1989).
39. Neil Gaiman, *Death: The High Cost of Living* (N.p.: DC Comics Vertigo, 1993).
40. This was true at the time of writing. More recently, however, a personification of Death has appeared in the final novel of the series, *Harry Potter and the Deathly Hallows* (London: Bloomsbury, 2007).
41. J. K. Rowling, *Harry Potter and the Philosopher's Stone* (London: Bloomsbury, 1997), 215.
42. Dylan Thomas, "Do Not Go Gentle into That Good Night," in *Collected Poems 1934–1952* (London: Dent, 1952), 116.
43. Roger Poole, "The Unknown Kierkegaard: Twentieth-Century Reception," in *The Cambridge Companion to Kierkegaard*, ed. Alistair Hannay and Gordin D. Marino (Cambridge: Cambridge University Press, 1998), 54–56.
44. Edith Kern, *Existential Thought and Fictional Technique: Kierkegaard, Sartre, Beckett* (New Haven, CT: Yale University Press, 1970).
45. Sartre was famously accused by Albert Camus of providing a moral cover for Stalinism, a charge he did not explicitly deny. For an analysis of this debate and an alternative view, see Ian H. Birchall, *Sartre against Stalinism* (New York: Berghahn, 2004).
46. Jean-Paul Sartre, *No Exit and Three Other Plays* (New York: Vintage, 1989).
47. Joseph F. Rutherford, *Millions Now Living Will Never Die* (Brooklyn, NY: International Bible Students Association, 1920).
48. James J. Hughes, "The Future of Death: Cryonics and the Telos of Liberal Individualism," *Journal of Evolution and Technology* 6 (2001): 1–23, http://jetpress.org/volume6/death.html.
49. Robert C.W. Ettinger and R. Michael Perry, *The Prospect of Immortality*, ed. Charles Tandy (1965; Palo Alto, CA: Ria University Press, 2005).
50. "Cryonics," Wikipedia, http://en.wikipedia.org/wiki/Cryonics (accessed May 2006).
51. John R. Wilmoth and Shiro Horiuchi, "Rectangularization Revisited: Variability of Age at Death within Human Populations," *Demography* 36 (1999): 475–95.
52. Jesse H. Ausubel, Perrin S. Meyer, and Iddo K. Wernick, "Death and the Human Environment: The United States in the 20th Century," *Technology in Society* 23 (2001): 134–46.
53. Ezekiel Kalipeni, ed., *HIV and AIDS in Africa: Beyond Epidemiology* (Malden, MA: Blackwell, 2004).

54. Roger Cooter, "The Dead Body," in *Companion to Medicine in the Twentieth Century*, ed. Roger Cooter and John Pickstone (London: Routledge, 2000), 479.
55. Elisabeth Kübler-Ross, *On Death and Dying* (New York: Simon and Schuster, 1969).
56. Shirley du Boulay, *Cicely Saunders: Founder of the Modern Hospice Movement* (London: Hodder and Stoughton, 1984).
57. Michelle Stanworth, "Reproductive Technologies and the Deconstruction of Motherhood," in *Reproductive Technologies: Gender, Motherhood and Medicine*, ed. Michelle Stanworth (Minneapolis: University of Minnesota Press, 1988), 10–35.
58. Sherry F. Colb, "What Is a Mother? The California 'Egg Donor' Case Gets It Wrong," *FindLaw Legal News and Commentary* (2004), http://writ.news.findlaw.com/colb/20040519.html (accessed April 2006).
59. Barbara Duden, *Disembodying Women: Perspectives on Pregnancy and the Unborn* (Cambridge, MA: Harvard University Press, 1993).
60. M. B. McNay and J. E. Fleming, "Forty Years of Obstetric Ultrasound 1957–1997: From A-Scope to Three Dimensions," *Ultrasound in Medicine and Biology* 25 (1999): 3–56.
61. Monica J. Casper, *The Making of the Unborn Patient: A Social Anatomy of Fetal Surgery* (New Brunswick, NJ: Rutgers University Press, 1998).
62. Rachel Roth, *Making Women Pay: The Hidden Costs of Fetal Rights* (Ithaca, NY: Cornell University Press, 2000).
63. Malcolm Nicolson, "Ian Donald, Diagnostician and Moralist," Web site of the Royal College of Physicians of Edinburgh, 2004, http://www.rcpe.ac.uk/library/donald/donald1.html.
64. Irvine Loudon, *Death in Childbirth: An International Study of Maternal Care and Maternal Mortality, 1800–1950* (Oxford: Clarendon, 1992).
65. Ministry of Health, *Report on an Investigation into Maternal Mortality*, Cmd. 5422 (London: HMSO, 1937).
66. Loudon, *Death in Childbirth*, 489.
67. Ibid.
68. R. Woods, "The Measurement of Historical Trends in Fetal Mortality in England and Wales," *Population Studies* 59 (2005): 147–62.
69. Ian Donald, *Practical Obstetric Problems* (London: Lloyd-Luke, 1969), 78.
70. Norman Del Mar, *Richard Strauss: A Critical Commentary on His Life and Works*, 3 vols. (London: Barrie & Rockliff, 1962–1972), 3:471.

Chapter 2

I would like to thank Eric Avery, Anne Kveim Lie, Cynde Moya, Irene Rafanell, Pablo Schyfter, and Rachel Spronk for their comments on this chapter, which very much improved the argument I am forwarding.

1. For the theoretical underpinning of this claim, see Michel Foucault, *Discipline and Punish: The Birth of the Prison*, trans. Alan Sheridan (New York: Random House, 1977), and *History of Sexuality I: The Will to Knowledge*, trans. Robert Hurley (London: Penguin, 1978).
2. I take "forms of life" from Michael M.J. Fischer, *Emergent Forms of Life and the Anthropological Voice* (Durham, NC: Duke University Press, 2003).

3. At least since Havelock Ellis people have known sexuality to be cultural anyway—see his *Evolution of Modesty* (Philadelphia: F. A. Davis, 1900).
4. See Lesley A. Hall, "Hauling Down the Double Standard: Feminism, Social Purity and Sexual Science in Late Nineteenth-Century Britain," *Gender and History* 16 (2004): 36–56. For an alternative account of the emergence of sexology in Britain, see Ivan Crozier, "British Medical Writing before Havelock Ellis: The Missing Story," *Journal for the History of Medicine and Allied Sciences* 63 (2008): 65–102.
5. The most prominent advocate of this view is Foucault, *History of Sexuality I*. For applications of Foucault's ideas, see Ian Hacking, "Making Up People," in *Reconstructing Individuality: Autonomy, Individuality and the Self in Western Thought*, ed. Thomas C. Heller, Morton Sosna, and David E. Wellbery (Stanford, CA: Stanford University Press, 1986), 222–36; for the specific example of homosexuality, see (among others) Harry Oosterhuis, *Step-Children of Nature* (Chicago: Chicago University Press, 1999).
6. I owe this point to a comment made by John Forrester at a day symposium on the World League for Sexual Reform held at Clare College, Cambridge, in November, 2000. On Freud's place in the wider reading public, see Dean Rapp, "The Early Discovery of Freud by the British General Educated Public, 1912–1919," *Social History of Medicine* 3 (1990): 217–43.
7. Particularly the 1885 Law Amendment Act in England and Wales. See F. B. Smith, "Labouchere's Amendment to the Criminal Law Amendment Act," *Historical Studies* 17 (1976): 16–73; for the way this act affected medical writing, see Ivan Crozier, "The Medical Construction of Homosexuality and Its Relation to the Law in Nineteenth-Century England," *Medical History* 45 (2001): 61–82. Of course, some acts, such as sodomy, had been illegal for much longer.
8. See the 2003 special edition of the *Journal for the History of Sexuality* for a number of articles exploring aspects of the World League for Sexual Reform in various contexts. See also Harry Cocks, "Saucy Stories: Pornography, Sexology and the Marketing of Sexual Knowledge in Britain, c. 1918–1970," *Social History* 29 (2004): 465–84.
9. Alfred Kinsey, Wardell Pomeroy, C. E. Martin, and P. H. Gebhart, *Sexual Behaviour in the Human Male* (New York: W. B. Saunders, 1948) and *Sexual Behaviour in the Human Female* (New York: W. B. Saunders, 1953); William Masters and Virginia Johnston, *Human Sexual Response* (Boston: Little, Brown, 1966); and Shere Hite, *The Hite Report on Female Sexuality* (New York: MacMillan, 1976). For a discussion of Kinsey's effect on the American public, see James Jones, *Kinsey: A Public/Private Life* (New York: W. W. Norton, 1997). For a more general discussion of sex surveys and their impacts in the twentieth century, see Angus McLaren, *Twentieth-Century Sexuality* (Oxford: Blackwells, 1999). There has of late been a backlash by the religious right against Kinsey and others who discuss sexuality in open ways. It is not entirely clear what they think will happen if people discuss sex openly or why they blame Kinsey personally for the changing tide in modern sexual morality—especially considering the material to be found in earlier sexological writings.
10. See, for example, Germaine Greer, "Lady, Love Your Cunt," in *The Mad Woman's Underclothes: Essays and Occasional Writings 1968–1985* (London: Picador, 1986), 74–77 (originally published in 1971 in the magazine *Suck*). For an alternative meditation on female genitalia by a very different brand of feminist thinker, see

the title essay in Luce Irigary's *The Sex Which Is Not One*, trans. Catherine Porter and Carolyn Burke (Ithaca, NY: Cornell University Press, 1985). An interesting nonacademic musing on female genitalia can be found at All About my Vagina, http://www.myvag.net.

11. This section owes everything to Irene Rafanell.
12. See, for example, Timothy F. Murphy, *Gay Science: The Ethics of Sexual Orientation Research* (New York: Columbia University Press, 1997). For strong constructivist perspectives, see David Halperin, *How to Do the History of Homosexuality* (Chicago: Chicago University Press, 2002); and Arnold Davidson, *The Emergence of Sexuality* (Cambridge, MA: Harvard University Press, 2001). While it is clear that in the nineteenth century, discussions of sexuality focused on the biology of sex as a way of escaping the scope of moral discourses, these biological arguments have—especially via eugenics and sociobiology—been more recently associated with conservative arguments. These uses of biology, while interesting in themselves and deserving of full attention, are sadly beyond the scope of this chapter.
13. A central position on this is Judith Butler, *Bodies That Matter: On the Discursive Limits of "Sex"* (New York: Routledge, 1993). See the complementary work of Elizabeth Grosz, *Volatile Bodies: Toward a Corporeal Feminism* (Bloomington: Indiana University Press, 1994).
14. Barry Barnes, "Social Life as Bootstrapped Induction," *Sociology* 17 (1983): 524–45.
15. In Pierre Bourdieu's sense of individual habituated practices. See Bourdieu, *Outline of a Theory of Practice* (Cambridge: Cambridge University Press, 1977). In this, I am in great debt to Irene Rafanell, who has consistently criticized models such as Bourdieu's that posit external controlling mechanisms to explain social actions. See her doctoral thesis, "The Sexed and Gendered Body as a Social Institution: A Critical Reconstruction of Two Social Constructionists' Models" (University of Edinburgh, 2003), on which she is building in profound ways in a series of forthcoming arguments.
16. The important, but neglected, role of the collective has been highlighted in the work of Irene Rafanell, "Durkheim and the Performative Model: Reconfiguring Social Objectivity," in *Sociological Objects: The Reconfiguration of Social Theory*, ed. Geoff Cooper, Andrew King, and Ruth Retti (London: Ashgate, 2009), 59–76.
17. For some discussion of this issue, see the papers in Karin Knorr-Cetina and Aaron Cicourel, eds., *Advances in Social Theory and Methodology: Toward an Integration of Micro- and Macrosociologies* (London: Routledge and Kegan Paul, 1981).
18. This is an extremely summarized account of a very complex analytic discussion found in Barnes, "Social Life as Bootstrapped Induction."
19. Barnes, "Social Life as Bootstrapped Induction," 524. For a sophisticated development of Barnes's work, see Martin Kusch, *Psychological Knowledge* (London: Routledge, 1998).
20. This point is developed at length by Butler in *Bodies That Matter*.
21. This model does not collapse into a radical idealism. Bodily actions are not entirely social, as they are limited by the "body inescapable"—see R. W. Connell, *Masculinities*, 2nd ed. (Berkeley: University of California Press, 1995)—but physicality is always mediated by culture. This conception of naturalness recognizes that practitioners, individuals, are reflexively able to position their bodies, to adapt them in subtle ways. See Lucy Bailey, "Refracted Selves? A Study of Changes in Self-Identity

in the Transition to Motherhood," *Sociology* 33 (1999): 335–52, who follows Anthony Giddens, *Modernity and Self-Identity: Self and Society in the Late Modern Age* (Cambridge, UK: Polity, 1991) and *The Transformation of Intimacy: Sexuality, Love and Eroticism in Modern Societies* (Cambridge, UK: Polity, 1992).

22. See Butler, *Gender Trouble: Feminism and the Subversion of Identity* (London: Routledge, 1990) and *Bodies That Matter*; and J. L. Austin, *How to Do Things with Words: The William James Lectures Delivered at Harvard University in 1955* (1955; Oxford: Clarendon, 1970).
23. Rafanell, "Durkheim and the Performative Model."
24. This idea of rules as social institutions derives from David Bloor's sophisticated account of rule following presented in *Wittgenstein: Rules and Institutions* (London: Routledge, 1997).
25. This argument is rehearsed in different ways by Michel Foucault, *Power/Knowledge: Selected Interviews and Other Writings, 1972–1977*, ed. Colin Gordon (Brighton, UK: Harvester, 1980). At this point Foucault meets theories of performativity. For an expansion of the link between performativity and power, see Butler, *Gender Trouble*, and Barry Barnes, *The Nature of Power* (Cambridge, UK: Polity, 1988). A new application of such performative theories of power is presented in Hugo Gorringe and Irene Rafanell, "The Embodiment of Caste: Oppression, Protest and Change," *Sociology* 41 (2007): 97–114.
26. A secrecy in which many do not put much store, considering the attention sex gets in conversations, gossip, the media, entertainment, art, and so on.
27. Foucault's model of episodic change derives from Gaston Bachelard's concept of scientific change as a dialectical process, channeled through Georges Canguilhem's notion of normalization. See Bachelard, *The New Scientific Spirit*, trans. Arthur Goldhammer (1934; Boston: Beacon Books, 1984); and Canguilhem, *The Normal and the Pathological*, trans. Caroline Fawcett (1943; New York: Zone Books, 1993). For more, see Gary Gutting, *Michel Foucault's Archaeology of Scientific Reason* (Cambridge: Cambridge University Press, 1989).
28. See, for example, Jennifer Terry, *An American Obsession: Science, Medicine, and Homosexuality in Modern Society* (Chicago: Chicago University Press, 2000), as one of many who, portraying themselves as following Foucault, have privileged sexological discourses in order to trace ideas of homosexuality in the twentieth century. This is not to say that legitimate study cannot be made of the history of sexology, but it is not the place to look for the history of homosexuality.
29. For the malleability of female bodies, see Victoria Pitts, *In the Flesh: The Cultural Politics of Body Modification* (New York: Palgrave, 2003), 51.
30. See, for example, Hera Cook, *The Long Sexual Revolution* (Oxford: Oxford University Press, 2004).
31. See Roy Porter and Lesley Hall, *The Facts of Life: The Creation of Sexual Knowledge in Britain, 1650–1950* (New Haven, CT: Yale University Press, 1995). See also Lutz L.D.H. Sauerteig and Roger Davidson, eds., *Shaping Sexual Knowledge: A Cultural History of Sex Education in Twentieth-Century Europe and North America* (London: Routledge, 2009).
32. See, for example, the article by Hilda Hutcherson, "5 Myths of Anal Sex Uncovered," iVillage, http://love.ivillage.com/lnssex/sextaboos/0,,jdgc,00.html (accessed August 11, 2006). Such sexual advice pieces on having anal sex, or how to give the perfect blowjob, and so on, are staple articles in glossy magazines such as *Cosmopolitan*. Other sexual activities, for example, fist-fucking, while becoming

much more prevalent in the twentieth century, have not completely moved over to broadly mainstream activities and remain in particular subcultures, for example, the sadomasochism community and certain homosexual communities. I thank Thomas Roeske for a discussion on this topic one evening in 2003 in Berlin.

33. See Havelock Ellis, *Sex in Relation to Society* (Philadelphia: F. A. Davis, 1910); and Marie Stopes, *Married Love*, ed. Ross McKibbin (1918; Oxford: Oxford University Press, 2004). These and other works are discussed in Lesley A. Hall, *Sex, Gender and Social Change in Britain since 1880* (London: MacMillan, 2000).
34. Giddens, *Transformation of Intimacy.*
35. Sodomy came under the 1885 Law Amendment Act's classification of "unnatural acts." It was, of course, treated as unnatural in the Judeo-Christian tradition, as shown in some of the writings of Paul: "Therefore, God handed them over to degrading passions. Their females exchanged natural relations for unnatural, and the males likewise gave up natural relations with females and burned with lust for one another. Males did shameful things with males and thus received in their own persons the due penalty for their perversity" (Romans 1:26–27).
36. For a discussion of medical texts addressing anal sex, see Ivan Crozier, "All the Parts Were Perfectly Natural," in *Body Parts: Critical Explorations in Corporeality*, ed. Christopher E. Forth and Ivan Crozier (Lanham, MD: Lexington Books, 2005), 65–84. Notable early pornographic representations include the Marquis de Sade's *Philosophy in the Bedroom* (1797) and Walter's *My Secret Life* (ca. 1888); the latter is available at http://www.my-secret-life.com/index.php.
37. http://www.medscape.com/viewarticle/532034 (accessed September 13, 2007).
38. William Saletan, "Ass Backwards: The Media's Silence about Rampant Anal Sex," Slate.com, September 20, 2005, http://www.slate.com/?id=2126643&nav=tap2/, quoting William D. Mosher, Anjani Chandra, and Jo Jones, "Sexual Behavior and Selected Health Measures: Men and Women 15–44 Years of Age, United States, 2002," Advance Data from Vital and Health Statistics, September 15, 2005, http://www.cdc.gov/nchs/data/ad/ad362.pdf (accessed September 13, 2007).
39. Although avoiding such technological determinism, Elizabeth Siegel Watkins, *On the Pill: A Social History of Oral Contraceptives, 1950–1970* (Baltimore, Johns Hopkins University Press, 1998), has argued that it was in fact the married American woman whose life was most changed by the pill, with unmarried women taking risks and/or other precautions as they delved into unmarried sexual liaisons. Cook also addresses this issue in her *Long Sexual Revolution.* For an important study of contraception in an earlier period, see Kate Fisher, *Birth Control, Sex and Marriage in Britain, 1918–1960* (Oxford: Oxford University Press, 2006).
40. Cook in particular argues that one of the impacts of reliable contraception is an alteration of family life and the role occupied by women when motherhood is a choice rather than a virtual inevitability.
41. See, for example, Roger Davidson and Lesley Hall, eds., *Sex, Sin and Suffering: Venereal Disease and European Society since 1870* (London: Routledge, 2001).
42. See Hall, *Sex, Gender and Social Change.*
43. Nineteenth-century advice for avoiding syphilis included using condoms, washing the penis after sex, and so on. This information was not as much part of a major health-advice movement like today, however, although specific advice was disseminated to soldiers and other risk groups.
44. I am reliably informed on this subject by G. B., a graduate student at the Science Studies Unit of Edinburgh University.

45. For an array of discussions about surgical manipulation of the body, see Angelika Taschen, ed., *Aesthetic Surgery* (Cologne: Taschen, 2005); and Sander L. Gilman, *Making the Body Beautiful: A Cultural History of Aesthetic Surgery* (Princeton, NJ: Princeton University Press, 1999). For recent feminist engagements with aesthetic surgery, see Ruth Holliday and Jacqueline Sanchez Taylor, "Aesthetic Surgery as False Beauty," *Feminist Theory* 7 (2006): 179–95; and, more interestingly, Victoria Pitts-Taylor, *Surgery Junkies: Wellness and Pathology in Cosmetic Culture* (New Brunswick, NJ: Rutgers University Press, 2007).
46. For an account of the problems surrounding male circumcision—and a damnation of this act—see Robert Darby, *A Surgical Temptation: The Demonisation of the Foreskin in Victorian Medicine* (Chicago: Chicago University Press, 2005).
47. Indeed, there is an increasing trend for men of all sexual persuasions to remove body hair, and increasingly this has come to mean pubic hair. As I edit this chapter, trips to Sauna Deco in Amsterdam reveal that a large proportion of men also remove their pubic hair—although this must be considered in light of the fact that nudity is in part a performative act in such spaces. As with the reconceptualizing of female pubic hair, this has been driven at least in part by the prevalence of pornographic images. It remains the case that men are less likely to put themselves in physically uncomfortable situations. See Nelly Oudshoorn, *The Male Pill* (Durham, NC: Duke University Press, 2003). My awareness of the extent of young heterosexual American men's acceptance of hair removal—and particularly shaved balls—was greatly enhanced by conversations held in a youth hostel in Austin, Texas (where I was based for a month researching Havelock Ellis's papers in 2002). I was especially struck by a response from a female interlocutor who said she "appreciated the effort" that her friend was willing to go to, although she did not require it as a prerequisite for sex. I am pleased to finally be able to acknowledge my debt to Sean and Amanda for this discussion.
48. Christine Hope, "Caucasian Female Body Hair and American Culture," *Journal of American Culture* 5 (1982): 93–99. See also Rebecca Herzig, "Removing Roots: 'North American Hiroshima Maidens' and the X-ray," *Technology and Culture* 40 (1999): 723–45.
49. "Walter," *My Secret Life*, vol. 7, chap. 5.
50. And, in the case of naturists, was considered normal and natural, and therefore desirable. See Michael Hau, *The Cult of Health and Beauty in Germany: A Social History, 1890–1930* (Chicago: Chicago University Press, 2003).
51. "Page Three Girls," MediaUK, December 6, 2004, http://www.mediauk.com/article/17252 (accessed August 29, 2006). For those interested in seeing more, numerous Web sites show scanned images of Liv Lindeland to which I refer, perhaps to their copyright law detriment.
52. Michael Castleman, "Porn-Star Secrets: Going Naked in Front of the Camera Necessitates Lots of Hair-Removal Tricks," Salon.com, 2000, http://dir.salon.com/sex/feature/2000/09/06/hair_removal/index.html (accessed August 30, 2006).
53. Ibid.
54. Cath Rapley, "Pubic Relations," *The Observer*, April 14, 2002, http://observer.guardian.co.uk/life/story/0,,683936,00.html (accessed August 30, 2006).
55. "Hair Removal Advice for Every Area," PubicShavingAdvice.com, 2008, http://pubicshave.stores.yahoo.net/ (accessed August 30, 2006).
56. "Shaving Your Pubic Area," PubicShavingAdvice.com, 2008, http://pubicshave.stores.yahoo.net/remhairfromy7.html (accessed August 30, 2006).

57. Ibid.
58. Ibid.
59. Hope, "Caucasian Female Body Hair," and Herzig, "Removing Roots."
60. Or perhaps worse—a fetish: numerous pornographic Web sites are devoted specifically to hairy genitalia.
61. Magdala Peixoto Labre, "The Brazilian Wax: New Hairlessness Norm for Women?" *Journal of Communication Inquiry* 26 (2002): 113.
62. reddog199200, Post subject: Shave it!, August 5, 2004, www.femalefirst.co.uk/board/ftopic6451.html (accessed March 23, 2005).
63. David2, Post subject: shaved, August 14, 2004, www.femalefirst.co.uk/board/ftopic6451.html (accessed March 23, 2005).
64. Stephen, August 28, 2004, www.femalefirst.co.uk/board/ftopic6451.html (accessed March 23, 2005).
65. Chelsea Summers, Pretty Dumb Things, http://prettydumbthings.typepad.com/chelsea girl/2005/05/index.html (accessed August 30, 2006). For a very different experience, see the enlightening online discussion "Pubic Hair and Bikini Styling," All About My Vagina, March 5, 2005, http://www.myvag.net/fur/ (accessed August 31, 2006). This site would repay further research in terms of modern embodiment and the construction of the gendered self. Very few historical sources are as detailed and reflexive.
66. In line with this comment, one of Ruth Barcan's respondents—an auteur of amateur erotic film—recalled the removal of pubic and body hair by one of his would-be "porn stars" over a period of three weeks as he conformed to pornographic industry standard. Barcan, "In the Raw: 'Home-Made' Porn and Reality Genres," *Journal of Mundane Behavior* 3 (2002): 13.
67. See Mike Featherstone, ed., *Body Modification* (London: Sage, 2000). For a detailed analysis of individual experiences of body modifiers, see Pitts, *In the Flesh*. For a more general approach to piercing, see Mariana Torgourvich, "Piercing," in *Late Imperial Culture*, ed. Román de la Capa, E. Ann Kaplan, and Michael Sprinker (London: Verso, 1995), who points out (on page 128) that genital piercings are comparatively rare outside the West, despite the fact that many of those who modify their bodies in such a way consider themselves to be somehow in touch with other premodern cultural values and practices.
68. See Pitts, *In the Flesh*, 64–65.
69. Elayne Angel, master piercer from New Orleans, e-mail message to author, March 23, 2005.
70. Elayne Angel, "Female Genital Piercings," Rings of Desire, http://ringsofdesire.com/femalegenital.html (accessed March 24, 2005). The research mentioned was published in V. S. Millner, B. H. Eichold, T. H. Sharpe, and S. Lynn, "First Glimpse of the Functional Benefits of Clitoral Hood Piercings," *American Journal of Obstetrics and Gynecology* 193 (2005): 675–76.
71. Angel, "Female Genital Piercings."
72. H. Ferguson, "Body Piercing," *British Medical Journal* 319 (1999): 1627–29.
73. A. Stirn, "Body Piercing: Medical Consequences and Psychological Motivations," *Lancet* 361 (2003): 1205–15. See also N. Buhrich, "The Association of Erotic Piercing with Homosexuality, Sadomasochism, Bondage, Festishes and Tattoos," *Archives of Sexual Behavior* 12 (1983): 161–71.
74. Stirn, "Body Piercing." See also F. E. Willmott, "Body Piercing: Lifestyle Indicator or Fashion Accessory?" *International Journal of STD and AIDS* 12 (2001): 358–60.

75. This issue is further discussed in Lois Bibbings and Peter Alldridge, "Sexual Expression, Body Alteration, and the Defence of Consent," *Journal of Law and Society* 20 (1999): 356–70.
76. See Giddens, *Transformation of Intimacy*. See also the papers on piercings in Featherstone, *Body Modification*.
77. Chelsea Summers, "spandex & lucite shoes: part 12, seeing stars and seeing stars," Pretty Dumb Things, June 20, 2005, http://prettydumbthings.typepad.com/chelseagirl/2005/06/spandex_lucite_1.html (accessed August 30, 2006).
78. John51, "Waited Too Long for My PA," Bmezine.com, July 2006, http://www.bmezine.com/pierce/09-male/pa/A60713/pawaited.html (accessed October 11, 2006).
79. This recalls Martin Kusch's extension of Barnes in *Psychological Knowledge*.
80. See Sigmund Freud, *Three Essays on the Theory of Sexuality*, in *Standard Edition*, trans. and ed. James Strachey (London: Hogarth, 1953): 125–245.
81. See Thomas Laqueur, "Amor Veneris, Vel Dulcedo Appeletur," in *Zone 3: Fragments for a History of the Human Body*, ed. Michel Feher, Ramona Naddaff, and Nadia Tazi (New York: Zone Books, 1989), 90–131.
82. The standard work on the relation of orgasm and generation before Freud is Thomas Laqueur, *Making Sex: The Body and Gender from the Greeks to Freud* (Cambridge, MA: Harvard University Press, 1990).
83. Alex Skene, "The Anatomy and Pathology of Two Important Glands of the Female Urethra," *American Journal of Obstetrics Diseases of the Woman and Child* 13 (1880): 265–70.
84. John Huffman, "The Development of the Periurethral Glands in the Human Female," *American Journal of Obstetrics and Gynecology* 46 (1946): 773–85; John Huffman, "The Detailed Anatomy of the Paraurethral Ducts in the Adult Human Female," *American Journal of Obstetrics and Gynecology* 55 (1948): 86–101; and John Huffman, "Clinical Significance of the Paraurethral Ducts and Glands," *Archives of Surgery* 62 (1951): 615–26. Huffman was joined in his investigations by Alfred Folsom and Harold O'Brien, "The Female Obstructing Prostate," *Journal of the American Medical Association* 121 (1943): 573–80; and by R. L. Deter, G. T. Caldwell, and A. I. Folsom, "A Clinical and Pathological Study of the Posterior Female Urethra," *Journal of Urology* 55 (1946): 651–62. More recently, Shelley Tepper, Jaishree Jagirdar, Desmond Heath, and Stephen Geller, "Homology between the Female Paraurethral (Skene's) Glands and the Prostate," *Archives of Pathology and Laboratory Medicine* 108 (1984): 423–25, confirmed this finding and used a chemical analysis to show that the ejaculate was not urine.
85. Havelock Ellis, *Psychology of Sex* (London: Heineman, 1933), 17.
86. Ernest Gräfenberg, "The Role of Urethra in Female Orgasm," *International Journal of Sexology* 3 (1950): 145–48.
87. "Orgasms, Female Ejaculation, and the G-Spot, Again," Go Ask Alice: Columbia University's Health Q&A Internet Service, August 11, 2006, http://www.goaskalice.columbia.edu/1267.html (accessed October 3, 2006).
88. F. Addiego, E. G. Belzer, J. Comolli, W. Moger, J. D. Perry, and B. Whipple, "Female Ejaculation: A Case Study," *Journal of Sex Research* 17 (1981): 13–21. More recently, see Whipple, "Beyond the G Spot: New Research on Human Female Sexual Anatomy and Physiology," *Scandinavian Journal of Sexology* 3 (2000): 35–42; and Whipple and B. R. Komisaruk, "Beyond the G Spot: Recent Research on Female Sexuality," *Medical Aspects of Human Sexuality* 1 (1998): 19–23.

89. "Cytherea," Wikipedia, http://en.wikipedia.org/wiki/Cytherea_%28erotic_actress%29 (accessed October 3, 2006). One should recall that Cytherea is a synonym for Aphrodite—a rare classical reference in the adult industry.
90. Louise Achille and Catherine Wilkinson, "Female Ejaculation: Research Contrary to BBFC Ruling," Feminists Against Censorship, http://www.fiawol.demon.co.uk/FAC/femejac.htm (accessed October 3, 2006).
91. "Sex Tutor Topics: Female Ejaculation," www.sextutor.com, http://www.sextutor.com/ejaculation.shtml (accessed October 3, 2006).
92. Thomas Laqueur, *Solitary Sex: A Cultural History of Masturbation* (New York: Zone Books, 2003).
93. See Maren Möhring, "Working Out the Body's Boundaries: Physiological, Aesthetic, and Psychic Dimensions of the Skin in German Nudism, 1890–1930," in *Body Parts: Critical Explorations in Corporeality*, ed. Christopher Forth and Ivan Crozier (Lanham, MD: Lexington Books, 2005), 229–46. The Freikörperkultur is still going strong. See http://fkk-online.de/ (accessed October 11, 2006)
94. For interesting work on nudism in the Australian context, as well as attention to European and American material, see Ruth Barcan, "Regaining What Mankind Has Lost through Civilisation: Early Nudism and Ambivalent Moderns," *Fashion Theory: The Journal of Dress, Body and Culture* 8 (2004): 1–20, and "The Moral Bath of Bodily Unconsciousness: Female Nudism, Bodily Exposure and the Gaze," *Continuum: Journal of Media and Cultural Studies* 15 (2001): 305–19.
95. See the Web site of the new attempt: Richard Collins, http://www.nakedwalk.org/ (accessed October 11, 2006).
96. Freikörperkultur Online, http://fkk-online.de/ (accessed October 11, 2006).
97. See "Natural Family," FKK Online, http://fkkonline.free.fr/img0401/natfam.jpg (accessed October 11, 2006) for a depiction of the idealized German naturist family.
98. See Marisa Torrieri, "Naked Yoga: Do It Downward-Doggie-Style at These Clothing-Free Workouts," Metromix, NY, August 17, 2007, http://newyork.metromix.com/events/article/naked-yoga/156195/content (accessed September 13, 2007) and Isis Phoenix's site Naked Yoga NYC, http://www.nakedyoganyc.com/ (accessed September 13, 2007).
99. For my favorite, Saunabad at 10 Rykestr., Prenzlauer Berg, Berlin, see http://www.saunabad-berlin.de (accessed October 11, 2006).
100. Gay saunas do, however, often involve public sex and mutual masturbation. Multiple reviews on the Internet give interested readers a good insight into what happens behind the doors of such establishments.
101. Gail Weiss, *Body Images: Embodiment as Intercorporeality* (New York: Routledge, 1999), 3.
102. See Rachel Shtier, *Striptease: The Untold Story of the Girlie Show* (Oxford: Oxford University Press, 2004). For Roland Barthes's essay, see *Mythologies*, trans. Annette Lavers (New York: Hill and Wang, 1984).
103. See Matt Houlbrook, *Queer London: Perils and Pleasures in the Sexual Metropolis, 1918–1957* (Chicago: Chicago University Press, 2005).
104. See Ruth Barcan, "In the Raw."
105. For an analysis of self-representations of nude bodies in public, see Ruth Barcan, "Home on the Rage: Nudity, Celebrity, and Ordinariness in the Home Girls/Blokes Pages," *Continuum: Journal of Media and Cultural Studies* 14 (2000):

145–58; and Katherine Albury, "Homie-Erotica: Heterosexual Female Desire in *The Picture*," *Media International Australia* 84 (1997): 19–27.

106. See Foucault, *History of Sexuality*. See also James Miller, *The Passion of Michel Foucault* (New York: Simon and Schuster, 1993).
107. See the chapters in this collection by Christopher Forth and Dan O'Connor.

Chapter 3

1. Roger Cooter and John Pickstone, "Introduction," in *Companion to Medicine in the Twentieth Century*, ed. Roger Cooter and John Pickstone (London: Routledge, 2000), xiii–xix; and Deborah Lupton, "The Social Construction of Medicine and the Body," in *The Handbook of Social Studies in Health and Medicine*, ed. Gary L. Albrecht, Ray Fitzpatrick, and Susan C. Scrimshaw (London: Sage, 2000), 50–63.
2. Thomas Schlich, "The Emergence of Modern Surgery," in *Medicine Transformed: Health, Disease and Society in Europe, 1800–1939*, ed. Deborah Brunton (Manchester, UK: Manchester University Press, 2004), 61–91.
3. Christopher Lawrence, "Democratic, Devine and Heroic: The History and Historiography of Surgery," in *Medical History and Surgical Practice: Studies in the History of Surgery*, ed. Christopher Lawrence (London and New York: Routledge, 1992), 15.
4. Owsei Temkin, "The Role of Surgery in the Rise of Modern Medical Thought," *Bulletin for the History of Medicine* 25 (1951): 248–59.
5. On the notion of control in surgery, see Thomas Schlich, "Surgery, Science and Modernity: Operating Rooms and Laboratories as Spaces of Control," *History of Science* 45 (2007): 231–56.
6. Emil Theodor Kocher, "Concerning Pathological Manifestations in Low-Grade Thyroid Diseases (Nobel Lecture, December 11, 1909)," in *Nobel Lectures: Physiology or Medicine, 1901–1921*, ed. Nobel Foundation (Amsterdam: Elsevier, 1967), 330–83, see 330–31.
7. On the redefinition of appendicitis as a surgical disease as a major step in the transformation of the public and professional understanding of surgery's place in medicine, see Dale C. Smith, "Appendicitis, Appendectomy, and the Surgeon," *Bulletin for the History of Medicine* 70 (1996): 414–44.
8. Ibid., 428–29.
9. Henry E. Sigerist, *American Medicine* (English translation of the 1933 German original; New York: W. W. Norton, 1934), 270.
10. Roy Porter, *The Greatest Benefit of Mankind: A Medical History of Humanity* (New York: W. W. Norton, 1999), 233.
11. Thomas Schlich, *Surgery, Science and Industry: A Revolution in Fracture Care, 1950–1990s* (Basingstoke, UK: Palgrave, 2002).
12. Thomas Schlich, "Trauma Surgery and Traffic Policy in Germany in the 1930s: A Case-Study in the Co-evolution of Modern Surgery and Society," *Bulletin for the History of Medicine* 80 (2006): 73–94. This is also true for other industrialized countries, such as the United States, a country that one surgical author characterized as being afflicted by the "modern monstrous *disease* of trauma," thus emphasizing the medical nature of the problem. Sawnie R. Gaston, "The Role of Leadership in the Quality of Fracture Care," *Bulletin for the American College of Surgeons* 60 (November 1975): 16–23.

13. In a similar way, surgery is also an integral part of modern warfare. To the same extent that modern war technologies have provided more efficient means of injuring and killing people, modern war surgery has developed new ways of treating injuries caused by those technologies. The promise to alleviate the effects of war on individual bodies makes war more acceptable to the fighting soldiers as well as to the societies from which they came. In this sense, surgery is actually built into the option of fighting a war. Like automobile traffic and trauma surgery, modern warfare and war medicine can be seen as different aspects of the development of modern societies. Roger Cooter and Steve Sturdy, "Of War, Medicine and Modernity: Introduction," in *War, Medicine and Modernity*, ed. Roger Cooter, Mark Harrison, and Steve Sturdy (Phoenix Mill, UK: Sutton, 1998), 1–21.
14. Roger Cooter and Bill Luckin, "Accidents in History: An Introduction," in *Accidents in History: Injuries, Fatalities and Social Relations*, ed. Roger Cooter and Bill Luckin (Amsterdam and Atlanta: Rodopi, 1997), 1–16, see 7. On naturalizing the social through medicine, see Patrick Joyce, *The Rule of Freedom: Liberalism and the Modern City* (London and New York: Verso, 2003), especially 62–93.
15. Andrew Barry, *Political Machines: Governing a Technological Society* (London and New York: Athlone, 2001), 7–8.
16. Lawrence, "Democratic, Devine and Heroic," 32.
17. Alice Domurat Dreger, *Hermaphrodites and the Medical Invention of Sex* (Cambridge, MA: Harvard University Press, 1998), 187.
18. Karl Figlio, "The Historiography of Scientific Medicine: An Invitation to the Human Sciences," *Comparative Studies in Society and History* 19 (1977): 262–86, see 266.
19. James L. Stone, "Dr. Gottlieb Burckhardt—The Pioneer of Pyschosurgery," *Journal of the History of the Neurosciences* 10 (2001): 79–92.
20. Jack D. Pressman, *Last Resort: Psychosurgery and the Limits of Medicine* (Cambridge: Cambridge University Press, 1998), 48–50.
21. Ibid., 8–10.
22. Ibid., 156–89.
23. Ibid., 138.
24. Ibid., 185.
25. David Shutts, *Lobotomy: Resort to the Knife* (New York: Van Nostrand Reinhold, 1982), 238.
26. Ibid., 215–30.
27. Joan Busfield, "Mental Illness," in Cooter and Pickstone, *Companion to Medicine*, 633–51, see 650.
28. Sander L. Gilman, *Making the Body Beautiful: A Cultural History of Aesthetic Surgery* (Princeton, NJ: Princeton University Press, 1999), 3–8.
29. 1.7 million were surgical procedures in the strict sense, with liposuction, nose reshaping, breast augmentation, eyelid surgery, and face-lift as the top five; see the Web site of the American Society of Plastic Surgeons, May 19, 2005, http://www.plasticsurgery.org/news_room/press_releases/2004-overall-statistics.cfm
30. Elizabeth Haiken, "The Making of the Modern Face: Cosmetic Surgery," *Social Research* 67 (2000): 90.
31. Elizabeth Haiken, "Plastic Surgery and American Beauty at 1921," *Bulletin for the History of Medicine* 68 (1994): 433. According to Gilman (*Making the Body Beautiful*, 10), the rise of aesthetic surgery at the end of the sixteenth century was

rooted in the appearance of epidemic syphilis and the stigma that went along with the syphilitic nose

32. Haiken, "Making of the Modern Face," 84.
33. Ibid., 88.
34. Ibid., 89.
35. Haiken, "Plastic Surgery," 431.
36. Haiken, "Making of the Modern Face," 83. For a more detailed analysis of the culture of self-representation in relation to cosmetic surgery, see Virginia L. Blum, *Flesh Wounds: The Culture of Cosmetic Surgery* (Berkeley: University of California Press, 2003). However, in virtually all nonliterate societies, identity and status are derived from visual self-representation, decoration, and so on; see, for example, Terence Turner, "The Social Skin," in *Not Work Alone: A Cross-Cultural View of Activities Superfluous to Survival*, ed. Jeremy Cherfas and Roger Lewin (Beverly Hills, CA: Sage, 1980), 112–39. I thank Margaret Lock for making me aware of this fact.
37. Haiken, "Making of the Modern Face," 89.
38. Bernice L. Hausman, *Changing Sex: Transsexualism, Technology, and the Idea of Gender* (Durham, NC, and London: Duke University Press, 1995), 65.
39. Gilman, *Making the Body Beautiful*, 85–118 (on the "racial nose"). Gilman sees the racialization of the nineteenth century as the offspring of aesthetic surgery. For the present, see, for example, Eugenia Kaw, "Medicalization of Racial Features: Asian-American Women and Cosmetic Surgery," in *The Politics of Women's Bodies: Sexuality, Appearance, and Behavior*, ed. Rose Weitz (New York and Oxford: Oxford University Press, 1998), 167–83.
40. Haiken, "Making of the Modern Face," 93.
41. Hausman, *Changing Sex*, 50.
42. Haiken, "Making of the Modern Face," 90.
43. Hausman, *Changing Sex*, 58.
44. Dreger, *Hermaphrodites*, in particular 167–201.
45. Ibid., 186.
46. Alice Domurat Dreger, *One of Us: Conjoined Twins and the Future of the Normal* (Cambridge, MA: Harvard University Press, 2004).
47. Dwight B. Billings and Thomas Urban, "The Socio-medical Construction of Transsexualism: An Interpretation and Critique," *Social Problems* 19 (1982): 267–68. Technically, many surgical procedures for sex assignment and reassignment actually came from plastic surgery. In addition, cosmetic surgery also provided the discursive framework for performing operations on demand without a strictly medical justification. Cosmetic surgeons created the notion that it made sense to treat unhappiness by surgically changing the body. Billings and Urban, "Sociomedical Construction," 268; Hausman, *Changing Sex*, 50, 63.
48. Stefan Hirschauer, "Performing Sexes and Genders in Medical Practices," in *Differences in Medicines: Unraveling Practices, Techniques, and Bodies*, ed. Marc Berg and Annemarie Mol (Durham, NC, and London: Duke University Press, 1998), 13–27, see 14; for more details see Stefan Hirschauer, *Die soziale Konstruktion der Transsexualität* (Frankfurt am Main: Suhrkamp, 1993).
49. See, for example, Joanne Meyerowitz, *How Sex Changed: A History of Transsexuality in the United States* (Cambridge, MA: Harvard University Press, 2002), 51–167.
50. Hausman, *Changing Sex*, 2.

51. Billings and Urban, "Socio-medical Construction," 266.
52. Meyerowitz, *How Sex Changed*, 4–5. Meyerowitz stresses the early twentieth-century origins, whereas Hausman, in *Changing Sex*, emphasizes the 1950s as the crucial period.
53. Hausman, *Changing Sex*, 3, 110. See also Billings and Urban, "Socio-medical Construction," 271, 275. In the reality of biographical experiences, the boundaries between transsexualism and other sexual deviances are not as clear. Instead, transsexuals appear to be a heterogeneous group, the common identity of which is being artificially constructed around surgery; see, for example, Billings and Urban, "Socio-medical Construction," 276.
54. Hausman, *Changing Sex*, 117. This view is being questioned by Joanne Meyerowitz, who holds that new technologies do not explain why sex-change surgery took root in some instances and not in others. She sees it as neither a necessary nor a sufficient cause of the emergence of transsexuality. Instead, cultural and political factors such as a new understanding of sex were decisive; see Meyerowitz, *How Sex Changed*, 21.
55. Hausman, *Changing Sex*, 14.
56. Billings and Urban, "Socio-medical Construction,"277.
57. Ibid., 276.
58. Ibid., 277.
59. Meyerowitz, *How Sex Changed*, 148.
60. Figlio, "Historiography," 277.
61. Stefan Hirschauer, "The Manufacture of Bodies in Surgery," *Social Studies of Science* 21 (1991): 279–319.
62. Figlio, "Historiography," 266. Of course, there has always been a tension between the body as an assembly of elements and the body as a unified whole, situated in a specific context. This tension has never been completely resolved; see, for example, Christopher E. Forth and Ivan Crozier, "Introduction: Parts, Wholes, and People," in *Body Parts: Critical Explorations in Corporeality*, ed. Christopher E. Forth and Ivan Crozier (Lanham, MD: Lexington Books, 2005), 1–16. What is important, however, is that the technological-fix strategy as analyzed here seems to tip the balance in the direction of fragmentation.
63. Thomas Schlich, *The Origins of Organ Transplantation: Surgery and Laboratory Science, 1880s–1930s.* (Rochester, NY: The University of Rochester Press, 2010).
64. Renée C. Fox and Judith P. Swazey, *Spare Parts: Organ Replacement in American Society* (New York and Oxford: Oxford University Press, 1992), 3–94; and Michael A. Bos, *The Diffusion of Heart and Liver Transplantation across Europe* (London: King's Fund Centre, 1991). On the early period of renal transplants, see Renée C. Fox and Judith P. Swazey, *The Courage to Fail: A Social View of Organ Transplants and Dialysis* (Chicago: University of Chicago Press, 1974).
65. For more on this, see Thomas Schlich, "Die Konstruktion der notwendigen Krankheitsursache: Wie die Medizin Krankheit beherrschen will," in *Anatomien medizinischen Wissens*, ed. Cornelius Borck (Frankfurt am Main: Fischer, 1996), 201–29.
66. Fox and Swazey, *Spare Parts*, xv.
67. Ibid., 204.
68. Paul Ramsey, *The Patient as Person: Explorations in Medical Ethics*, 2nd ed. (New Haven, CT: Yale University Press, 2002), 238.

69. Paul Rabinow, "Severing the Ties: Fragmentation and Dignity in Late Modernity," *Knowledge and Society: The Anthropology of Science and Technology* 9 (1992): 185.
70. Fox and Swazey, *Spare Parts*, 207.
71. Lesley A. Sharp, "The Commodification of the Body and Its Parts," *Annual Revue of Anthropology* 29 (2000): 287–328, see 306.
72. Ibid., 306; see also Margaret Lock, *Twice Dead: Organ Transplants and the Reinvention of Death* (Berkeley: University of California Press, 2002), 347–62.
73. Sharp, "Commodification," 288.
74. See, for example, Douglas Starr, *Blood: An Epic History of Medicine and Commerce* (New York: Alfred A. Knopf, 1998).
75. Sharp, "Commodification," 289. On that topic see also Nancy Scheper-Hughes and Loic Wacquant, eds., *Commodifying Bodies* (London: Sage, 2002).
76. Donald Joralemon, "Organ Wars: The Battle for Body Parts," *Medical Anthropology Quarterly*, n.s., 9 (1995): 335–56. In detail, the reasons for this are complex, and to a certain extent variable in different countries, but they are basically linked to the problem of the relationship between the body and personhood.
77. On this more generally, see, for example, Rabinow, "Severing the Ties," 169–87.
78. The association between the failure of transplants from one individual to another and immune mechanisms originated in the early twentieth-century quest for an immune therapy for cancer; for a detailed account, see Thomas Schlich, *Die Erfindung der Organtransplantation: Erfolg und Scheitern des chirurgischen Organersatzes (1880–1930)* (Frankfurt: Campus, 1998), 305–20.
79. Thomas Schlich, "Körper und Person: Kultur, Chirurgie und persönliche Identität," *Zeitschrift für medizinische Ethik* 48 (2002): 237–45.
80. Emily Martin, *Flexible Bodies: Tracking Immunity in American Culture—From the Days of Polio to the Age of AIDS* (Boston: Beacon, 1994).
81. Martin, cited in Lupton, "Social Construction," 54.
82. Martin, *Flexible Bodies*.
83. Sharp, "Commodification," 309.
84. Patricia A. Baird, "Identification of Genetic Susceptibility to Common Diseases," *Perspectives in Biology and Medicine* 45 (2002): 516–28, see 518.
85. Surgery is sometimes explicitly used as an analogy and a model. Thus, when the beta-blockers were developed for treatment of high blood pressure, they were seen as a pharmaceutical analogous to a surgical intervention, sympathectomy. Sympathectomy lowers blood pressure by removing parts of the sympathic nerve system. The inventors of propanolol explicitly equated the purity of its chemical blockade to the surgical knife; see Carsten Timmermann, "To Treat or Not to Treat: Drug Research and the Changing Nature of Essential Hypertension," in *The Risks of Medical Innovation: Risk Perception and Assessment in Historical Context*, ed. Thomas Schlich and Ulrich Tröhler (London and New York: Routledge, 2005), 133–47.
86. D. J. Weatherall, "Introduction," *Journal of Internal Medicine* 247 (2000): 3–5, see 3–4.
87. Ibid., 4.
88. Haiken, "Making of the Modern Face," 94.
89. Baird, "Identification of Genetic Susceptibility," 526.
90. Cf. Hans-Jörg Rheinberger, "Molekulare Medizin als Paradigma? Gentechnologie im Blick von Wissenschaftstheorie und medizinischer Ethik," in *Meilensteine der Medizin*, ed. Heinz Schott (Dortmund, Germany: Harenberg, 1996), 555–61.

91. Cornelius Borck, "Anatomien medizinischer Erkenntnis: Der Aktionsradius der Medizin zwischen Vermittlungskrise und Biopolitik," in *Anatomien medizinischen Wissens: Medizin—Macht—Moleküle*, ed. Cornelius Borck (Frankfurt am Main: Fischer, 1996), 9–52.
92. Margaret Lock, "Death in Technological Time: Locating the End of Meaningful Life," *Medical Anthropology Quarterly*, n.s., 10 (1996): 575–600. On the problem of investigating the history of brain death in the cultural context of different definitions of death and different ways of diagnosing it, see Thomas Schlich, "Tod, Geschichte, Kultur," in *Hirntod: Zur Kulturgeschichte der Todesfeststellung*, ed. Thomas Schlich and Claudia Wiesemann (Frankfurt am Main: Suhrkamp, 2001), 9–42. The culture dependence of brain death becomes particularly striking by intercultural comparison; see Lock, *Twice Dead*. See also Malcolm Nicolson's chapter in this volume.
93. See, for example, Lock, *Twice Dead*, on Japan, and on Germany, Linda F. Hogle, *Recovering the Nation's Body: Cultural Memory, Medicine, and the Politics of Redemption* (New Brunswick, NJ, and London: Rutgers University Press, 1999).
94. Schlich, *Transplantation*.
95. Peter Galison, "Die Ontologie des Feindes: Norbert Wiener und die Vision der Kybernetik," in *Räume des Wissens, Repräsentation, Codierung, Spur*, ed. Hans-Jörg Rheinberger, Michael Hagner, and Bettina Wahrig-Schmidt (Berlin: Akademie Verlag, 1997), 281–324.
96. Frederic Pohl and Hans Moravec, "Souls in Silicon," in *The American Body in Context: An Anthology*, ed. Jessica R. Johnston (1993; Wilmington, DE: Scholarly Resources, 2001), 49–59.
97. Ray Kurzweil, *The Age of Intelligent Machines* (Cambridge, MA: MIT Press, 1990).
98. Sharp, "Commodification," 297.
99. Nancy Scheper-Hughes, "Bodies for Sale—Whole or in Parts," in Scheper-Hughes and Wacquant, *Commodifying Bodies*, 1–8, see 3.
100. Joralemon, "Organ Wars," 347.
101. Ibid.

Chapter 4

1. Important here is Charles Rosenberg, "What Is Disease? In Memory of Owsei Temkin," *Bulletin of the History of Medicine* 77 (2003): 491–505.
2. For example, on smallpox, see Sanjoy Bhattacharya, *Expunging Variola: The Control and Eradication of Smallpox in India, 1947–1977* (New Delhi and London: Orient Longman, 2006); on tuberculosis, see Linda Bryder, *Below the Magic Mountain: A Social History of Tuberculosis in Twentieth-Century Britain* (Oxford: Clarendon, 1988); on AIDS, see Steven Epstein, *Impure Science: AIDS, Activism, and the Politics of Knowledge* (Berkeley: University of California Press, 1996); on cancer, see David Cantor, ed., *Cancer in the Twentieth Century* (Baltimore: Johns Hopkins University Press, 2008); and on disease history, see Michael Worboys, *Spreading Germs: Disease Theories and Medical Practice in Britain, 1865–1900* (New York: Cambridge University Press, 2000). This list is far from exhaustive.
3. Michel Foucault, "Nietzsche, Genealogy, History," in *The Foucault Reader*, ed. Paul Rabinow (London: Penguin, 1984), 87.

4. Lisa Downing, "The Measure of Sexual Dysfunction: A Plea for Theoretical Limitlessness," in "Regions of Sexuality," ed. Iain Morland and Wendy O'Brien, special issue, *Transformations: Region, Culture, Society* 8 (2004), http://transformations.cqu.edu.au/journal/issue_08/article_02.shtml#return2 (accessed July 2008).
5. David Cantor, "The Diseased Body," in *Companion to Medicine in the Twentieth Century*, ed. Roger Cooter and John Pickstone (London: Routledge, 2000), 349.
6. See David Armstrong, "The Temporal Body," in Cooter and Pickstone, *Companion to Medicine*, 247–59; J.D.N. Turney and Brian Balmer, "The Genetic Body," in Cooter and Pickstone, *Companion to Medicine*, 399–415; and Charis Thompson, *Making Parents: The Ontological Choreography of Reproductive Technologies* (2005; Cambridge, MA: MIT Press, 2007).
7. See Stephen J. Cross and Randall Albury, "Walter B. Cannon, L. J. Henderson, and the Organic," *Osiris* 3 (1987): 165–92; and Christopher Lawrence and George Weiss, eds., *Greater than the Parts: Holism in Biomedicine, 1920–1950* (New York and Oxford: Oxford University Press, 1998).
8. See Olga Amsterdamska and Anja Hiddinga, "The Analyzed Body," in Cooter and Pickstone, *Companion to Medicine*, 417–33.
9. Annemarie Mol, *The Body Multiple: Ontology in Medical Practice* (Durham, NC: Duke University Press, 2002). See also Annemarie Mol, "Lived Reality and the Multiplicity of Norms: A Critical Tribute to George Canguilhem," *Economy and Society* 27 (1998): 274–84.
10. This argument about bodily multiplicities was thought out a long time ago by Judith Butler, *Bodies That Matter: On the Discursive Limits of "Sex"* (New York: Routledge, 1993); Elizabeth Grosz, *Space, Time and Perversion: Essays on the Politics of Bodies* (New York: Routledge; Sydney: Allen and Unwin, 1995); and Moira Gatens, *Imaginary Bodies: Ethics, Power and Corporeality* (London and New York: Routledge, 1996). In medical history these insights have been used by Thomas Laqueur, *Making Sex: The Body and Gender from the Greeks to Freud* (Cambridge, MA: Harvard University Press, 1990); and Nelly Oudshoorn, *Beyond the Natural Body: An Archaeology of Sex Hormones* (London: Routledge, 1994).
11. See Anne Kveim Lie, "Stories of Origin and the Norwegian Radesyge," *Social History of Medicine* 20 (2007): 563–79.
12. If, as has been argued, diseases are multiple and based in varying practices both diachronically and synchronically, then it is important to consider the roles played by medical practices to appreciate how the diseased body is made anew with specific developments in health care and associated practices. For this reason, this chapter focuses on the ways in which disease is managed through different forms of medical bodily control. For examples of other ways of representing diseased bodies in the twentieth century, see the introduction and practically all other chapters in this volume where art is the focus, especially the chapters by Ana Carden-Coyne, Ivan Crozier, Michael Fischer, and Anna Cole and Anna Haebich. This attention is mostly with reference to different conceptions and representations of sexually transmitted diseases.
13. Terence Ranger and Paul Slack, eds., *Epidemics and Ideas: Essays on the Historical Perception of Pestilence* (Cambridge: Cambridge University Press, 1992).
14. William F. Bynum, "Policing Hearts of Darkness: Aspects of the International Sanitary Conferences," *History of Philosophy and Life Sciences* 15 (1993): 421–34; and João Rangel de Almeida, "The International Sanitary Conferences (1851–1866):

A Social Locale for the Negotiation of Knowledge" (master's thesis, University of Edinburgh, 2006).

15. Krista Maglen, "A World Apart: Geography, Australian Quarantine, and the Mother Country," *Journal of the History of Medicine and Allied Sciences* 60 (2005): 196–217; and Alison Bashford, *Medicine at the Border: Disease, Globalization and Security, 1850 to the Present* (Basingstoke, UK: Palgrave Macmillan, 2006).
16. Mary Spongberg, *Feminizing Venereal Disease* (New York: New York University Press, 1997); Frank Mort, *Dangerous Sexualities: Medico-moral Politics in England since 1830* (1987; London: Routledge, 2000); Judith Walkowitz, *Prostitution and Victorian Society: Women, Class and the State* (Cambridge: Cambridge University Press, 1980); and Frances Finnegan, *Poverty and Prostitution* (Cambridge: Cambridge University Press, 1979). For developments since then, see Roger Davidson and Lesley Hall, eds., *Sex, Sin and Suffering: Venereal Disease and European Society since 1870* (London: Routledge, 2001).
17. See William F. Bynum, *Science and the Practice of Medicine in the Nineteenth Century* (Cambridge: Cambridge University Press, 1994); Worboys, *Spreading Germs*.
18. Such normalizing practices are, of course, analyzed by Michel Foucault, especially in *Discipline and Punish: The Birth of the Prison*, trans. Alan Sheridan (New York: Pantheon, 1977). See also Georges Canguilhem, *The Normal and the Pathological*, trans. Caroline Fawcett (1943; New York: Zone Books, 1993).
19. See, for example, World Health Organization, "Avian Influenza," http://www.who.int/csr/disease/avian_influenza/en/ (accessed August 2008).
20. European Centre for Disease Control, "About ECDC," http://ecdc.europa.eu/About_ECDC.html (accessed July 2008). There are clearly resonances here with Susan Sontag's two studies: *Illness as Metaphor* (New York: Vintage Books, 1978) and *AIDS and Its Metaphors* (New York: Farrar, Straus, Giroux, 1989).
21. David A. Koplow, *Smallpox: The Fight to Eradicate a Global Scourge* (Berkeley: University of California Press, 2003).
22. Vaccine Resource Library, PATH, http://www.path.org/vaccineresources.
23. Nadja Durbach, *Bodily Matters: The Anti-vaccination Movement in England, 1853–1907* (Durham, NC, and London: Duke University Press, 2005).
24. A. J. Wakefield, S. H. Murch, A. Anthony, J. Linnell, D. M. Casson, M. Malik, M. Berelowitz, et al., "Ileal-Lymphoid-Nodular Hyperplasia, Non-Specific Colitis, and Pervasive Developmental Disorder in Children," *Lancet* 351 (1998): 637–41.
25. Susan Mayor, "Authors Reject Interpretation Linking Autism and MMR Vaccine," *British Medical Journal* 328 (2004): 602.
26. Ana Carden-Coyne and Christopher E. Forth, *Cultures of the Abdomen: Diet, Digestion and Fat in the Modern World* (New York: Palgrave, 2005).
27. See Robert Nye, "The Evolution of the Concept of Medicalization in the Late Twentieth Century," *Journal of History of the Behavioral Sciences* 39 (2003): 115–29. See also Canguilhem, *Normal and Pathological*.
28. James H. Mills and Satadru Sen, eds., *Confronting the Body: The Politics of Physicality in Colonial and Post-colonial India* (London: Anthem, 2004); and Anna Crozier, "Sensationalising Africa: British Medical Impressions of Sub-Saharan Africa 1890–1939," *Journal of Imperial and Commonwealth History* 35 (2007): 393–415.

29. David Arnold, *Colonizing the Body: State Medicine and Epidemic Disease in Nineteenth-Century India* (Berkeley: University of California Press, 1993).
30. David Arnold, ed., *Imperial Medicine and Indigenous Societies* (Manchester, UK: Manchester University Press, 1988); Arnold, *Colonizing the Body*; John Farley, *Bilharzia: A History of Imperial Tropical Medicine* (Cambridge: Cambridge University Press, 1991); Maryinez Lyons, *The Colonial Disease: A Social History of Sleeping Sickness in Northern Zaire, 1900–1940* (Cambridge: Cambridge University Press, 1992); and Megan Vaughan, *Curing Their Ills: Colonial Power and African Illness* (Cambridge, UK: Polity, 1991).
31. Kirk Arden Hoppe, *Lords of the Fly: Sleeping Sickness Control in British East Africa, 1900–1960* (Westport, CT: Praeger, 2003); and Christoph Gradmann, "'It Seemed About Time to Try One of Those Modern Medicines': Animal and Human Experimentation in the Chemotherapy of Sleeping Sickness 1905–1908," in *Twentieth Century Ethics of Human Subjects Research: Historical Perspectives on Values, Practices, and Regulations*, ed. Giovanni Maio and Volker Roelcke (Stuttgart, Germany: Steiner, 2004), 83–97.
32. Arnold, *Colonizing the Body*; Arden Hoppe, *Lords of the Fly*; Myron Echenberg, *Black Death, White Medicine: Bubonic Plague and the Politics of Public Health in Colonial Senegal, 1914–1945* (Portsmouth, NH: Heinemann; Oxford: James Currey; and Cape Town: David Philips, 2002); David Gordon, "A Sword of Empire? Medicine and Colonialism at King William's Town, Xhosaland, 1856–91," in *Medicine and the Colonial Identity*, ed. Mary P. Sutphen and Bridie Andrews (London: Routledge, 2003), 41–60; Roy Macleod and Milton Lewis, eds., *Disease, Medicine and Empire: Perspectives on Western Medicine and the Experience of European Expansion* (New York: Routledge, 1988); Obioma Nnaemeka, ed., *Female Circumcision and the Politics of Knowledge: African Women in Imperialist Discourses* (Westport, CT, and London: Praeger, 2005); and Michael Tuck, "Venereal Disease, Sexuality and Society in Uganda," in Davidson and Hall, *Sex, Sin and Suffering*, 191–203.
33. For example, this enforced and violent treatment sometimes continued beyond colonialism; see, for example, James H. Mills, "Body as Target, Violence as Treatment: Psychiatric Regimes in Colonial and Post-colonial India," in Mills and Sen, *Confronting the Body*, 80–101.
34. Warwick Anderson, "The Third-World Body," in Cooter and Pickstone, *Companion to Medicine*, 236. See also Warwick Anderson, "Excremental Colonialism: Public Health and the Poetics of Pollution," *Critical Enquiry* 21 (1995): 640–69.
35. Mark Harrison, "'The Tender Frame of Man': Disease, Climate, and Racial Difference in India and the West Indies, 1760–1860," *Bulletin of the History of Medicine* 70 (1996): 68–93; Dane Kennedy, "The Perils of the Midday Sun: Climatic Anxieties in the Colonial Tropics," in *Imperialism and the Natural World*, ed. John M. MacKenzie (Manchester, UK: Manchester University Press, 1990); and Warwick Anderson, *Colonial Pathologies: American Tropical Medicine, Race and Hygiene in the Philippines* (Durham, NC: Duke University Press, 2006), 130–57.
36. John Stephens and Rickard Christophers, "The Malaria Infection of Native Children," in *Reports to the Malaria Committee*, ed. Royal Society, 3rd ser. (London: Royal Society, 1900), 4–14; John Stephens and Rickard Christophers, "The Segregation of Europeans," in *Reports to the Malaria Committee*, 21–24; and John Stephens and Rickard Christophers, "The Native as a Prime Agent in the Malaria

Infection of Europeans," in *Further Reports to the Malaria Committee*, ed. Royal Society (London: Royal Society, 1900), 3–19.

37. E. A. Reeves, *Hints to Travellers*, 11th ed. (London: Royal Geographical Society, 1935–1938), 1:1–2; 2:380 and 413.
38. The Crown Agents for the Colonies, *Hints on the Preservation of Health in Tropical Africa* (1938; London, The Crown Agents for the Colonies, 1943), 17, 28, and 24; and Charles J. Ryan, *Health Preservation in West Africa* (London: John Bale, Sons & Danielsson, 1914), 11 and 46.
39. Specific examples are given in Crozier, "Sensationalising Africa."
40. Dane Kennedy, "Diagnosing the Colonial Dilemma: Tropical Neurasthenia and the Alienated Briton," in *Decentering Empire: Britain, India, and the Transcolonial World*, ed. Dane Kennedy and Durba Ghosh (Hyderabad: Orient Longman, 2006), 157–81; and Warwick Anderson, "The Trespass Speaks: White Masculinity and Colonial Breakdown," *American Historical Review* 102 (1997): 1343–70.
41. Anna Crozier, "What Was Tropical about Tropical Neurasthenia? The Utility of the Diagnosis in the Management of British East Africa," *Journal of the History of Medicine and Allied Sciences* 64 (2009): 518–48.
42. For an outline of some of the policy implications of development and health care, see Maureen Mackintosh and Meri Koivusalo, eds., *Commercialization of Health-care: Global and Local Dynamics and Policy Responses* (Basingstoke, UK: Palgrave Macmillan, 2005), which builds on Walt Gill and Lucy Gilson, "Reforming the Health Sector in Developing Countries: The Central Role of Policy Analysis," *Health Policy and Planning* 9 (2004): 353–70.
43. Frantz Fanon, *Black Skin, White Masks*, trans. Charles Lam Markmann (New York: Grove, 1967), 113.
44. Timothy Burke, *Lifebouy Men and Lux Women: Commodification, Consumption, and Cleanliness in Modern Zimbabwe* (Durham, NC: Duke University Press, 1996).
45. Paula Treicher, "AIDS and HIV Infection in the Third World: A First World Chronicle," in *Remaking History*, ed. Barbara Kruger and Phil Mariani (Seattle, WA: Bay Press, 1989), 31–86.
46. See Bertrand Taithe, "The Rise and Fall of European Syphilization: The Debates on Human Experimentations and Vaccination of Syphilis, c. 1845–1870," in *Sexual Cultures in Europe, 1700–1995*, ed. Franz Eder, Lesley Hall, and Gert Hekma (Manchester, UK: Manchester University Press, 1998), 2:34–57.
47. See Walkowitz, *Prostitution and Victorian Society*; Spongberg, *Feminzing Venereal Disease*.
48. Ivan Crozier, "William Acton and the History of Sexuality: The Professional and Medical Contexts," *Journal of Victorian Culture* 5 (2000): 1–27; and Bernhard Witkop, "Paul Ehrlich and His Magic Bullets—Revisited," *Proceedings of the American Philosophical Society* 143 (1999): 540–57. For the place of Ehrlich's work in the history of antimicrobial chemotherapy, see David Greenwood, *Antimicrobial Drugs* (Oxford: Oxford University Press, 2008), esp. 57–61.
49. See Michel Foucault, *History of Sexuality I: The Will to Knowledge*, trans. Robert Hurley (London: Penguin, 1978); Ivan Crozier, "Introduction: Havelock Ellis, John Addington Symonds and the Construction of *Sexual Inversion*," in *Sexual Inversion: A Critical Edition*, ed. Ivan Crozier (Basingstoke, UK, and New York: Palgrave Macmillan, 2008), 1–86. For more on the medico-moral dimension, see Frank Mort, *Dangerous Sexualities*.

50. Steve Epstein, *Impure Science.*
51. For an alternative look at modern African sexual health (and identity) that does not reduce to a discussion of African AIDS in need of external Western control, see Rachel Spronk, "Ambiguous Pleasures: Sexuality and New Self-Definitions in Nairobi" (PhD diss., University of Amsterdam, 2006).
52. For example, Charles Rosenberg, "The Bitter Fruit: Heredity, Disease and Social Thought in Nineteenth Century America," *Perspectives in American History* 8 (1974): 189–235; John Waller, " 'The Illusion of an Explanation': The Concept of Hereditary Disease, 1770–1870," *Journal of the History of Medicine and Allied Sciences* 57 (2002): 410–48; and Daniel Pick, *Faces of Degeneration* (Cambridge: Cambridge University Press, 1989).
53. Elizabeth Ettorre, "Reproductive Genetics, Gender and the Body: 'Please Doctor, May I Have a Normal Baby?' " *Sociology* 34 (2000): 403–20.
54. See Thompson, *Making Parents*; see also Elizabeth Ettorre, *Reproductive Genetics, Gender and the Body* (London: Routledge, 2002).
55. David Armstrong, S. Michie, and T. Marteau, "Revealed Identity: A Study in the Process of Genetic Counselling," *Social Science and Medicine* 47 (1998): 1653–58.
56. Nina Hallowell, C. Foster, R. Eeles, A. Ardern-Jones, and M. Watson, "Accommodating Risk: Women's Responses to BRCA1/2 Genetic Testing Following a Cancer Diagnosis," *Social Science and Medicine* 59 (2004): 553–65; and Nina Hallowell, "Reconstructing the Body or Reconstructing the Woman? Perceptions of Prophylactic Mastectomy for Hereditary Breast Cancer Risk," in *Ideologies of Breast Cancer: Feminist Perspectives*, ed. L. Potts (London: Macmillan, 2000), 153–80.
57. For basic introductory information on stem cells, see http://stemcells.nih.gov/info/basics/basics1.asp (accessed July 2008). Stem-cell research is a burgeoning topic in science and technology studies. For a general discussion of these new technologies, see Andrew Webster, *Health Technology and Society: A Sociological Critique* (Basingstoke, UK: Palgrave Macmillan, 2007). A lot of this work seems to focus on public participation in science policy surrounding stem cells rather than technical debates in the field. Other fields of biotechnology, especially modern genetics, have been given broader treatment from philosophical and social studies of science. See Paul Griffiths, Karola Stotz, and Rob Knight, "How Scientists Conceptualise Genes: An Empirical Study," *Studies in History and Philosophy of Biological and Biomedical Sciences* 35 (2004): 647–73.
58. Michael Mulkay, *The Embryo Research Debate: Science and the Politics of Reproduction* (Cambridge, UK: Cambridge University Press, 1997).
59. Laury Oaks, "Antiabortion Positions and Young Women's Life Plans in Contemporary Ireland," *Social Science and Medicine* 56 (2003): 1973–86. For a defense of the practice, see A. Kero, U. Högberg, L. Jacobsson, and A. Lalos, "Legal Abortion: A Painful Necessity," *Social Science and Medicine* 53 (2001): 1481–90.
60. See Matthew Hill, "Ukraine Babies in Stem Cell Probe," BBC News, December 12, 2006, http://news.bbc.co.uk/1/hi/world/europe/6171083.stm (accessed July 2008).
61. See Christopher E. Forth and Ivan Crozier, "Introduction: Parts, Wholes and People," in *Body Parts: Critical Explorations in Corporeality*, ed. Christopher E. Forth and Ivan Crozier (Lanham, MD: Lexington Books, 2005), 1–16.
62. See Mol, *The Body Multiple.*

Chapter 5

1. David Wallechinksy, *The Complete Book of the Olympics* (London: Penguin Books, 1984), 132.
2. *Times*, August 11, 1983.
3. "Suspicion Surrounds Flo-Jo's Death," BBC News: Sport, September 23, 1998, http://news.bbc.co.uk/1/hi/sport/177433.stm.
4. For an excellent extrapolation of the practices involved in this term, see Linda F. Hogle, "Enhancement Technologies and the Body," *Annual Review of Anthropology* 34 (2005): 695–716.
5. I am grateful to Michael Rabinovich, who supplied all the track-and-field statistics for this article. See Track and Field Statistics Web site, http://trackandfield.brinkster.net.
6. *New York Times*, "US Track Team Leads in Moscow," July 21, 1963.
7. *New York Times*, "The Red-Faced Reds," October 23, 1964.
8. *Times*, "Kiev Ban on Top Woman Sprinter," September 15, 1967.
9. *New York Times*, "Sex Test Disqualifies Athlete," September 16, 1967.
10. Steven Ungerleider, *Faust's Gold: Inside the East German Doping Machine* (New York: St. Martin's Press, 2001).
11. John Hoberman, *Darwin's Athletes: How Sport Has Damaged Black America and Preserved the Myth of Race* (New York: Houghton Mifflin, 1997).
12. Dorothy Porter, "The Healthy Body," in *Companion to Medicine in the Twentieth Century*, ed. Roger Cooter and John Pickstone (London: Routlege, 2000), 211.
13. Barbara Duden, *The Woman beneath the Skin: A Doctor's Patients in Eighteenth Century Germany*, trans. Thomas Dunlap (Cambridge, MA: Harvard University Press, 1991), 1–50.
14. See Leon S. Roudiez, "Translator's Note," in *The Powers of Horror: An Essay on Abjection*, by Julia Kristeva, trans. Leon S. Roudiez (New York: Columbia University Press, 1982), viii.
15. See Susan Reynolds Whyte and Benedicte Ingstad, "Introduction," in *Disability and Culture*, ed. Susan Reynolds Whyte and Benedicte Ingstad (Berkeley: University of California Press, 1995), 8.
16. Joanna Bourke, *Dismembering the Male: Men's Bodies, Britain, and the Great War* (London: Reaktion, 1996).
17. Donna Haraway, *Simians, Cyborgs, and Women: The Reinvention of Nature* (London: Routlege, 1991).
18. *The Six Million Dollar Man*, The Internet Movie Database, http://www.imdb.com/title/tt0071054/.
19. MKULTRA was renamed MKSearch in 1962.
20. Elizabeth Haiken, *Venus Envy: A History of Cosmetic Surgery* (Baltimore: Johns Hopkins University Press, 1997); Sander L. Gilman, *Making the Body Beautiful: A Cultural History of Aesthetic Surgery* (Princeton, NJ: Princeton University Press, 1999).
21. For more on the American legal belief in corporeal privacy, see Alan Hyde, *Bodies of Law* (Princeton, NJ: Princeton University Press, 1997).
22. Janet L. Golden, *Message in a Bottle: The Making of Fetal Alcohol Syndrome* (Cambridge, MA: Harvard University Press, 2005).
23. Aaron Sorkin, "What Kind of Day Has It Been?" *The West Wing*, season 1, episode 22, broadcast May 17, 2000, by NBC.

24. See Ad Hoc Committee of Harvard Medical School, "A Definition of Irreversible Coma," *Journal of the American Medical Association* 210 (1968): 337–40.
25. David Armstrong, "The Temporal Body," in Cooter and Pickstone, *Companion to Medicine*, 249–50.
26. B. Jennet, "Brain Death and the Vegetative State," in *Oxford Textbook of Medicine*, ed. D. J. Weatherall *et al.* (Oxford: Oxford University Press, 1966), 3933, cited in Roger Cooter, "The Dead Body," in Cooter and Pickstone, *Companion to Medicine*, 469.
27. In 1997, the state of Oregon ratified the "Death with Dignity" Act, legalizing physician-assisted suicide. Euthanasia was also made legal around this time in both the Netherlands and Switzerland.
28. A full copy of the Hays Code is available at "Hays Code," artsreformation.com, http://www.artsreformation.com/a001/hays-code.html.
29. David Serlin, "Broadcasting the Body: Performing Live Surgery on Television and the Internet since 1945," in *Imagining Illness: Public Health and Visual Culture*, ed. David Serlin (Minneapolis: University of Minnesota Press, 2007).

Chapter 6

1. Anthony Giddens, *Modernity and Self-Identity: Self and Society in the Late Modern Age* (Stanford, CA: Stanford University Press, 1991), 178.
2. Joanna Bourke, *Dismembering the Male: Men's Bodies, Britain and the Great War* (London: Reaktion, 1996), 212–14; Elizabeth Haiken, *Venus Envy: A History of Cosmetic Surgery* (Baltimore: Johns Hopkins University Press, 1997), 29; Michael Hau, *The Cult of Health and Beauty in Germany: A Social History, 1890–1930* (Chicago: University of Chicago Press, 2003), 178; Sophie Delaporte, *Gueules cassées de la grande guerre* (Paris: Agnès Viénot, 1996), 32–33, 141–54; and Sander L. Gilman, *Making the Body Beautiful: A Cultural History of Aesthetic Surgery* (Princeton, NJ: Princeton University Press, 1999), 168.
3. Anthony Synnott, *The Body Social: Symbolism, Self and Society* (London: Routledge, 1993), 78–100; and Londa Schiebinger, *Nature's Body: Gender in the Making of Modern Science* (Boston: Beacon, 1993).
4. Warren I. Susman, " 'Personality' and the Making of Twentieth-Century Culture," in *Culture as History: The Transformation of American Society in the Twentieth Century* (New York: Pantheon, 1984), 271–85.
5. Ute Frevert, *Women in German History: From Bourgeois Emancipation to Sexual Liberation*, trans. Stuart McKinnon Evans (Oxford: Berg, 1989), 181–82.
6. Kathy Peiss, *Hope in a Jar: The Making of America's Beauty Culture* (New York: Metropolitan Books, 1998), 152; and Joan Jacobs Brumberg, *The Body Project: An Intimate History of American Girls* (New York: Random House, 1997), 71–72, 79–82.
7. Maxine Leeds Craig, *Ain't I a Beauty Queen? Black Women, Beauty, and the Politics of Race* (Oxford: Oxford University Press, 2002).
8. Peiss, *Hope in a Jar*, 160, 264–65; Kenon Breazeale, "In Spite of Women: *Esquire* Magazine and the Construction of the Male Consumer," *Signs* 20, no. 1 (1994): 1–22; and Tom Pendergast, *Creating the Modern Man: American Magazines and Consumer Culture, 1900–1950* (Columbia: University of Missouri Press, 2000).
9. Gilman, *Making the Body Beautiful*, 132–36, 169–71; Haiken, *Venus Envy*, 175–227; Kathy Davis, *Reshaping the Female Body: The Dilemma of Cosmetic Surgery* (London: Routledge, 1995), 6; and Kathy Davis, " 'A Dubious Equal-

ity': Men, Women and Cosmetic Surgery," *Body and Society* 8, no. 1 (2002): 58–60.

10. Valerie Steele, *The Corset: A Cultural History* (New Haven, CT: Yale University Press, 2001), 143, 148, 152; Marilyn Yalom, *A History of the Breast* (New York: Knopf, 1997), 176–78.
11. Bess M. Mensendieck, *"It's Up to You"* (New York: Mensendieck System, 1931), 19.
12. Quoted in Carolyn Ward Comiskey, " 'I Will Kill Myself . . . If I Have to Keep My Fat Calves!': Legs and Cosmetic Surgery in Paris in 1926," in *Body Parts: Critical Explorations in Corporeality*, ed. Christopher E. Forth and Ivan Crozier (Lanham, MD: Lexington Books, 2005), 251.
13. Yalom, *History of the Breast*, 246.
14. Valerie Steele, *Fashion and Eroticism: Ideals of Feminine Beauty from the Victorian Era to the Jazz Age* (New York: Oxford University Press, 1985), 171–72; Tamar Garb, *Bodies of Modernity: Figure and Flesh in Fin-de-Siècle France* (London: Thames and Hudson, 1998), 145–77.
15. Mary Lynn Stewart, *For Health and Beauty: Physical Culture for Frenchwomen, 1880s–1930s* (Baltimore: Johns Hopkins University Press, 2001), 166–67.
16. Victoria de Grazia, *How Fascism Ruled Women: Italy, 1922–1945* (Berkeley: University of California Press, 1992), 211–14.
17. Haiken, *Venus Envy*, 243–44; Fredric Wertham, *Seduction of the Innocent* (New York: Rinehart, 1953), 199–210.
18. Haiken, *Venus Envy*, 235–42, 256.
19. Cf. Jana Evans Braziel and Kathleen LeBesco, eds., *Bodies Out of Bounds: Fatness and Transgression* (Berkeley: University of California Press, 2001).
20. Steele, *Corset*, 154.
21. Ina Zweiniger-Bargielowska, "The Body and Consumer Culture," in *Women in Twentieth-Century Britain*, ed. Ina Zweiniger-Bargielowska (London: Pearson, 2001), 183–97.
22. Martin Voracek and Maryanne L. Fisher, "Shapely Centrefolds? Temporal Change in Body Measures: Trend Analysis," *British Medical Journal* 325 (2002): 1447–48.
23. Hau, *Cult of Health*, 15, 33, 46–49, 200.
24. Todd Samuel Presner, " 'Clear Heads, Solid Stomachs, and Hard Muscles': Max Nordau and the Aesthetics of Jewish Regeneration," *Modernism/Modernity* 10, no. 2 (2003): 269–96.
25. Candid Jane, "Our Search for Beauty: Men Are the Guilty Ones," *Australian Women's Weekly*, July 8, 1933, 7.
26. Kenneth R. Dutton, *The Perfectible Body: The Western Ideal of Male Physical Development* (New York: Continuum, 1995), 130–36.
27. Gershoy Legman, *Love and Death: A Study in Censorship* (1949; New York: Hacker Art Books, 1963), 43.
28. Wertham, *Seduction*, 217. On the role of such ads for proletarian and lower-middle-class men, see Erin A. Smith, *Hard-Boiled: Working-Class Readers and Pulp Magazines* (Philadelphia: Temple University Press, 2000).
29. Richard Dyer, *White* (London: Routledge, 1997), 169.
30. Richard I. Jobs, "Tarzan under Attack: Youth, Comics, and Cultural Reconstruction in Postwar France," *French Historical Studies* 26, no. 4 (2003): 705–13; Richard Dyer, *White* (London: Routledge, 1997), 169; and Maggie Günsberg, *Italian Cinema: Gender and Genre* (New York: Palgrave Macmillan, 2005), 102.

31. Dutton, *Perfectible Body*, 143–47; David Forrest, "'We're Here, We're Queer, and We're Not Going Shopping': Changing Gay Male Identities in Contemporary Britain," in *Dislocating Masculinity: Comparative Ethnographies*, ed. Andrea Cornwall and Nancy Lindisfarne (London: Routledge, 1994), 97–110; Harrison G. Pope, Jr., Katharine A. Phillips, and Roberto Olivardia, *The Adonis Complex: The Secret Crisis of Male Body Obsession* (New York: Free Press, 2000), 40–47; and Richard A. Leit, Harrison G. Pope, Jr., and James J. Gray, "Cultural Expectations of Muscularity in Men: The Evolution of Playgirl Centerfolds," *International Journal of Eating Disorders* 29 (2001): 90–93.
32. Susan Jeffords, *The Remasculinization of America: Gender and the Vietnam War* (Bloomington: Indiana University Press, 1989); Alan M. Klein, *Little Big Men: Bodybuilding Subculture and Gender Construction* (Albany: State University of New York Press, 1993), 242–50; Sam Fussell, *Muscle: Confessions of an Unlikely Bodybuilder* (London: Sphere Books, 1991); and Murray J.N. Drummond, "Men's Bodies: Listening to the Voices of Young Gay Men," *Men and Masculinities* 7, no. 3 (2005): 270–90.
33. F. T. Marinetti, "Multiplied Man and the Reign of the Machine," in *Marinetti: Selected Writings*, trans. R. W. Flint and Arthur A. Coppotelli (New York: Farrar, Straus and Giroux, 1971), 91; and Stephen Kern, *The Culture of Time and Space, 1880–1918* (Cambridge, MA: Harvard University Press, 1983), 109–30.
34. John M. Hoberman, *Mortal Engines: The Science of Performance and the Dehumanization of Sport* (New York: Free Press, 1992), 103.
35. Kevin White, *The First Sexual Revolution: The Emergence of Male Heterosexuality in Modern America* (New York: New York University Press, 1993), 156–57.
36. Joanna Bourke, *Working-Class Cultures in Britain, 1890–1960* (London: Routledge, 1994), 35.
37. Frevert, *Women in German History*, 191.
38. Rachel P. Maines, *The Technology of Orgasm: "Hysteria," the Vibrator, and Women's Sexual Satisfaction* (Baltimore: Johns Hopkins University Press, 1999), 117–18; Stewart, *For Health and Beauty*, 14–115; Frevert, *Women in German History*, 190–92.
39. Atina Grossmann, *Reforming Sex: The German Movement for Birth Control and Abortion Reform, 1920–1950* (New York: Oxford University Press, 1995), 34.
40. White, *First Sexual Revolution*, 84, 99.
41. Lesley A. Hall, *Hidden Anxieties: Male Sexuality, 1900–1950* (Cambridge, UK: Polity, 1991), 71, 120–24.
42. Stewart, *For Health and Beauty*, 140; Jennifer Terry, *An American Obsession: Science, Medicine, and Homosexuality in Modern Society* (Chicago: University of Chicago Press, 1999), 162; and Chandak Sengoopta, "Glandular Politics: Experimental Biology, Clinical Medicine, and Homosexual Emancipation in Fin-de-Siècle Central Europe," *Isis* 89 (1998): 445–73.
43. Thomas R. Cole, *The Journey of Life: A Cultural History of Aging in America* (Cambridge: Cambridge University Press, 1992), 198, 224.
44. Barbara L. Marshall and Stephen Katz, "Forever Functional: Sexual Fitness and the Ageing Male Body," *Body and Society* 8, no. 4 (2002): 43–70; and Susan Bordo, *The Male Body: A New Look at Men in Public and in Private* (New York: Farrar, Straus and Giroux, 1999), 59–61.

45. Bonnie Smith, *Changing Lives: Women in European History since 1700* (Toronto: D. C. Heath, 1989), 436.
46. Jessamyn Neuhaus, "The Joy of Sex Instruction: Women and Cooking in Marital Sex Manuals, 1920–1963," in *Kitchen Culture in America: Popular Representations of Food, Gender, and Race*, ed. Sherrie A. Inness (Philadelphia: University of Pennsylvania Press, 2001), 109.
47. Anson Rabinbach, *The Human Motor: Energy, Fatigue, and the Origins of Modernity* (Berkeley: University of California Press, 1990); and Cecelia Tichi, *Shifting Gears: Technology, Literature, and Culture in Modernist America* (Chapel Hill: University of North Carolina Press, 1987).
48. Victoria E. Bonnell, *Iconography of Power: Soviet Political Posters under Lenin and Stalin* (Berkeley: University of California Press, 1997), 39.
49. Quoted in Jackson Lears, *Fables of Abundance: A Cultural History of Advertising in America* (New York: Basic Books, 1994), 167; see also Rabinbach, *Human Motor*, 292–300.
50. Linda M. Blum and Nena F. Stracuzzi, "Gender in the Prozac Nation: Popular Discourse and Productive Femininity," *Gender and Society* 18, no. 3 (2004): 269–86; and John M. Hoberman, *Testosterone Dreams: Rejuvenation, Aphrodisia, Doping* (Berkeley: University of California Press, 2005), 30–32.
51. Nikolas Rose, "The Neurochemical Self and Its Anomalies," in *Risk and Morality*, ed. Richard V. Ericson and Aaron Doyle (Toronto: University of Toronto Press, 2003), 419.
52. Frank Becker, "Sportsmen in the Machine World: Models for Modernization in Weimar Germany," *International Journal of the History of Sport* 12, no. 1 (1995): 153–68.
53. Susan K. Cahn, *Coming on Strong: Gender and Sexuality in Twentieth-Century Women's Sport* (New York: Free Press, 1994), 26–27, 173–81, 239, 262–63; Stewart, *For Health and Beauty*, 168–72.
54. Zygmunt Bauman, *Life in Fragments: Essays in Postmodern Morality* (Oxford: Blackwell, 1995), 163.

Chapter 7

1. Frank 1997: 103.
2. Thomas Csordas, *The Sacred Self: A Cultural Phenomenology of Charismatic Healing* (Berkeley: University of California Press, 1994).
3. Donna Haraway, *Simians, Cyborgs, and Women: The Reinvention of Nature* (London: Routledge, 1991), 198.
4. Vicki Kirby, *Telling Flesh: The Substance of the Corporeal* (London: Routledge, 1997), 2.
5. Elizabeth Grosz, *Volatile Bodies: Towards a Corporeal Feminism* (Bloomington: Indiana University Press, 1994), 144.
6. "Armless Iraqi Boy Bears No Grudges for U.S. Bombing," Reuters, August 11, 2003, http://www.iraqbodycount.org/analysis/reference/ibc-in-the-media/reuters_11aug2003.php. Iraq Body Count estimated that between 16,000 and 20,000 civilians were wounded in the war based on media reports and the findings of independent investigators. Britain and the United States say it is impossible to give any accurate figures.

7. See Ana Carden-Coyne's chapter in this volume for more on Lapper.
8. Alison Lapper (with Guy Feldman), *My Life in My Hands* (New York: Simon and Schuster, 2005).
9. The fourth plinth, Trafalgar Square, London, designed by Charles Barry, was built in 1841, but the funds to top it with an equestrian statue never materialized.
10. Rachel Cooke, "Bold, Brave, Beautiful," *The Observer*, September 18, 2005, Review section.
11. Mark Quinn, quoted in London Authority press release, "Alison Lapper Pregnant and Hotel for the Birds Selected for 4th Plinth," March 15, 2004, Official Web site for the Mayor of London and the Greater London Authority, http://www.london.gov.uk/view_press_release.
12. Susan Benson, "Inscriptions of the Self: Reflections on Tattooing and Piercing in Contemporary Euro-America," in *Written on the Body: The Tattoo in European and American History*, ed. Jane Caplan (London: Reaktion Books, 2000), 234.
13. Ibid.
14. Kirby, *Telling Flesh*, 3–4.
15. Ibid., 3.
16. Stanley Porteus, *The Psychology of a Primitive People: A Study of the Australian Aborigines* (Sydney: Books for Libraries Press, 1931), 122.
17. See Anna Cole and Anna Haebich, "Corporal Punishments and Corporeal Publics: A Cross-Cultural Historical Approach to the Techniques and Cosmologies of Body Modification" (paper presented at Body Modification, Mark II, Macquarie University, Sydney, April 21–23, 2005).
18. Csordas, *Sacred Self*, 3; Elaine Scary, *The Body in Pain: The Making and Unmaking of the World* (Oxford: Oxford University Press, 1985); Jean Jackson, *Women in Pain* (Philadelphia: University of Pennsylvania Press, 1994); and Jean Jackson, *"Camp Pain": Talking with Chronic Pain Patients* (Philadelphia: University of Pennsylvania Press, 2000).
19. Jennifer Biddle, "Country, Skin, Canvas: The Intercorporeal Art of Kathleen Petyarre," *Australian and New Zealand Journal of Art* 4, no. 1 (2003): 61–76.
20. Deborah Bird-Rose, *Dingo Makes Us Human: Life and Land in an Australian Aboriginal Culture* (Cambridge: Cambridge University Press, 1992), 58–59.
21. Christine Watson, "Touching the Land: Towards an Aesthetic of Balgo Contemporary Painting," in *From the Land: Dialogues with the Kluge-Ruhe Collection of Australian Aboriginal Art*, ed. H. Morphy and M. Smith-Bowles (Charlottesville: University of Virginia Press, 1999), 4.
22. Biddle, "Country, Skin, Canvas," 8.
23. Ibid., 61.
24. Ibid., 65.
25. Ibid., 69.
26. Jennifer Biddle, "Inscribing Identity: Skin as Country in the Central Desert," in *Thinking Through the Skin*, ed. S. Ahmed and J. Stacey (London: Routledge, 2001), 17–19.
27. Ibid., 17.
28. Biddle, "Country, Skin, Canvas"; see also M. J. Meggit, *Desert People: A Study of the Walbiri Aborigines of Central Australia* (Sydney: Angus and Robertson, 1971), 270–80.
29. Biddle, "Country, Skin, Canvas," 74.
30. Grosz, *Volatile Bodies*, 144.

31. *Report on the Work of the Horn Scientific Expedition to Central Australia* (Melville, Mullen and Slade, 1896; fascimile, Bundaberg, Australia: Corkwood, 1994), 29–31.
32. Porteus, *Psychology of a Primitive People*, 122.
33. Henry Louis Gates, "A Liberalism of Heart and Spine," *New York Times*, March 27, 1994, cited in Christine Walley, "Searching for Voices: Feminism, Anthropology, and the Global Debate over Female Genital Operations," *Cultural Anthropology* 12, no. 3 (1997): 422.
34. Nawal El Saadawi, *The Hidden Face of Eve* (London: Zed Books, 1980).
35. Frances Althaus, "Female Circumcision: Rite of Passage or Violation of Rights?" *International Family Planning Perspectives* 23, no. 3 (1997): 130–33.
36. Walley, "Searching for Voices," 418.
37. Leonard J. Kouba and Judith Muasher, "Female Circumcision in Africa: An Overview," *African Studies Review* 28, no. 1 (1985): 103.
38. Penelope Hetherington, "Female Circumcision and the Writing of History," *Indian Ocean Studies Review* 5, no. 3 (1992): 7.
39. Ibid.; Vicki Kirby, "On the Cutting Edge: Feminism and Clitoridectomy," *Australian Feminist Studies* 5 (1987): 35–55.
40. Hetherington, "Female Circumcision," 7.
41. Walley, "Searching for Voices," 420.
42. Hetherington, "Female Circumcision," 10.
43. See, for example, Leila Ahmed, *Women and Gender in Islam: Roots of an Historical Debate* (New Haven, CT: Yale University Press, 1992); and Lata Mani, "Contentious Traditions: The Debate on Sati in Colonial India," in *Recasting Women: Essays in Indian Colonial History*, ed. K. Sangari and V. Sudesh (New Brunswick, NJ: Rutgers University Press, 1990), 88–126.
44. Susan Pederson, "National Bodies, Unspeakable Acts; the Sexual Politics of Colonial Policy Making," *Journal of Modern History* 63 (1991): 647–80; and Jocelyn Murray, "The CMS and the Female Circumcision Issue in Kenya, 1929–1932," *Journal of Religion in Africa* 8 (1976): 92–104.
45. Penelope Hetherington, "The Politics of the Clitoris: Contaminated Speech, Feminism and Female Circumcision," *African Studies Review and Newsletter* 19, no. 1 (1997): 4–9; see Tim Schultz and Robert Feldman, "New Moves in Sex Surgery," *Cleo* 77 (March 1979): 50–59, cited in Hetherington, "Politics of the Clitoris," 8.
46. Hetherington, "Politics of the Clitoris," 6.
47. Ibid., 7.
48. Walley, "Searching for Voices," 412.
49. Walter Goldschmidt, *The Sebei: A Study in Adaptation* (New York: Holt, Rinehart & Winston, 1986).
50. Janice Boddy, "Womb as Oasis: The Symbolic Context of Pharaonic Circumcision in Rural Northern Sudan," *American Ethnologist* 9 (1982): 688.
51. Hanny Lightfoot-Klein, *Prisoners of Ritual: An Odyssey into Female Genital Circumcision in Africa* (New York: Haworth, 1989).
52. Nahid Toubia, "Female Circumcision as a Public Health Issue," *New England Journal of Medicine* 331, no. 11 (1994): 712–16.
53. Hetherington, "Female Circumcision," 6.
54. Chris Shilling, *The Body and Social Theory* (London: Sage, 1993).
55. Anthony Giddens, *Modernity and Self-Identity: Self and Society in the Late Modern Age* (Stanford, CA: Stanford University Press, 1991).

56. Grosz, *Volatile Bodies*, 196, for an overview of this approach to the body.
57. See for discussion Alfred Gell, *Wrapping in Images: Tattooing in Polynesia* (Oxford: Oxford University Press, 1993), 10.
58. Ibid., 12. See also the writings of criminologist Cesare Lombroso for an example of such an approach.
59. Margo DeMello, *Bodies of Inscription: A Cultural History of the Modern Tattoo Community* (Durham, NC: Duke University Press, 2000), 2.
60. Karl Broome, "Kids Love Ink: London's Youth Tattoo Milieux" (PhD diss., Goldsmiths College, University of London, 2006).
61. Ibid., 65.
62. Ibid., 123.
63. Les Back, *Inscriptions of Love* (London: Polity, 2004).
64. Victoria Pitts, *In the Flesh: The Cultural Politics of Body Modification* (New York: Palgrave/MacMillan, 2003), 67.
65. Ibid.
66. Ibid., 41.
67. Julia Kristeva, *Powers of Horror: An Essay on Abjection*, trans. Leon S. Roudiez. (New York: Columbia University Press, 1982).
68. Pitts, *In the Flesh*, 42.
69. Mikhail Bakhtin, *Problems of Dostoevsky's Politics* (Ann Arbor: University of Michigan Press, 1984).
70. Judith Butler, *Bodies That Matter: On the Discursive Limits of "Sex"* (New York: Routledge, 1993); see also Donald Morton, ed., *The Material Queer: A Lesbigay Cultural Studies Reader* (Boulder, CO: Westview, 1996).
71. Pitts, *In the Flesh*, 68.
72. Gell, *Wrapping in Images*, 204.

Chapter 8

1. Michel Foucault, *Discipline and Punish*, trans. Alan Sheridan (New York: Pantheon, 1977), 29.
2. Francis Barker, *The Tremulous Private Body: Essays on Subjection* (Ann Arbor: University of Michigan Press, 1995), vi–viii.
3. Walter Benjamin, "To the Planetarium," in *One Way Street*, in *Walter Benjamin: Selected Writings*, Vol. 1, *1913–1926*, ed. Marcus Bullock and Michael W. Jennings (Cambridge, MA: Belknap Press of Harvard University Press, 1996), 486.
4. On this reading of Benjamin, see Gertrud Koch, "Cosmos in Film: On the Concept of Space in Walter Benjamin's 'Work of Art' Essay," in *Walter Benjamin's Philosophy*, ed. Andrew Benjamin and Peter Osborne (London and New York: Routledge, 1994), 205–15.
5. Simian immunodeficiency syndrome in chimpanzees, thought to have mutated into HIV-1 human immunodeficiency syndrome, may itself be a hybrid of monkey viruses. Stefan Lovgren, "HIV Originated with Monkeys, Not Chimps, Study Finds," *National Geographic*, June 12, 2003, http://news.nationalgeographic.com/news/2003/06/0612_030612_hivvirusjump.html.
6. Emergence is modeled in "artificial life," or a-life, experiments with genetic algorithms, agent-based flocking rules, bottom-up robotics, cellular autonomata, metal alloys, and biochemical reactions; see P. Cariani, "Emergence and Artificial Life,"

in *Artificial Life II*, ed. C. G. Langton, C. Taylor, J. D. Farmer, and S. Rasmussen (Redwood, CA: Addison-Wesley, 1992), 775–97; Stefan Helmreich, *Silicon Second Nature: Culturing Artificial Life in a Digital World* (Berkeley: University of California Press, 1998); Mitchell Resnick, *Turtles, Termites, and Traffic Jams: Explorations in Massively Parallel Microworlds* (Cambridge, MA: MIT Press, 1994); and Mitchell Whitelaw, *Metacreation* (Cambridge, MA: MIT Press, 2004). Emergence is thought out in philosophies of relations between the physical, chemical, and biological, as in Slavoj Žižek, *Organs without Bodies: On Deleuze and Consequences* (New York: Routledge, 2004). Philosophically, both Whitelaw and Žižek draw on the discussions of causality initiated by G. H. Lewes and J. S. Mill, proceeding through Hegel and Kant, to Nagel (for Whitelaw) and via Weismann, Freud, Lacan, Deleuze, and bodies without organs (for Žižek). Keith Ansell-Pearson attempts to trace out a Nietzschean lineage in *Viroid Life: Perspectives on Nietzsche and the Transhuman Condition* (New York: Routledge, 1997). Both Whitelaw and Žižek comment on the idea that emergence has to do with an accumulation of causes that can be understood only retrospectively as a form of feedback loop, or causes whose relations surpass those of actuality (and thus can be understood as a form of virtuality). I would add to this the Wittgensteinian notion of the social consequences of decisions and actions in the evolution of technoscientific worlds; see Michael M.J. Fischer, "Cultural Critique with a Hammer, Gouge, and Woodblock: Art and Medicine in the Age of Social Retraumatization," in *Emergent Forms of Life and the Anthropological Voice* (Durham, NC: Duke University Press, 2003), 90–144; and Michael M.J. Fischer, "Technoscientific Intrastructures and Emergent Forms of Life: A Commentary," *American Anthropologist* 107 (2005): 55–61.

7. Structural in the sense of Levi Strauss or Noam Chomsky's generative syntactic structures, tropes in the sense of Hayden White or Paul de Man, ritual process in the sense of Victor Turner, and oedipal-psychosocial in the sense of Freud or Lacan (all with their nineteenth-century precursors, as Levi-Strauss remarks, in Lyell's geology, Marx's modes of production, and Darwin's evolutionary ecologies).
8. Victor Turner, *Forest of Symbols: Aspects of Ndembu Ritual* (Ithaca, NY: Cornell University Press, 1967) and *Drums of Affliction: A Study of Religious Processes among the Ndembu* (Oxford: Clarendon, 1968); Vincent Crapanzano, "Rite of Return: Circumcision in Morocco," in *Psychoanalytic Study of Society*, ed. Warner Meunsterberger and L. Bryce Boyer (New York: Library of Psychological Anthropology, 1981), 9:15–36; Hélène Cixous, *Portrait of Jacques Derrida as a Young Saint* (New York: Columbia University Press, 2004); Gananath Obeyesekere, *Medusa's Hair: An Essay on Personal Symbols and Religious Experience* (Chicago: University of Chicago Press, 1981).
9. Victor Turner, "Bodily Marks," in *The Encyclopedia of Religion*, ed. Mircea Eliade (New York: Macmillan Free Press, 1978), 2:269–75.
10. Ibid.
11. Michael Mejia, *Forgetfulness* (Tuscaloosa: University of Alabama Press, 2005), 13.
12. See also Haraway's play with techno-eyes in "Situated Knowledges," in *Simians, Cyborgs, and Women: The Reinvention of Nature* (London: Routlege, 1991).
13. Bataille, with André Breton, founded the anti-Fascist group Contre-Attaque in 1935. For a slightly different but complementary reading of Bataille and Hans Bellmer's work on the eros/thanatos thematics of Fascism, see Laura Frost, *Sex Drives: Fantasies of Fascism in Literary Modernism* (Ithaca, NY: Cornell University

Press, 2002), chap. 3. Bataille's obsession with the eye, the grossness of life, bodily loss of control, and deathly transition is plausibly rooted in his experiences as the child of a syphilitic (blind and paralytic) father and suicidal, eventually demented mother. Michel Surya, *Georges Bataille: An Intellectual Biography*, trans. E. T. Krzysztof Fijalkowski and Michael Richardson (New York: Verso, 1992). The pornography of the eye has to do not only with sex but also with the stigmata of the syphilitic: "The weirdest thing was certainly the way he looked while pissing . . . His pupils . . . pointed up into space, shifting under the lids . . . with a completely stupefying expression of abandon and aberration." Georges Bataille, *Story of the Eye* (1928; Harmondsworth, UK: Penguin Books, 1982), 72; Surya, *Georges Bataille*, chap. 1. The loss of inhibition around sex is a commonplace of dementia as well as a highly charged (anxious nervous system) transgression of the sacred constituted as a (nonstable) function of the social.

Bataille's own body and corpus, thus, are also suggestive in relation to bodily marks: a Nietzschean, post-Catholic (converting to Catholicism and later abandoning it), depressive sufferer of at least two bouts of tuberculosis and, toward the end of his life, cerebral arteriosclerosis. Other key influences were his interest in dreaming and the photographs of the execution of Fu Chou Li by being cut to pieces while alive, injected with opium to extend the torture. The pictures that he saw in 1925 were still occupying him when he published them in 1961, and they have a hallucinatory quality, with the eyes rolled back in an undecidable expression of either pain or demented ecstasy. Not only involved with (and against) the surrealists, Bataille founded a number of journals (*Documents*, *La critique sociale*, *Acéphale*, *Actualité*, *Critique*) and was an early publisher of Barthes, Foucault, Derrida, and others. With Michel Leiris and Roger Callois he founded in 1939 the Collège de Sociologie. Pablo Picasso, Max Ernst, and Juan Miro auctioned paintings to raise money for him when he was in financial straits toward the end of his life. His first wife, the actress Sylvia Makles Bataille, later married the psychoanalyst Jacques Lacan, who himself was a friend; their daughter Julie took her mother's name, Bataille.

14. Michael Grace, a biologist at the Florida Institute of Technology, is a leader in this work, reported by Lee Dye, "How Snakes See Two Ways: How Snake Eyes Could Lead to Smarter Missiles and Stop Cancer," ABCnews.com, January 9, [2008?], http://abcnews.go.com/Technology/story?id=98115&page=1.
15. See Sophia Roosth, "Sonic Eukaryotes: Sonocytology, Cytoplasmic Milieu, and the *Temps Intèrieur*" (manuscript, 2006), on Jim Gimzewski's use of atomic tunneling microscopes to listen to the vibrations of yeast cells and his hopes for this in cancer detection, since cancerous cells metabolize adenosine triphosphate more quickly and would therefore vibrate at a higher frequency than healthy cells. Roosth describes this as a series of transductions: "The yeast/atomic force microscope/human assemblage that performs sonocytology is a series of vibrations traveling through different material media and converted by mediating transducers into sound and signals. The kinetic motion of motor proteins becomes a cytoplasmic rumble that vibrates the cell wall, which exerts pressure on a cantilever, causing the piezoelectric crystal to convert the deflection into an electrical output, creating a graphic trace of its deflection, which is then converted using a computer program into an electrical signal that exits a pair of speakers as mechanical wave oscillation, creating a periodic turbulence in the air that vibrates the tympanum, that vibrates the ossicles, that vibrates the fluid of the cochlea, that triggers hair cells to send electrical signals

to nerves that travel to the brain, in which each time the signal travels from one neuron to another it must be transduced from electrical to chemical energy while traveling through the intercellular synapse. . . . Sound triangulates between space and time . . . If traditional light microscopy with its staining and fixing techniques offers a vision of flat surfaces frozen in time, and microcinematography animates these cellular landscapes, then sonocytology promises a volumized science . . . creating an acoustic space" (25–26).

16. On the work of Mrganka Sur at the Massachusetts Institute of Technology, see Sandra Brakeslee, "'Rewired' Ferrets Overturn Theories of Brain Growth," *New York Times*, April 25, 2000, http://web.mit.edu/msur/www/nytimes.html; and J. A. Sharma, A. Angelucci, and M. Sur, "Induction of Visual Orientation Modules in Auditory Cortex," *Nature* 404 (2000): 841–47.
17. Peter B.L. Meijer, "Augmented Reality for the Totally Blind," Seeing with Sound, http://www.seeingwithsound.com/voice.htm; and Alison Motluk, "Seeing with Your Ears," *New York Times Magazine*, December 11, 2005, http://www.nytimes.com/2005/12/11/magazine/11ideas_section3-14.html?scp=19&sq=voice&st=nyt.
18. Lisa Cartwright and Brian Goldfarb, "On the Subject of Neural and Sensory Prostheses," in *The Prosthetic Impulse*, ed. Marquard Smith and Joanne Morra (MIT Press, 2006), 141–43; and James Geary, *The Body Electric: An Anatomy of the New Bionic Senses* (London: Weidenfeld and Nicholson, 2002), chap. 2, describe current research on retinal implants that also rely on an external camera to send signals to electrode arrays that activate retinal cells to send signals to the optic nerve. With somewhat lower image quality, William Dobelle experimented with putting the electrode arrays in the visual cortex. A man with a Dobelle eye has had the implant since 1978. Both epiretinal and subretinal implants have been tried. Much still needs to be learned about nerve conduction and nanomanufacturing, and there are hopes for "wet chips" that match the body's conduction of chemical charges. Experiments in nanoknitting by Rutledge Ellis-Behnke and colleagues at the Massachusetts Institute of Technology have shown promise in restoring sight in rats: Nanofibers are injected into damaged portions of the brain and act as scaffolds for axons to regrow. Carey Goldberg, "Ultra-Tiny Knitting Thread Helps Restore Brain Function," *Boston Globe*, March 20, 2006.
19. See Steve Mann, *Cyborg: Digital Destiny and Human Possibility in the Age of the Wearable Computer* (New York: Random House Doubleday, 2001).
20. Geary, *The Body Electric*, chap. 2.
21. On the history of microcinematography, see Hannah Landecker, "New Times for Biology: Nerve Cultures and the Advent of Cellular Life in Vitro," *Studies in History and Philosophy of Biological and Biomedical Sciences* 33 (2002): 667–94; and Landecker, *Culturing Life: How Cells Became Technologies* (Cambridge, MA: Harvard University Press, 2007).
22. The classic film, based on H. G. Wells's 1896 story "The Island of Dr. Moreau," was made in 1927 (*Island of Lost Souls*, with Charles Laughton as Dr. Moreau). It focused on the use of surgery and blood-transfusion technology to create humans from lower animals; remakes under the original title, *The Island of Dr. Moreau*, in 1977 (with Burt Lancaster) and 1996 (with Marlion Brando) shifted the technologies to genetics and neuroscience.
23. Although a number of biologists and computer scientists have publicly invoked this fantasy, it is attributed originally to Harvard's Walter Gilbert in his enthusiasm for the information lingua franca of genetics and genomics. See Lily Kay, *Who Wrote*

the Book of Life: A History of the Genetic Code (Stanford, CA: Stanford University Press, 1999), on the way the "code of life" metaphor came to be inserted into the language of biology, its misrecognitions, and its functionality as a lingua franca across disciplines.

24. John Howard Griffin, *Black Like Me* (Boston: Houghton Mifflin, 1961), 45, 52–53.
25. Haraway, "Situated Knowledges." See also Paul N. Edwards, *The Closed World: Computers and the Politics of Discourse in Cold War America* (Cambridge, MA: MIT Press, 1996).
26. Griffin, *Black Like Me*, 45, 52–53.
27. Michael Westlake, *Imaginary Women* (Winnipeg, MB: Paladin Books, 1989).
28. Victoria Pitts, *In the Flesh: The Cultural Politics of Body Modification* (New York: Palgrave, 2003), 1–2.
29. Born in Dallas in 1920, Griffin was educated in France, doing a medical internship at the Asylum in Tours, experimenting with music therapy for the criminally insane, studying at the Conservatoire de Fontainebleau (with Nadia Boulanger, Robert Casadesus, and Jean Batalla) as well as studying Gregorian chant with Benedictines at the Abbey of Solesmes. At nineteen he became a medic with the French Resistance, evacuating Austrian Jews to the port of St. Nazaire. He served thirty-nine months with the U.S. Army Air Corps in the South Pacific, where he was injured. From 1946–1957 he lost his sight but wrote five novels in that time. He converted to Catholicism in 1952 and became a Third Order Carmelite. Elizabeth Griffin-Bonazzi, "Griffin, John Howard," The Handbook of Texas Online, http://www.tshaonline.org/handbook/online/articles/GG/fgr99.html.
30. Susan Ryan, "New FX Series Has Families Trading Races," *Boston Globe*, March 4, 2006.
31. Without chemically altering his skin, Israeli David Grossman performed a Griffin-like exploration of the lives of Palestinians in the occupied territories in *Yellow Wind* (1987). In an interview reflecting on his childhood, he comments on the Nazi tattoos: "In my neighborhood, you saw people with numbered tattoos. People used to cry out in their nightmares, and we heard them. But what was strange was that people did not talk about it. In those years, it was as if they didn't want to interfere with the momentum of building up the myth of a strong state. As children we got all the contradictory radiation of strength and fragility. This manic-depressive wave went through us all the time." Grossman, interview in "Reflections on the Body Politic" by Mark Sorkin, *The Nation*, July 11, 2005, http://www.thenation.com/doc/20050711/grossman.
32. A Brazilian cosmetics company distributed temporary "Justice for Jessica" tattoos in India in March 2006 to support a protest campaign against the acquittal of an alleged (and confessed) murderer of a barmaid-model, Jessica Lal. A less entangled mixing of cosmetics, commerce, and politics, but of art action is Jenny Holzer's 1993 *Lustmord-Zyklus* of thirty color photographs about rape and murder in the former Yugoslavia, with sudden shifts of perspective between perpetrators, victims' families, and victims. Texts are written on the skin with a felt pen, photographed, and enlarged, filling the visual space so as to be "menacingly close to the viewer." Claudia Benthien, *Skin: On the Cultural Border between Self and the World*, trans. Thomas Dunlap (New York: Columbia University Press, 2002), 3.
33. Nikki Sullivan, *Tattooed Bodies: Subjectivity, Textuality, Ethics and Pleasure* (Westport, CT: Praeger, 2001), 179.

34. I follow here the lovely reading of Dennis Patrick Slattery, *The Wounded Body* (Albany: State University of New York Press, 2000), chap. 9.
35. From a letter of O'Connor, cited in Slattery, *Wounded Body*, 199.
36. For the notion that late nineteenth-century American Protestants had so deprived themselves of multisensory ritual that in their tourism to Europe they sought out cathedrals and Catholic ritual, see W. Lloyd Warner's *Family of God: A Symbolic Study of Christian Life in America* (New Haven, CT: Yale University Press, 1961); and T. J. Jackson Lears's *No Place of Grace: Antimodernism and the Transformation of American Culture, 1880–1920* (New York: Pantheon Books, 1981).
37. As William Turner correctly notes, Maimonides explained this proscription against tattoos and marks as one against the temptation of idolatry, and other rabbis also explained that since man is created in the image of God, that perfection should not be profaned or marred. Turner, "Teaching of Moses Maimonides," *The Catholic Encyclopedia*, Vol. 9 (New York: Robert Appleton, 1910), available at http://www.newadvent.org/cathen/09540b.htm. Both explanations are used by Muslims as well and in the proscription by the Roman Catholic Church. On the paradoxes of circumcision rites, see below.
38. Many Muslims in North Africa and elsewhere have traditional tattoos, as people notice especially during the hajj, when people from many different areas come together. Temporary henna tattooing is part of weddings and some circumcision ceremonies, but even permanent tattoos are not *haram* (forbidden) if done in ignorance or before one has converted and need not be removed. It is a topic discussed on a number of Muslim Web sites today (e.g., "New Converts Having Tattos and Tending [sic] to Perform Hajj," online discussion on Islamonline.net, January 18, 2004, http://www.islamonline.net/servlet/Satellite?cid=1119503543310&pagename=IslamOnline-English-Ask_Scholar/FatwaE/FatwaEAskTheScholar).

    Among Jews, the famed Moskowitz family of tattoo artists—beginning with Willie in 1918, an immigrant from Russia, and continuing on to his grandson Marvin (who retired in 2000)—kept the business going, despite a health department ban in Manhattan in the 1960s, by moving to Long Island. They watched the tattoo move from a sign of rebellion, to an insignia against the Holocaust ("never again" or Star of David tattoos), to gentrification.
39. Liminal periods in the ritual process, as Victor Turner elaborates, fuse together ("con-fuse") emotion (fear, anxiety), bodily pain, and bodily substances (blood, milk) together with the cognitive (ideology, rationale, explanation), moral (social affiliation, status), and symbolic. The most powerful symbols, Turner argues, have this bipolar structure, viscerally grounding social understandings and expectations, so that the socially obligatory is transformed into something internalized and viscerally sacralized, such that transgressions cause unnerving physical reactions. The transgressive thereby becomes itself ritually powerful, as in trance curing rituals in which Muslims (for instance, in the Zar cults of East Africa and the Persian Gulf) are made to drink prohibited blood and in which separation, drumming, and hyperventilation also make the body pliable and diets and prohibitions continue the cultic discipline. Victor Turner, *Dramas, Fields and Metaphors: Symbolic Action in Human Society* (Ithaca, NY: Cornell University Press, 1974) and *The Ritual Process: Structure and Anti-Structure* (Chicago: Aldine, 1969).
40. For a somewhat similar and brilliantly elaborated Jewish reading of circumcision, see Jacques Derrida, "Circonfession," in *Jacques Derrida, Collaboration with*

Geoffrey Bennington (Chicago: University of Chicago Press, 1991); and Cixous, *Portrait of Jacques Derrida*, drawing on an array of Jewish and Catholic, North African and French, male and female thematics of "too young to sign, he could only bleed," the stigmata of 1940 (and other dates of Passovers, transfers, expulsions, naturalizations, de-citizenships, exinclusions, blacklistings, doors slammed in your face, . . . the archives of what he calls "my nostalgeria" and that I call my "algeriance"), scarification/mortification ("what happens to the skin and flesh of the text, incision, graft of a fragment lifted from another segment of time"), the hidden name of substitution/superstition of new identity "dissociated from this initial hallmarking" (Elie, qu'on elit, the Elie's of the dead, the name of election), male names with female endings (Jackie, Elie), North African inflections (Ali Baba, Baba Elie), a prince whose parentage is provisionally concealed to keep him alive (dead elder brother Paul Moses), Oedipus of El-Biar, and more. (The quotations are from Derrida, "Circonfession.")

41. "With milk dripping from her breasts, her back in bloom, her womb about to give birth, and her feet wrapped in nature's substance, Sethe is reminiscent of the great goddess of fertility, of the earth goddess, of life itself insisting on finding an aperture into the world." Slattery, *Wounded Body*, 215.
42. I draw here on and tweak Paula Willoquet-Maricondi's reading of the film in "Fleshing the Text: Greenaway's *Pillow Book* and the Erasure of the Body," *Postmodern Culture* 9 (1999): 52–82, http://muse.jhu.edu/journals/postmodern_culture/v009/9.2willoquet.html. Hélène Cixous invokes the Greenaway film on the first page of her essay "The Laugh of the Medusa," *Signs* 4 (1976): 875–93. See also Alissa Bersin, "My Father's Pen: Writing, the Body, and Female Pleasure in Helene Cixous's *Inside* and Peter Greenaway's *The Pillow Book of Nogiko*" (bachelor's thesis, Harvard University, 2005), who traces the parallels and differences between Cixous's and Greenaway's search for the father as a source of writing. Cixous's "feminine" writing need not be done by a woman; Joyce's writing is feminine for Cixous. But Bersin argues that while Greenaway's version of Cixous's search for enabling feminine writing is instructive, it is not fully worked out in these terms.
43. Both quotations are from Giuliana Bruno, "M Is for Mapping: Art, Apparel, Architecture Is for Peter Greenaway," in *Atlas of Emotion: Journeys in Art, Architecture, and Film* (New York: Verso, 2002), 327.
44. See Bruno's comments on the "assemblage of three screen ratios—one widescreen and color; one of smaller dimensions in black and white; and the last a tiny videographic window screen." "M Is for Mapping," 286.
45. Pitts, *In the Flesh*.
46. This and the following quotations in this paragraph are from ibid., 11, 54, 58, and 59.
47. V. Vale, *Modern Primitives: An Investigation of Contemporary Adornment and Ritual* (San Francisco: Research Publications, 1989).
48. Quoted in Pitts, *In the Flesh*, 44.
49. Ibid., 153.
50. Hélène Mialet, "Do Angels Have Bodies: Two Stories about Subjectivity in Science, The Cases of William X and Mr. H," in *The Philosophy of Expertise*, ed. E. Selinger and R. P. Crease (New York: Columbia University Press, 2006), 246–79.
51. John Locke, quoted by Hélène Mialet, "Do Angels Have Bodies? Two Stories about Subjectivity in Science: The Cases of William X and Mister H," *Social Studies of Science* 29 (1999): 563.

52. Antonin Artaud, *To Have Done by the Judgement of God*, trans. Clayton Eshleman and Norman Glass (1948; Los Angeles: Black Sparrow Press, 1975).
53. Jean-François Lyotard, *Libidinal Economy* (1974; London: Athlone, 1993), 32–33, 41.
54. For more examples of prosthetic and sensory rewiring experiments, organized by the five senses, see Geary, *The Body Electric*. On the early exploration of the body electric (that is, the stimulation of muscles by electrical impulses), see Robert O. Becker and Gary Selden, *The Body Electric: Electromagnetism and the Foundation of Life* (New York: Morrow, 1985).
55. The term *bioavailability* was coined by Lawrence Cohen for how state policies impact bodies, for instance, in regulations for markets in transplantation or requirements for sterilization before organ donation is allowed, and the pattern of surgeries that tends to form. Cohen, "Where It Hurts: Indian Material for an Ethics of Organ Transplantation," *Daedelus* 128 (1999): 135–65.
56. Gunther von Hagens, *Body Worlds: The Anatomical Exhibition of Real Human Bodies* (Heidelberg, Germany: Institute for Plastination, 2004).
57. On November 27, 2005, Isabelle Dinoire, age thirty-eight, underwent the world's first partial facial transplant in Amiens, France, with tissues, muscle, arteries, and veins from a brain-dead donor in Lilles. She had lost her nose, lips, and chin from a dog bite. "Woman Has First Face Transplant," BBC News, November 30, 2005, http://news.bbc.co.uk/1/hi/health/4484728.stm.
58. Mialet, "Do Angels Have Bodies?" (1999 and 2006 versions).
59. The term *material-semiotic objects* is taken from Donna Haraway, *Modest_Witness@Second_Millennium.FemaleMan©Meets _OncoMouse: Feminism and Technoscience* (New York: Routledge, 1997). These are objects that not only reconfigure the material world (sensors, transducers, newly created objects) but also simultaneously configure semantic relationships.
60. Aslihan Sanal, "Flesh Yours, Bones Mine: The Making of the Biomedical Body in Turkey" (PhD diss., Massachusetts Institute of Technology, 2005).
61. Renée C. Fox and Judith P. Swazey, *The Courage to Fail: A Social View of Organ Transplants and Dialysis* (Chicago: University of Chicago Press, 1974); and Renée C. Fox and Judith P. Swazey, *Spare Parts: Organ Replacement in American Society* (New York and Oxford: Oxford University Press, 1992).
62. Sanal, "Flesh Yours, Bones Mine." Sanal recounts how in one dramatic, heavily media-covered incident, a young woman who had shot her husband and then herself was redeemed by having her organs distributed to save the lives of six others. In a subsequent case, amid anxieties about young women committing suicide, a young woman left a suicide note forbidding her organs to be transplanted as a transgressive rejection of the patriarchal and Islamic constraints on women.
63. Cohen, "Where It Hurts," 135–65.
64. Jane Goodall, "The Will to Evolve," in *Stelarc, the Monograph*, ed. Marquard Smith (Cambridge, MA: MIT Press, 2005), 12.
65. On these and similar technologies, see Geary, *The Body Electric*, chap. 6.
66. Biocybernetics was launched with Defense Department funding in the 1970s and now is funded also by the National Institute of Neurological Disorders and Stroke. Some devices are implanted in the brain. In 2003, Miguel Nicolelis, a neurobiologist at Duke, put eighty-six microwires into a monkey brain, taught him to use a joystick, and then unplugged the joystick; the monkey learned to do the same tasks merely with his brain waves, dropping the joystick and not using his hands.

Cyberkinetics Neurotechnology Systems, Inc. of Foxborough, Massachusetts, has a "BrainGate" device in a four-person clinical trial and hopes to have the device on the market in two to three years. BrainGate is a two-millimeter-by-two-millimeter chip with an array of a hundred electrodes implanted into the motor cortex that samples patterns of neuron firings and feeds them through a wire to a small platform on the patient's head and thence to a computer. The chip was adapted by John P. Donoghue (chairman of Brown University's neuroscience department) and Dick Normann from a microarray chip Normann designed in Utah (where the chips are still manufactured) to send signals into the brain as part of a visual prosthetic; Donoghue realized it could also be an uplink. They tested it first on a cat, then twenty-two monkeys, before asking the Food and Drug Administration for permission to run a human clinical trial. Nicholas Hatsopoulos, formerly at Brown (and CalTech) and now at the University of Chicago, and a founder and director of Cyberkinetics, explains that although the 100 electrodes massively undersample the 1.6 million neurons under the chip, prediction of motion seems to work because to some degree the cells are redundant and for that reason need to be studied collectively rather than cell by cell.

Donoghue is convinced that implants will be required to get the signal clarity to transform noisy signals into something that a patient can use. But others are experimenting with noninvasive methods. Jonathan Wolpaw, with the support of engineers from the Altran Foundation for Innovation (and its subsidiary, Cambridge Consultants in Boston), is testing a cap with sixty-four electrodes, electrogel, and an amplifier that does not require implantation into the brain. It is being tested with a medical scientist with ALS, "who is losing control of his eyes, the last part of his body he can move," to "replace an older system that let him use a computer through eye movement." He is able to use it to send email messages. (The small market for the device, well under 170,000, is a constraint.) Stephen Heuser, "A Case of Mind over Matter: Mind-Reading Devices Show Promise," *Boston Globe*, April 2, 2006; Richard Martin, "Mind Control: Matthew Nagle is Paralyzed. He's Also a Pioneer in the New Science of Brain Implants," *Wired* 13 (2005), http://www.wired.com/wired/archive/13.03/brain.html?pg=2&topic=brain&topic_set; *University of Chicago Magazine*, June 2006, pp. 20–21.

67. Heuser, "A Case of Mind over Matter."
68. See also Helmreich, *Silicon Second Nature*.
69. Martin, "Mind Control," 1.
70. John E. Sarno, *The Divided Mind: The Epidemic of Mindbody Disorders* (New York: Reagan Books, 2006).
71. Michael Gershon, *The Second Brain* (New York: Harper Collins, 1998); and Elizabeth Wilson, "The Brain in the Gut," in *Psychosomatic: Feminism and the Neurological Body* (Durham, NC: Duke University Press, 2004), chap. 2.
72. The phrase BwO comes from Antonin Artaud's radio play, *To Have Done with the Judgement of God*. The sense in which it is discussed here comes from Gilles Deleuze and Félix Guttari, "November 18, 1947: How Do You Make Yourself a Body Without Organs," in *A Thousand Plateaus: Capitalism and Schizophrenia*, trans. Brian Massumi (Minneapolis: University of Minnesota Press, 1987), 149–66.
73. Robert Ayers, "Serene and Happy and Distant: An Interview with Orlan," *Body and Society* 5 (1999): 183.
74. Ibid., 183.

75. This and the following two quotations are from ORLAN, "Carnal Art Manifesto," http://www.dundee.ac.uk/transcript/volume2/issue2_2/orlan/orlan.htm.
76. ORLAN Offical Website, http://www.orlan.net/.
77. Ayers, "Serene and Happy and Distant," 183.
78. ORLAN, "Carnal Art Manifesto." The revolt against Catholic corporeal theology and iconography included posing iconically as the Virgin Mary with one breast exposed, the use of crosses and skulls, bloodied facial imprints on gauze cloths (like the Shroud of Turin), and reliquaries of her fat. The first of the surgical series was called *The Reincarnation of Saint Orlan.*
79. Julie Clarke, "The Sacrificial Body of Orlan," *Body and Society* 5 (1999): 185–207.
80. Blood was also part of the facial imprints on gauze, and it partakes of ORLAN's intertwined symbolism (blood/wine and grapes) of both Christianity and Dionysus/Bacchus.
81. Adams, *The Emptiness of the Image* (London: Routledge, 1996). She takes Holbein's *The Ambassadors* as a classic anamorphosis, where depending on the angle you see a blur or a skull (produced by drawing the image projected from a cylinder mirror). Adams then suggests that isomorphically to subject any oppositional pair (male/female, mind/body, subject/object, essence/appearance) to an anamorphic process "is to reveal the extent to which each term of the pair is *not* in contradiction to the other term," but "the relations between them . . . are strewn with strange thresholds and hybrid forms" (142), a point made long ago by Claude Levi-Strauss in arguing that the generativity of category and mythic oppositions lies in the thirds that they produce as mediators, each mediation in turn producing more. In Lacanian terms, this is also generated in the gap between signifier and signified, one that here, she suggests, is also invested with a foundational anxiety.
82. Ibid., 142, 154, and 156.
83. Ibid., 154.
84. Ibid. The word *circus—carnival* might be better—reminds of Ray Bradbury's *The Illustrated Man* and the role of tattoos in carnivals.
85. Invoking the metaphor of a door that opens but cannot be entered, as well as the unhinging of a door, a body part, or a mind.
86. Jacques Lacan, *The Seminar of Jacques Lacan II: The Ego in Freud's Theory and in the Technique of Psychoanalysis, 1954–1955*, ed. Jacques-Alain Miller, trans. Sylvana Tomaselli (New York: W. W. Norton, 1988), cited in Adams, *Emptiness of the Image*, 154.
87. Adams, *Emptiness of the Image*, 145.
88. On the dissociations experienced in anatomy dissection classes, see Byron Good, *Medicine, Rationality and Experience* (Cambridge: Cambridge University Press, 1994), chap. 3; and Avery in Michael Fischer, "Cultural Critique." For a denial of this, see Eugene A. Arnold, "Autopsy: The Final Diagnosis," in *Images of the Corpse: From the Renaissance to Cyberspace*, ed. Elizabeth Klaver (Madison: University of Wisconsin Press, 2004).
89. On the film *Journey to Qandahar* (or *Kandahar*) by Mohsen Makhmalbaf and its powerful image of prostheses dropped by Red Cross helicopters to Afghan men running on crutches to receive them, see Michael Fischer, *Mute Dreams, Blind Owls, and Dispersed Knowledges* (Durham, NC: Duke University Press, 2004). On the silencing of survivors of the massacres and rapes on Bali in 1965 and their continuing effects, see Leslie Dwyer and Degung Santikarma, "When the World

Turned to Chaos: 1965 and Its Aftermath in Bali, Indonesia," in *The Spector of Genocide: Mass Murder in Historical Perspective*, ed. Robert Gellately and Ben Kiernan (Cambridge: Cambridge University Press, 2003), 298–99. Walter Benjamin observed that soldiers returning from World War I were impoverished rather than empowered in storytelling abilities, and indeed the trauma of war has silenced generations of soldiers. There is a transposition of the public inscriptions of regimes of "truth" on the body of torture victims in the premodern period, described generally by Georg Rusche and Otto Kirchheimer in *Punishment and Social Structure* (New York: Columbia University Press, 1939), for France by Foucault in *Discipline and Punish*, and for Iran by Michael M.J. Fischer, in "Legal Postulates in Flux: Justice, Wit and Hierarchy in Iran," in *Law and Politics in the Middle East*, ed. D. Dwyer (New York: J. F. Bergin, 1989) and Darius Rejali, in *Torture and Modernity: Self, Society and State in Modern Iran* (Boulder, CO: Westview, 1994), to a hidden disciplinary and rehabilitation system in the modern period, with today a reversion to public torture and rape but on a population basis as in Bali in 1965, the Balkans and Rwanda in the 1990s, and today in Darfur and elsewhere. Frantz Fanon, in *The Wretched of the Earth* (New York: Grove, 1963), pointed out the psychological damage inflicted on the torturers as well as the tortured; today we have the effects of Agent Orange and posttraumatic stress disorder on Vietnam veterans and Gulf War Syndrome (chemical and possible radiation effects) among veterans from the first Gulf War.

90. Quoted in Smith, *Stelarc*, viii.
91. Made by the Israeli company Given Imaging, this capsule endoscopy was available in nearly 300 hospitals by 2002. Marilyn Chase, "To Avoid Surgery, Eat This Camera," *Wall Street Journal*, August 15, 2002, available at www.cpmc.org/advanced/endoscopy/wsj_article_aug2002.html.
92. Goodall, "The Will to Evolve," 8.
93. Ibid., 17.
94. Smith, *Stelarc*, 98.
95. Amelia Jones, "Stelarc's Technological Transcendence/Stelarc's Wet Body: The Insistent Return of the Body," in Smith, *Stelarc*, 87–124. See Adams's analysis earlier in this chapter for a more astute feminist take, one grounded in the splits of the speech act or enunciative effects, indeed one that rejects a simple masculine-feminine split, asserting instead in ORLAN's case a woman-woman inscription of sexual difference.
96. Brian Massumi, "The Evolutionary Alchemy of Reason," in Smith, *Stelarc*, 160.
97. Ibid., 167.
98. Stelarc and Marquard Smith, "Animating Bodies, Mobilizing Technologies: Stelarc in Conversation," in Smith, *Stelarc*, 216.
99. Massumi, "Evolutionary Alchemy of Reason," 180.
100. Geary, *The Body Electric*, 39–41.
101. Stelarc and Smith, "Animating Bodies, Mobilizing Technologies," 321.
102. Geary, *The Body Electric*, chap. 2.
103. Ibid., 222.
104. Ibid., 229.
105. Clynes and Kline coined the word *cyborg* in 1960 in an article, "Cyborgs and Space," published in the journal *Astronautics*. Stelarc proposed to work on cyborgs at the U.S. National Aeronautics and Space Administration.

106. Stelarc and Ross Farnell, "In Dialogue with 'Posthuman' Bodies: Interview with Stelarc," in *Body Modification*, ed. Mike Featherstone (London: Sage, 2000), 130 and 133.
107. Melanitis Yiannis, "Stelarc Interview, Web site of Melanitis Yiannis/Artist, http://melanitis.com/StelarcInterview.htm.
108. Lyotard, *Libidinal Economy*, 42.
109. Žižek, *Organs without Bodies*, 132–33.
110. Gilles Deleuze and Felix Guattari, *A Thousand Plateaus: Capitalism and Schizophrenia* (1980; Minneapolis: University of Minnesota Press, 1987), 151.
111. The compilation of criminal tattoos from the Soviet period by Danzig Batlaev, Alexei Plutser-Sarno, and Sergei Vesiliev, in *Russian Criminal Tattoo Encyclopedia* (Saint Petersburg, Russia: Steidl Fuel, 2003), suffers precisely from an insufficiently described account of which tattoos are voluntary, which forcibly imposed, and under what conditions, and how particular meanings of a more general code are negotiated.
112. See Aryn Martin, "Can't Anybody Count? Counting as an Epistemic Theme in the History of Human Chromosomes," *Social Studies of Science* 34 (2004): 923–48, on the making visible and counting of chromosomes ("colored bodies") by staining and, after the 1950s, by other means of generating micrographs.
113. See Geary, *The Body Electric*, chaps. 3 and 7, for a review of sociable and affective robotics technologies.
114. The film *Gattaca* (1997), written and directed by Andrew Niccol, describes a world in which DNA identification can be done from any scrap of the body (a strand of hair, shed skin cells) and in which probabilistic predispositions to illness read from the genome disqualify one for jobs that require educational and other investment. As a result, parents strive for genetically perfect mixes of sperm and eggs preimplantation (making "love children" both antiquated and disadvantaged), and a surveilled caste society develops. One understands the intricacies of the implications through the story of a love child with tremendous will power and intelligence who tries to game the system by using a genetically superior sibling who lacks the same drive to provide him with biological covers.
115. "Will you still need me, will you still feed me, when I'm sixty-four." Beatles, "When I'm Sixty-Four," *Seargent Pepper's Lonely Hearts Club Band*, London, Parlaphone Records, 1967.
116. *Catacoustis*, the inner echo "of a musical order," the return, in very precise circumstances, of a melodic fragment, like "the psychopathology of everyday life," of a "tune in one's head" that "keeps coming back," like the Kol Nidre. Philippe Lacoue-Labarthe, "The Echo of the Subject," in *Typography: Mimesis, Philosophy, Politics* (Stanford, CA: Stanford University Press, 1998).
117. One thinks of the diaries of pioneers into the American West, who, unable to articulate directly what they saw, drew on more familiar tropes of castles and paradise. Or more to the point of the body is the art of physician-artist Eric Avery, who sees in an autopsied skull the face of Edward Munch's *The Scream*, in a patient being lifted by nurses *Lazarus Arising from the Dead*, in the fingers of the surgeon reattaching a hand Michelangelo's image of God's hand touching man, and the stigmata of Saint Sebastian in the 1981 Rio Sumpul Massacre in El Salvador (see Fischer, "Cultural Critique").

118. Emmanel Levinas's formulation of ethics as the response to the face of the other has become a useful shorthand in contemporary discussions of ethics, as well as a counter to the metaphysics of being in philosophy.
119. Jacques Derrida's *Given Time* [*Donner le temps*] (University of Chicago Press, 1991) is both a contribution to the growing literature spawned by Marcel Mauss's 1925 *Essai sur le don* (*The Gift*) and a meditation on the time given before death; volume two was titled *Given Death* [*Donner le mort*] (University of Chicago Press, 1995).
120. Jonathan Eller and William F. Touponce, *Ray Bradbury: The Life of Fiction* (Kent, OH: Kent State University Press, 2004), 383.
121. Joseph Dumit, *Drugs for Life* (Durham, NC: Duke University Press, forthcoming).
122. Merleau-Ponty, *Phenomenology of Perception* (New York: Humanities Press, 1962), 13; Henri Bergson, *Matter and Memory*, trans. Nancy Margaret Paul and W. Scott Palmer (London: Allen and Unwin, 1911); Taussig, *Mimesis and Alterity: A Particular History of the Senses* (New York: Routledge, 1993), 57; Feld, "Places Sensed, Senses Placed," in *Empire of the Senses*, ed. David Howes (New York: Berg, 2005), 180–81; Classen, "McLuhan in the Rainforest: The Sensory Worlds of Oral Cultures," in Howes, *Empire of the Senses*; and Sacks, "The Mind's Eye: What the Blind See," in Howes, *Empire of the Senses*.
123. "I had only to look at a picture or an anatomical specimen, and its image would remain both vivid and stable . . . for hours. I could mentally project the image onto the paper before me—it was as clear and distinct as if projected by a camera lucida—and trace its outlines with a pencil. My drawings were not elegant, but they were, everyone agreed, very detailed and accurate. . . . I had only to think of a face, a place, a picture, a paragraph in a book to see it vividly in my mind." Sacks, "The Mind's Eye," 39.
124. Quoted in Žižek, *Organs without Bodies*, 125.
125. Žižek, *Organs without Bodies*, 126.

Chapter 9

1. Sue Malvern, *Modern Art, Britain and the Great War: Witnessing, Testimony and Remembrance* (New Haven, CT: Yale University Press, 2004).
2. Amelia Jones, *Irrational Modernism: A Neurasthenic History of New York Dada* (Cambridge, MA: MIT Press, 2004), 16–20.
3. Julia Kristeva, *Powers of Horror: An Essay on Abjection*, trans. Leon S. Roudiez (New York: Columbia University Press, 1982).
4. See Elizabeth Cowling and Jennifer Mundy, eds., *On Classic Ground: Picasso, Leger, de Chirico and the New Classicism, 1910–1930* (London: Tate Gallery, 1990).
5. Bernd Huppauf, "Languemarck, Verdun and the Myth of a New Man in Germany after the First World War," *War and Society* 6, no. 2 (1988): 70–103; and Tim Armstrong, *Modernism, Technology and the Body: A Cultural Study* (Cambridge: Cambridge University Press, 1998), 77.
6. Paul O'Keefe, "Art, Action and the Machine," in *Dynamism: The Art of Modern Life before the Great War*, ed. Penelope Curtis (Liverpool: Tate Gallery, 1991), 49 (an exhibition catalog).

7. Peter de Francia, *Fernand Léger* (New Haven, CT: Yale University Press, 1983), 30.
8. Kenneth Silver, *Esprit de Corps: The Art of the Parisian Avant-Garde and the First World War, 1914–1925* (Princeton, NJ: Princeton University Press, 1989), 107.
9. Michel Seuphor, "In Defence of an Architecture," in *The Tradition of Constructivism*, ed. Stephen Bann (London: Thames and Hudson, 1974), 185.
10. Romy Golan, *Modernity and Nostalgia: Art and Politics between the Wars* (New Haven, CT: Yale University Press, 1995).
11. Herbert Read, *A Concise History of Modern Sculpture* (London: Thames and Hudson, 1971), 101.
12. Vladimir Tatlin, T. Shapiro, I. Meyerzon, and Pavel Vinogradov, "The Work Ahead of Us" (December 31, 1920, Moscow), in Bann, *Tradition of Constructivism*, 12.
13. Clive Bell, "The Rise and Decline of Cubism," *Vanity Fair*, 1923, reprinted in Susan Noyes Platt, *Modernism in the 1920s: Interpretations of Modern Art in New York from Expressionism to Constructivism* (Ann Arbor, MI: UMI Research Press, 1983), 81.
14. Herbert Read, *Art Now: An Introduction to the Theory of Modern Painting and Sculpture* (1933; London: Faber and Faber, 1960). 77.
15. See the depiction of woman's abundant fertility in Picasso's *The Source* (1921), which is related to *Three Women at the Spring* (1921). A similar idealization of the sexually ready and fertile nude is depicted in Arturo Martini's terra-cotta sculpture of a *Reclining Woman* (1930–1931). See also Joaquim Sunyer, *Maternity* (1921); Achille Funi, *Maternity* (1921); and Mario Sironi, *Maternity* (1923). These paintings are reproduced in Cowling and Mundy, *On Classic Ground*, 173, 255, 107, 244.
16. Golan, *Modernity and Nostalgia*, 20–21.
17. See Mary Louise Roberts, *Civilisation without Sexes: Reconstructing Gender in Postwar France, 1917–1927* (Chicago and London: University of Chicago Press, 1994).
18. "British Painting," *The Studio* (89), 1931, 150–51; "Contemporary Figure Painters," *The Studio* (103), 1925.
19. Marie Stopes, *Married Love* (London: A. C. Fifield, 1918); Marie Stopes, *Radiant Motherhood: A Book for Those Who Are Creating the Future* (London: G. P. Putnam and Sons, 1926), 216; and Ettie Hornibrook, *Restoration Exercises for Women* (London: William Heinemann, 1931), ix. See also Roy Porter and Lesley Hall, " 'Good Sex': The New Rhetoric of Conjugal Relations," in *The Facts of Life: The Creation of Sexual Knowledge in Britain, 1650–1950* (New Haven, CT: Yale University Press, 1995), 202–23.
20. For similar works, see *Verdun: The Trench Diggers* (1916) and *Two Soldiers* (1915), where man has become machine. See also *Verdun* (1916) and *Le Blesse* (1917). This aesthetic attitude is carried over in a work such as *The Typographer* (1918), where it is purposefully difficult to decipher the machine from the figure. These forms are all rather jagged when compared with the more curvilinear and sparse geometry of his 1920s work, such as *Three Comrades* (1920), *Le Mechanicien* (1920), *Woman with a Vase* (1924–1927), and *Three Women on a Red Ground* (1927).
21. John Golding and Christopher Green, eds., *Léger and Purist Paris* (London: Tate Gallery, 1970), 10 (an exhibition catalog).
22. Golan, *Modernity and Nostalgia*, 72.

23. Cited in de Francia, *Fernand Léger*, 60.
24. See Edouard Jeanneret [Le Corbusier] and Amédée Ozenfant, *Après le Cubisme* (Paris: Édition des Commentaires, 1918). Reprinted in Silver, *Esprit de Corps*, 230.
25. Quoted from in Cowling and Mundy, *On Classic Ground*, 126–27.
26. Fernand Léger, "Art and the People," *Arts de France* (1946), reprinted in *Fernand Léger, 1881–1955*, ed. Robert T. Buck, Edward F. Fry, and Charlotta Kotik (New York: Abbeville, 1982), 146–47.
27. Cited in Matthias Eberle, *World War One and the Weimar Artists: Dix, Grosz, Beckmann, Schlemmer* (New Haven, CT, and London: Yale University Press, 1985), 108n10.
28. Theo van Doesburg, "Meditations at the Frontier," *De Avondpost*, 1915, reprinted in *Art in Theory, 1900–1990: An Anthology of Changing Ideas*, ed. Charles Harrison and Paul Wood (Oxford: Blackwell, 1995), 292.
29. Kasimir Malevich, "The Question of Imitative Art" (Smolensk, 1920), reprinted in part in Harrison and Wood, *Art in Theory*, 292.
30. Piet Mondrian, "Neo-Plasticism: The General Principle of Plastic Equivalence" (January 1921), in Harrison and Wood, *Art in Theory*, 288; see also Fernand Léger, "The New Realism Goes On," *Art Front* 3, no. 1 (February 1937): 493–96; and "A New Realism—the Object," in *Theories of Modern Art*, ed. Herschel B. Chipp (Berkeley: University of California Press, 1968), 279.
31. Naum Gabo, "The Constructive Idea in Art," in *Circle: International Survey of Constructive Art*, ed. J. L. Martin, B. Nicholson, and N. Gabo (London: Praeger, 1937), 1–10.
32. Cited in Read, *Concise History of Modern Sculpture*, 96–98.
33. Earle Powell III and James Wood, Foreword, in *Degenerate Art: The Fate of the Avant-Garde in Nazi Germany*, ed. Los Angeles County Museum of Art (New York: Abrams, 1991), 6.
34. Annegret Jandra, "The Fight for Modern Art: The Berlin Nationalgalerie After 1933," in Los Angeles County Museum of Art, *Degenerate Art*, 105.
35. Stephanie Barron, "Modern Art and Politics in Prewar Germany," in Los Angeles County Museum of Art, *Degenerate Art*, 9–23.
36. Klaus Theweleit, *Male Fantasies*, Vol. 2, *Male Bodies: Psychoanalyzing the White Terror* (Minneapolis: University of Minnesota Press, 1989).
37. Arnd Krüger, "Breeding, Rearing and Preparing the Aryan Body," *International Journal of the History of Sport* 16 (1999): 56. See also Arnd Krüger and Dietrich Ramba, "Sparta or Athens? The Reception of Greek Antiquity in Nazi Germany," in *The Olympic Games through the Ages: Greek Antiquity and Its Impact on Modern Sport*, ed. Roland Renson and Manfred Lammer (Athens, Greece: Hellenic Sports Research Institute, 1991), 345–56. See also Wilfried van der Will, "The Body and the Body Politic as Symptom and Metaphor in the Transition of German Culture to National Socialism," in *The Nazification of Art: Art, Design, Music, Architecture and Film in the Third Reich*, ed. Brandon Taylor and Wilfried van der Will (Winchester, UK: Winchester School of Art Press, 1990), 52–75.
38. Taylor and van der Will, *Nazification of Art*.
39. George Mosse, "Beauty without Sensuality: The Exhibition *Entartete Kunst*," in Powell and Wood, *Degenerate Art*, 25ff.
40. Jean Clair, *Identity and Alterity: Figures of the Body 1895–1995*, Venice Bienale (Venice: Marsilio, 1995), 309.

41. Karen Pinkus, *Bodily Regimes: Italian Advertising under Fascism* (Minneapolis: University of Minnesota Press, 1995), 16.
42. Pontus Hulten, "The Blind Lottery of Reputation or the Duchamp Effect," in *Marcel Duchamp* (London: Thames and Hudson, 1993), 18.
43. Christine Jarvis, *The Male Body at War: American Masculinity during World War II* (Dekalb: University of Northern Illinois Press, 2004), 14.
44. M. Darsie Alexander, Mary Chan, and Starr Figura, *Body Language* (New York: Museum of Modern Art and Harry Abrams, 1999) 62–63.
45. Henry Geldzahler, *Pop Art 1955–70* (New York: Museum of Modern Art, in conjunction with the International Cultural Corporation of Australia, 1985), 20 (an exhibition catalog).
46. James E.B. Breslin, "Out of the Body: Mark Rothko's Painting," in *The Body Imaged: The Human Form and Visual Culture since the Renaissance*, ed. Kathleen Adler and Marcia Pointon (Cambridge: Cambridge University Press, 1993), 43, 47.
47. An image is reproduced in *Angry Women*, ed. Andrea Juno and V. Vale (San Francisco: Re/Search, 1991), 188.
48. Laura Mulvey, "Visual Pleasure and Narrative Cinema," *Screen* 16, no. 3 (Autumn 1975): 6–18.
49. Thelma Golden, "My Brother," in *Black Male: Representations of Black Masculinity in Contemporary American Art*, ed. Thelma Golden (New York: Whitney Museum of Art and Abrams, 1994), 26.
50. Golden, "My Brother," 32.
51. Elizabeth Alexander, " 'Can You Be BLACK and Look at This?': Reading the Rodney King Video(s)," in Golden, *Black Male*, 108.
52. Maurice Merleau-Ponty, "The Body as Expression and Speech," in *The Phenomenology of Perception* (London: Routledge, 2002), 202–34.
53. Charles Merewether, "The Unspeakable Condition of Figuration," in *Body*, (Sydney, Australia: Art Gallery of New South Wales and Bookman, 1997), 153.
54. Amelia Jones and Andrew Stephenson, eds., *Performing the Body, Performing the Text*, exhibition catalogue compiled by Tony Bond (London and New York: Routledge, 1999), 1.
55. Joanna Frueh, "The Body through Women's Eyes," in *The Power of Feminist Art: The American Movement of the 1970s, History and Impact*, ed. Norma Broude and Mary D. Garrard (New York: Abrams, 1994), 190, 191.
56. The image is reproduced in Juno and Vale, *Angry Women*, 75 (image courtesy of Carolee Schneemann, photo Anthony McCall). See also Carolee Schneemann, *Imaging Her Erotics: Essays, Interviews, Projects* (Cambridge, MA: MIT Press, 2002).
57. Laura Cottingham, "The Feminist Continuum: Art after 1970," in Broude and Garrard, *Power of Feminist Art*, 276.
58. Francesca Alfano Miglietti, *Extreme Bodies: The Use and Abuse of the Body in Art* (Milan, Italy: Skira, 2003), 170–71.
59. Richard Cork, "British Sculpture in the Late Twentieth Century," in *Breaking the Mould: British Art of the 1980s and 1990s: The Weltkunst Collection* (London: Lund Humphries, 1997), 14.
60. Cork, "British Sculpture," 24.
61. John Pultz, *The Body and the Lens: Photography 1893 to the Present* (New York: Abrams, 1995), 144–47.

62. *Don't Leave Me This Way: Art in the Age of AIDS*, exhibition catalogue compiled by Ted Gott (Melbourne: National Gallery of Australia and Thames and Hudson, 1994), 91ff.
63. Cottingham, "Feminist Continuum," 285.
64. Robert Atkins and Thomas W. Sokolowski, *From Media to Metaphor: Art about AIDS* (New York: Independent Curators, 1992), 17.
65. Douglas Crimp, "Cultural Analysis/Cultural Activism," in *Melancholia and Moralism: Essays on AIDS and Queer Politics* (Cambridge, MA: MIT Press, 2002), 27–41.
66. Atkins and Sokolowski, *From Media to Metaphor*, 65.
67. Susan Sontag, *Illness as Metaphor and Aids and Its Metaphors* (New York: St. Martin's, 2001).
68. ORLAN, "What's All This Body Art?" *Flash Art*, 26, no. 168 (January/February 1993): 50–53, quoted in Kate Ince, *Orlan: Millenial Female* (Oxford: Berg, 2000), 102.
69. Miglietti, *Extreme Bodies*, 197–99.
70. Martin Kemp and Marina Wallace, *Spectacular Bodies: The Art and Science of the Human Body from Leonardo to Now* (Berkeley: University of California Press, 2000), 155.
71. Norman Rosenthal, "The Blood Must Continue to Flow," in *Young British Artists from the Saatchi Collection* (London: Royal Academy of the Arts, 1997), 11.
72. Wu Hung, "Between Past, Present and Future: A Brief History of Contemporary Chinese Photography," in *Between Past, Present and Future: New Photography and Video from China*, ed. Wu Hung and Christopher Phillips (Göttingen, Germany: Steidl, 2004), 30.
73. Kemp and Wallace, *Spectacular Bodies*, 157.
74. Amelia Jones, *Body Art: Performing the Subject* (Minneapolis: University of Minnesota Press, 1998), 197ff.
75. Donna Haraway, "A Cyborg Manifesto: Science, Technology, and Socialist-Feminism in the Late Twentieth Century," in *Simians, Cyborgs and Women: The Reinvention of Nature* (London: Routledge, 1991), 149–81, quoted in Jones, *Body Art*, 204.
76. Dominick La Capra, "Trauma, Absence, Loss," *Critical Inquiry* 25 (1999): 696–727.
77. Jones, *Body Art*, 238–40.

Chapter 10

1. From an interview with Douglas Keay, "Aids, Education and the Year 2000!" *Women's Own Magazine*, October 31, 1987, 8–10.
2. E. J. Hundert, "The European Enlightenment and the History of the Self," in *Rewriting the Self: Histories from the Renaissance to the Present*, ed. Roy Porter (London: Routledge, 1997), 72–83; and Thomas Hobbes, *The Elements of Law Natural and Politic*, ed. F. Toennies and M. M. Goldsmith (London: Simpkin, Marshall, 1889).
3. For a summary of prison reform, and bibliographic notes, see Daniel M. Vyleta, "The Cultural History of Crime," in *A Companion to Nineteenth Century Europe*, ed. Stephan Berger (London: Blackwell, 2005), chap. 27.
4. For all the similarities to Enlightenment models of selfhood, one should, however, point out that psychoanalytic theory was distinct in positing a self whose con-

scious, rational component was ignorant (and in strenuous denial) of the basic primordial drives that governed its desires. This aspect of the theory gained further prominence with the publication of *The Ego and the Id* in 1923; here, the introduction of the superego also made explicit the importance of "socialization" for the Freudian theory of the self. Further, one should note that Freud consistently allowed for "constitutive" (i.e., biological) factors to influence and even shape the psychic development of a given individual. These aspects of his theory partake in a biological materialism that was an important strand of nineteenth-century narratives of the self (see below).

5. In a Foucauldian view the theoretical articulation of socially penetrated selves lagged considerably behind the creation of a "disciplinary machinery" (in schools, the military, hospitals, etc.) that socially configured bodies. Cf. Michel Foucault, *Discipline and Punish: The Birth of the Prison*, trans. Alan Sheridan (New York: Vintage, 1977).
6. Georg Simmel, "Die Grossstädte und das Geistesleben," *Die Grossstädte: Vorträge und Aufsätze zur Städteausstellung; Jahrbuch der Gehe-Stiftung zu Dresden* 9 (1902–1903): 185–206.
7. Émile Durkheim, *The Elementary Forms of Religious Life*, trans. Carol Cosman (1912; Oxford: Oxford University Press, 2001), 11–21, 335–43.
8. For Durkheim, see the comments in ibid., xxvi. For Marx, see especially Karl Marx and Friedrich Engels, *Die deutsche Ideologie* [*The German Ideology*] (Moscow: David Riazanov, 1932). For Simmel, note the Hegelian overtones of the final sections in his essay on "The Metropolis and Mental Life."
9. For an introduction to the topic of degeneracy within two national contexts, see Robert A. Nye, *Crime, Madness and Politics in Modern France: The Medical Concept of National Decline* (Princeton, NJ: Princeton University Press, 1984); and Daniel Pick, *Faces of Degeneration* (Cambridge: Cambridge University Press, 1989). For a clinical history, see Edward Shorter, *A History of Psychiatry* (New York: Basil Wiley, 1997), 93–99.
10. From Alfred E. Hoche, *Krieg und Seelenleben* (Leipzig, Germany: Speyer & Koerner, 1915), as quoted in Paul Lerner, *Hysterical Men, War, Psychiatry, and the Politics of Trauma in Germany, 1890–1930* (Ithaca, NY: Cornell University Press, 2003), 51.
11. Cf. Tim Armstrong, *Modernism, Technology, and the Body: A Cultural Study* (Cambridge: Cambridge University Press, 1998); and John Jervis, *Exploring the Modern* (Oxford: Blackwell, 1998), 202ff.
12. On the theorizations of "digestive efficiency" and their implementation in Russia and elsewhere, see Douglas Greenfield, "Constipation and Utopia" (paper presented at the MLA Annual Convention, San Diego, December 2003).
13. Karel Čapek, *R.U.R.*, trans. Claudia Novack (London: Penguin, 2004), xii.
14. Yevgeny Zamyatin, *We*, trans. Clarence Brown (London: Penguin, 1993), 13.
15. Ibid., 99.
16. Ibid., 124.
17. On the making of and critical reception of *Metropolis*, see Michael Minden and Holger Bachmann, eds., *Fritz Lang's* Metropolis: *Cinematic Visions of Technology and Fear* (Rochester, NY: Camden House, 2000).
18. Ibid., 54–56.
19. It might be noted, though, that Freud's own contributions to this development, most notably *Das Unbehagen in der Kultur* [*Civilisation and Its Discontents*]

(Vienna: Psychoanalytischer Verlag, 1930), painted a more ambivalent picture, in which civilizatory progress and feelings of individual disquiet were necessary correlatives.

20. Otto Gross, *Von geschlechtlicher Not zur sozialen Katastrophe* (Hamburg: Edition Nautilus, 2000). Cf. Emanuel Hurwitz, *Otto Gross, Paradies Sucher zwischen Jung und Freud, Leben und Werk* (Zurich: Suhrkamp, 1979), 36ff.; Jacques Le Rider, "Hans Gross, criminologue, et son fils Otto Gross, 'délinquant sexuel' et psychoanalyste," in *Kulturwissenschaft im Vielvölkerstaat: Zur Geschichte der Ethnologie und verwandter Gebiete in Österreich ca. 1780–1918*, ed. Britta Rupp-Eisenreich und Justin Stegl (Vienna: Böhlau, 1995), 229–40.
21. Concluding remarks by Hess, made at the closing of the party congress. *Triumph of the Will*, DVD, directed by Leni Riefenstahl (1935; Detroit: Synapse Films, 2001).
22. Detlev J.K. Peukert, "The Genesis of the 'Final Solution' from the Spirit of Science," reprinted in *Nazism and German Society, 1933–1945*, ed. David F. Crew (London: Routledge, 1994), 274–99; Paul Weindling, *Health, Race and German Politics between National Unification and Nazism, 1870–1945* (Cambridge: Cambridge University Press, 1989); Stefan Kühl, *Die Internationale der Rassisten: Aufstieg und Niedergang der internationalen Bewegung für Eugenik und Rassenhygiene im 20. Jahrhundert* (Frankfurt am Main: Campus, 1997); and Peter Weingart, Jürgen Kroll, and Kurt Bayertz, *Rasse, Blut und Gene: Geschichte der Eugenik und Rassenhygiene in Deutschland* (Frankfurt am Main: Suhrkamp, 1988).
23. For a strong formulation of this thesis, see Michael Burleigh, *The Third Reich: A New History* (London: Pan, 2000). See also Michael Burleigh and Wolfgang Wippermann, *The Racial State: Germany 1933–1945* (Cambridge: Cambridge University Press, 1991).
24. Richard Wetzell, *Inventing the Criminal: A History of German Criminology, 1880–1945* (Chapel Hill: University of North Carolina Press, 2000), 256ff. See also Nikolaus Wachsmann, *Hitler's Prisons: Legal Terror in Nazi Germany* (New Haven, CT: Yale University Press, 2004).
25. Claudia Koonz, *Mothers in the Fatherland* (New York: St. Martin's, 1987). For a look at Nazi "antinatalism," aimed against the hereditarily tainted, see Gisela Bock, "Antinatalism, Maternity and Paternity in National Socialist Racism," in Crew, *Nazism and German Society*, 110–40.
26. Thomas Koch, *Zwangssterilisation im Dritten Reich: Das Beispiel der Uni-Frauenklinik Göttingen* (Frankfurt am Main: Mabuse, 1994); and Christian Ganssmüller, *Die Erbgesundheitspolitik des Dritten Reiches* (Cologne: Böhlau, 1987).
27. Michael Zimmermann, *Rassenutopie und Genozid: Die nationalsozialistische "Lösung der Zigeunerfrage"* (Hamburg, Germany: Wallstein, 1996), 139–46.
28. Wolfgang Ayass, *"Asoziale" im Nationalsozialismus* (Stuttgart, Germany: Klett-Cotta, 1995).
29. Ernst Klee, *Euthanasie im NS Staat* (Frankfurt am Main: Fischer, 1985).
30. Raul Hilberg, *Die Vernichtung der europäischen Juden*, rev. ed. (Frankfurt am Main: Fischer, 1990); and Christopher Browning, *The Origins of the Final Solution: The Evolution of Nazi Jewish Policy, September 1939 to March 1942* (Lincoln: University of Nebraska Press, 2004).
31. Cf. Ian Kershaw, *Hitler, 1889–1936: Hubris* (London: Penguin, 1999), 569–73.
32. Quoted in Burleigh, *Third Reich*, 660–61.
33. Georg Lilienthal, *Der "Lebensborn e.V."* (Frankfurt am Main: Fischer, 2003).

34. Eric A. Johnson, *Nazi Terror: The Gestapo, Jews, and Ordinary Germans* (New York: Basic Books, 1999), 18–21; Klaus-Michael Mallmann and Gerhard Paul, "Omniscient, Omnipotent, Omnipresent? Gestapo, Society and Resistance," in Crew, *Nazism and German Society*, 166–96; and Robert Gellatelly, "Denunciations in Twentieth-Century Germany," in *Accusatory Practices: Denunciation in Modern European History, 1789–1989*, ed. Sheila Fitzpatrick and Robert Gellately (Chicago: University of Chicago Press, 1997), 185–221.
35. On the question of a "social revolution," the popular success of Nazi ideology, "atomization," the German attitude toward Jews, and so on, see Richard Bessel, ed., *Life in the Third Reich* (Oxford: Oxford Paperbacks, 2001); Detlev J.K. Peukert, *Inside Nazi Germany: Conformity and Opposition in Everyday Life* (New Haven, CT: Yale University Press, 1989); and David Bankier, *The Germans and the Final Solution: Public Opinion under Nazism* (London: Blackwell, 1996).
36. As quoted in Ronald Grigor Suny, *The Soviet Experiment: Russia, the USSR, and the Successor States* (Oxford: Oxford University Press, 1998), 207.
37. Sheila Fitzpatrick, *The Commissariat of Enlightenment, Soviet Organisation of Education and the Arts under Lunacharsky* (Cambridge: Cambridge University Press, 1970); and Abbot Gleason, Peter Kenez, and Richard Stites, eds., *Bolshevik Culture: Experiment and Order in the Russian Revolution* (Bloomington: Indiana University Press, 1985).
38. Wendy Z. Goldman, *Women, the State, and Revolution: Soviet Family Policy and Social Life, 1917–1936* (Cambridge: Cambridge University Press, 1993); and Richard Stites, *The Women's Liberation Movement in Russia: Feminism, Nihilism, and Bolshevism, 1860–1930* (Princeton, NJ: Princeton University Press, 1978).
39. Richard Stites, *Russian Popular Culture: Entertainment and Society since 1900* (Cambridge: Cambridge University Press, 1992); James von Geldern and Richard Stites, eds., *Mass Culture in Soviet Russia* (Bloomington: Indiana University Press, 1995); Svetlana Boym, *Common Places: Mythologies of Everyday Life in Russia* (Cambridge, MA: Harvard University Press, 1994); and Peter Kenez, *Cinema and Soviet Society, 1917–1953* (London: I. B. Tauris, 2001).
40. Lewis H. Siegelbaum, *Stakhanovism and the Politics of Productivity in the USSR, 1935–1941* (Cambridge: Cambridge University Press, 1988); and Sheila Fitzpatrick, "Signals from Below: Soviet Letters of Denunciation of the 1930s," in Fitzpatrick and Gellately, *Accusatory Practices*, 85–120.
41. Sheila Fitzpatrick, "Ascribing Class: The Constitution of Social Identity in Soviet Russia," *Journal of Modern History* 65 (1993): 745–70; Stephen Kotkin, *Magnetic Mountain: Stalinism as a Civilisation* (Berkeley: University of California Press, 1995); and Peter Holquist, "To Count, to Extract, to Exterminate: Statistics and Population Politics in Late Imperial and Soviet Russia," in *A State of Nations: Empire and Nation-Making in the Age of Lenin and Stalin*, ed. T. Martin and R. Suny (Oxford: Oxford University Press, 2001), 111–44.
42. Jochen Hellbeck, "Fashioning the Stalinist Soul: The Diary of Stephan Podlubnyi (1931–1939)," *Jahrbücher für Geschichte Osteuropas* 44, no. 3 (1996): 344–73.
43. Kotkin, *Magnetic Mountain*, 198–237. Kotkin's position is actually more sophisticated, arguing for both internalization and opportunism. The phrase *speaking Bolshevik* is his. See also I. Halfin and J. Hellbeck, "Rethinking the Stalinist Subject: Stephan Kotkin's 'Magnetic Mountain' and the State of Soviet Historical Studies," *Jahrbücher für Geschichte Osteuropas* 44 (1996): 456–63.

44. Sheila Fitzpatrick, *Everyday Stalinism: Ordinary Life in Extraordinary Times, Soviet Russia in the 1930s* (Oxford: Oxford University Press, 1999); Daniel M. Vyleta, "City of the Devil, Bulgakovian Moscow and the Search for the Stalinist Subject," *Rethinking History* 4, no. 1 (2000): 37–53; and Slavoj Žižek, *The Sublime Object of Ideology* (London: Verso, 1989).
45. William Gibson, *Neuromancer* (New York: Berkeley Publishing Group, 1984). The term *cyberpunk* was coined in a 1983 short story by Bruce Bethke, and the movement was first defined in an article by Gardner Dozois in the same year.
46. Norman Spinrad, "The Neuromantics," *Isaac Asimov's Science Fiction Magazine*, May 1986, 186.
47. Stephenson, *The Diamond Age* (New York: Bantam, 1995).
48. Banks, *Consider Phlebas* (London: Macmillan, 1987); *The Player of Games* (London: Macmillan, 1988); *Use of Weapons* (London: Orbit, 1990); *The State of the Art* (London: Orbit, 1991); *Excession* (London: Orbit, 1996); *Inversions* (London: Orbit, 1998); *Look to Windward* (London: Orbit, 2000); and *Matter* (London: Orbit, 2008).

BIBLIOGRAPHY

Ad Hoc Committee of Harvard Medical School. "A Definition of Irreversible Coma." *Journal of the American Medical Association* 210 (1968): 337–40.

Adam, David. "What Is a Green Burial?" *The Guardian*, July 8, 2004.

Adams, Parveen. *The Emptiness of the Image*. London: Routledge, 1996.

Addiego, F., E. G. Belzer, J. Comolli, W. Moger, J. D. Perry, and B. Whipple. "Female Ejaculation: A Case Study." *Journal of Sex Research* 17 (1981): 13–21.

Ahmed, Leila. *Women and Gender in Islam: Roots of an Historical Debate*. New Haven, CT: Yale University Press, 1992.

Albury, Katherine. "Homie-Erotica: Heterosexual Female Desire in *The Picture*." *Media International Australia* 84 (1997): 19–27.

Albury, W. R. "Ideas of Life and Death." In *Companion Encyclopedia of the History of Medicine*, edited by William F. Bynum and Roy Porter, 249–80. London: Routledge, 1993.

Alexander, Elizabeth. " 'Can You Be BLACK and Look at This?' Reading the Rodney King Video(s)." In *Black Male: Representations of Black Masculinity in Contemporary American Art*, edited by Thelma Golden, 91–110. New York: Whitney Museum of Art and Abrams, 1994.

Alexander, M. Darsie, Mary Chan, and Starr Figura. *Body Language*. New York: Museum of Modern Art and Harry Abrams,.

Althaus, Frances. "Female Circumcision: Rite of Passage or Violation of Rights?" *International Family Planning Perspectives* 23, no. 3 (1997): 130–33.

Amsterdamska, Olga, and Anja Hiddinga. "The Analyzed Body." In Cooter and Pickstone, *Companion to Medicine*, 417–33.

Anderson, Warwick. *Colonial Pathologies: American Tropical Medicine, Race and Hygiene in the Philippines*. Durham, NC: Duke University Press, 2006.

Anderson, Warwick. "Excremental Colonialism: Public Health and the Poetics of Pollution." *Critical Enquiry* 21 (1995): 640–69.

Anderson, Warwick. "The Third-World Body." In Cooter and Pickstone, *Companion to Medicine*, 235–45.

Anderson, Warwick. "The Trespass Speaks: White Masculinity and Colonial Breakdown." *American Historical Review* 102 (1997): 1343–70.

Ansell-Pearson, Keith. *Viroid Life: Perspectives on Nietzsche and the Transhuman Condition*. New York: Routledge, 1997.

Arden Hoppe, Kirk. *Lords of the Fly: Sleeping Sickness Control in British East Africa, 1900–1960*. Westport, CT: Praeger, 2003.

Arditti, Rita. "Feminism and Science." In *Science and Liberation*, edited by Rita Arditti, Pat Brennan, and Steve Cavrak, 350–68. Boston: South End Press, 1979.

"Armless Iraqi Boy Bears No Grudges for U.S. Bombing." Reuters, August 11, 2003. http://www.iraqbodycount.org/analysis/reference/ibc-in-the-media/reuters_11aug2003.php.

Armstrong, David. "The Temporal Body." In Cooter and Pickstone, *Companion to Medicine*, 247–59.

Armstrong, David, S. Michie, and T. Marteau. "Revealed Identity: A Study in the Process of Genetic Counselling." *Social Science and Medicine* 47 (1998): 1653–58.

Armstrong, Tim. *Modernism, Technology, and the Body: A Cultural Study*. Cambridge: Cambridge University Press, 1998.

Arnold, David. *Colonizing the Body: State Medicine and Epidemic Disease in Nineteenth-Century India*. Berkeley: University of California Press, 1993.

Arnold, David, ed. *Imperial Medicine and Indigenous Societies*. Manchester, UK: Manchester University Press, 1988.

Arnold, David. *Warm Climates and Western Medicine: The Emergence of Tropical Medicine, 1500–1900*. Amsterdam: Rodopi, 1996.

Arnold, Eugene A. "Autopsy: The Final Diagnosis." In *Images of the Corpse: From the Renaissance to Cyberspace*, edited by Elizabeth Klaver. Madison: University of Wisconsin Press, 2004.

Artaud, Antonin. *To Have Done by the Judgement of God*. Translated by Clayton Eshleman and Norman Glass. 1948. Los Angeles: Black Sparrow Press, 1975.

Atkins, Robert, and Thomas W. Sokolowski. *From Media to Metaphor: Art about AIDS*. New York: Independent Curators, 1992.

Austin, J. L. *How to Do Things with Words: The William James Lectures Delivered at Harvard University in 1955*. Oxford: Clarendon, 1970.

Ausubel, Jesse H., Perrin S. Meyer, and Iddo K. Wernick. "Death and the Human Environment: The United States in the 20th Century." *Technology in Society* 23 (2001): 134–46.

Ayass, Wolfgang. *"Asoziale" im Nationalsozialismus*. Stuttgart, Germany: Klett-Cotta, 1995.

Ayers, Robert. "Serene and Happy and Distant: An Interview with Orlan." *Body and Society* 5 (1999): 171–84.

Bachelard, Gaston. *The New Scientific Spirit*. Translated by Arthur Goldhammer. 1934. Boston: Beacon Books, 1984.

Back, Les. *Inscriptions of Love*. London: Polity, 2004.

Baden-Powell, Robert. *Scouting for Boys: A Handbook for Instruction in Good Citizenship*. London: Horace Cox, 1908.

Bagheri, A., and S. Shoji. "The Model and Moral Justification for Organ Procurement in Japan." *Journal Internationale de Bioéthique* 16 (2005): 79–90, 194–95.

Bailey, Lucy. "Refracted Selves? A Study of Changes in Self-Identity in the Transition to Motherhood." *Sociology* 33 (1999): 335–52.

Baird, Patricia A. "Identification of Genetic Susceptibility to Common Diseases." *Perspectives in Biology and Medicine* 45 (2002): 516–28.

Bakhtin, Mikhail. *Problems of Dostoevsky's Politics*. Ann Arbor: University of Michigan Press, 1984.

Balfour, Andrew, and Henry H. Scott. *Health Problems of Empire*. London: W. Collins and Sons, 1924.

Ballantyne, Tony, and Antoinette Burton, eds. *Bodies in Contact*. Durham, NC: Duke University Press, 2005.

Bankier, David. *The Germans and the Final Solution: Public Opinion under Nazism*. London: Blackwell, 1996.

Banks, Ian M. *Consider Phlebas*. London: Macmillan, 1987.

Banks, Ian M. *Excession*. London: Orbit, 1996.

Banks, Ian M. *Inversions*. London: Orbit, 1998.

Banks, Ian M. *Look to Windward*. London: Orbit, 2000.

Banks, Ian M. *Matter*. London: Orbit, 2008.

Banks, Ian M. *The Player of Games*. London: Macmillan, 1988.

Banks, Ian M. *The State of the Art*. London: Orbit, 1991.

Banks, Ian M. *Use of Weapons*. London: Orbit, 1990.

Barcan, Ruth. "Home on the Rage: Nudity, Celebrity, and Ordinariness in the Home Girls/Blokes Pages." *Continuum: Journal of Media and Cultural Studies* 14 (2000): 145–58.

Barcan, Ruth. "In the Raw: 'Home-Made' Porn and Reality Genres." *Journal of Mundane Behavior* 3 (2002), http://www.mundanebehavior.org/issues/v3n1/barcan.htm (accessed September 28, 2007).

Barcan, Ruth. "The Moral Bath of Bodily Unconsciousness: Female Nudism, Bodily Exposure and the Gaze." *Continuum: Journal of Media and Cultural Studies* 15 (2001): 305–19.

Barcan, Ruth. "Regaining What Mankind Has Lost through Civilisation: Early Nudism and Ambivalent Moderns." *Fashion Theory: The Journal of Dress, Body and Culture* 8 (2004): 1–20.

Barker, Francis. *The Tremulous Private Body: Essays on Subjection*. Ann Arbor: University of Michigan Press, 1995.

Barnes, Barry. *The Nature of Power*. Cambridge, UK: Polity, 1988.

Barnes, Barry. "Social Life as Bootstrapped Induction." *Sociology* 17 (1983): 524–45.

Barnes, Barry. *T. S. Kuhn and Social Science*. London: MacMillan, 1982.

Barron, Stephanie. "Modern Art and Politics in Prewar Germany." In Los Angeles County Museum of Art, *Degenerate Art*, 9–23.

Barry, Andrew. *Political Machines: Governing a Technological Society*. London and New York: Athlone, 2001.

Barry, John M. *The Great Influenza: The Epic Story of the Deadliest Plague in History*. New York: Viking, 2004.

Barthes, Roland. *Mythologies*. Translated by Annette Lavers. New York: Hill and Wang, 1984.

Bashford, Alison. *Medicine at the Border: Disease, Globalization and Security, 1850 to the Present*. Basingstoke, UK: Palgrave Macmillan, 2006.

Bashford, Alison, and Carolyn Strange. "Public Pedagogy: Sex Education and Mass Communication in the Mid Twentieth Century." *Journal of the History of Sexuality* 13 (2004): 71–99.

Bataille, Georges. *Story of the Eye*. 1928. Harmondsworth, UK: Penguin Books, 1982.

Batlaev, Danzig, Alexei Plutser-Sarno, and Sergei Vesiliev. *Russian Criminal Tattoo Encyclopedia*. Saint Petersburg, Russia: Steidl Fuel, 2003.

Bauman, Zygmunt. *Life in Fragments: Essays in Postmodern Morality*. Oxford: Blackwell, 1995.

Becker, Frank. "Sportsmen in the Machine World: Models for Modernization in Weimar Germany." *International Journal of the History of Sport* 12, no. 1 (1995): 153–68.

Becker, Robert O., and Gary Selden. *The Body Electric: Electromagnetism and the Foundation of Life*. New York: Morrow, 1985.

Beecham, Linda. "BMA Opposes Any Change to Law on Assisted Suicide." *British Medical Journal* 325 (2002): 66.

Bell, Clive. "The Rise and Decline of Cubism." *Vanity Fair*, 1923. Reprinted in *Modernism in the 1920s: Interpretations of Modern Art in New York from Expressionism to Constructivism*, by Susan Noyes Platt. Ann Arbor, MI: UMI Research Press, 1983.

Benjamin, Walter. "To the Planetarium." In *One Way Street*. In *Walter Benjamin: Selected Writings*. Vol. 1, *1913–1926*, edited by Marcus Bullock and Michael W. Jennings. Cambridge, MA: Belknap Press of Harvard University Press, 1996.

Benson, Susan. "Inscriptions of the Self: Reflections on Tattooing and Piercing in Contemporary Euro-America." In *Written on the Body: The Tattoo in European and American History*, edited by Jane Caplan, 234–55. London: Reaktion Books, 2000.

Benthien, Claudia. *Skin: On the Cultural Border between Self and the World*. Translated by Thomas Dunlap. New York: Columbia University Press, 2002.

Bergson, Henri. *Matter and Memory*. Translated by Nancy Margaret Paul and W. Scott Palmer. London: Allen and Unwin, 1911.

Bersin, Alissa. "My Father's Pen: Writing, the Body, and Female Pleasure in Helene Cixous's *Inside* and Peter Greenaway's *The Pillow Book of Nogiko*." Bachelor's thesis, Harvard University, 2005.

Bessel, Richard, ed. *Life in the Third Reich*. Oxford: Oxford Paperbacks, 2001.

Bhattacharya, Sanjoy. *Expunging Variola: The Control and Eradication of Smallpox in India, 1947–1977*. New Delhi and London: Orient Longman, 2006.

Bibbings, Lois, and Peter Alldridge. "Sexual Expression, Body Alteration, and the Defence of Consent." *Journal of Law and Society* 20 (1999): 356–70.

Biddle, Jennifer. "Country, Skin, Canvas: The Intercorporeal Art of Kathleen Petyarre." *Australian and New Zealand Journal of Art* 4, no. 1 (2003): 61–76.

Biddle, Jennifer. "Inscribing Identity: Skin as Country in the Central Desert." In *Thinking Through the Skin*, edited by S. Ahmed and J. Stacey, 17–27. London: Routledge, 2001.

Billings, Dwight B., and Thomas Urban. "The Socio-medical Construction of Transsexualism: An Interpretation and Critique." *Social Problems* 19 (1982): 266–82.

Birchall, Ian H. *Sartre against Stalinism*. New York: Berghahn, 2004.

Bird-Rose, Deborah. *Dingo Makes Us Human: Life and Land in an Australian Aboriginal Culture*. Cambridge: Cambridge University Press, 1992.

Bloch, Iwan. *Beiträge zur Aetiologie der Psychopathia Sexualis*. 2 vols. Dresden, Germany: H. R. Dohrn, 1902–1903.

Bloch, Marc. *The Royal Touch: Sacred Monarchy and Scrofula in England and France*. Translated by J. E. Anderson. 1924. London: Routledge and Kegan Paul, 1973.

Bloor, David. *Wittgenstein: Rules and Institutions*. London: Routledge, 1997.

Blum, Linda M., and Nena F. Stracuzzi. "Gender in the Prozac Nation: Popular Discourse and Productive Femininity." *Gender and Society* 18, no. 3 (2004): 269–86.

Blum, Virginia L. *Flesh Wounds: The Culture of Cosmetic Surgery.* Berkeley: University of California Press, 2003.

Bock, Gisela. "Antinatalism, Maternity and Paternity in National Socialist Racism." In Crew, *Nazism and German Society*, 110–40.

Boddy, Janice. "Womb as Oasis: The Symbolic Context of Pharaonic Circumcision in Rural Northern Sudan." *American Ethnologist* 9 (1982): 682–98.

Bonnell, Victoria E. *Iconography of Power: Soviet Political Posters under Lenin and Stalin.* Berkeley: University of California Press, 1997.

Borck, Cornelius. "Anatomien medizinischer Erkenntnis: Der Aktionsradius der Medizin zwischen Vermittlungskrise und Biopolitik." In *Anatomien medizinischen Wissens: Medizin—Macht—Moleküle*, edited by Cornelius Borck, 9–52. Frankfurt am Main: Fischer, 1996.

Bordo, Susan. *The Male Body: A New Look at Men in Public and in Private.* New York: Farrar, Straus and Giroux, 1999.

Bordo, Susan. *Unbearable Weight: Feminism, Western Culture and the Body.* Berkeley: University of California Press, 1993.

Bos, Michael A. *The Diffusion of Heart and Liver Transplantation across Europe.* London: King's Fund Centre, 1991.

Bourdieu, Pierre. *Outline of a Theory of Practice.* Cambridge: Cambridge University Press, 1977.

Bourke, Joanna. *Dismembering the Male: Men's Bodies, Britain and the Great War.* London: Reaktion, 1996.

Bourke, Joanna. *An Intimate History of Killing: Face-to-Face Killing in Twentieth Century Warfare.* London: Granta, 1999.

Bourke, Joanna. *Working-Class Cultures in Britain, 1890–1960.* London: Routledge, 1994.

Boym, Svetlana. *Common Places: Mythologies of Everyday Life in Russia.* Cambridge, MA: Harvard University Press, 1994.

Brakeslee, Sandra. " 'Rewired' Ferrets Overturn Theories of Brain Growth." *New York Times*, April 25, 2000. http://web.mit.edu/msur/www/nytimes.html.

Braziel, Jana Evans, and Kathleen LeBesco, eds. *Bodies Out of Bounds: Fatness and Transgression.* Berkeley: University of California Press, 2001.

Breazeale, Kenon. "In Spite of Women: *Esquire* Magazine and the Construction of the Male Consumer." *Signs* 20, no. 1 (1994): 1–22.

Breslin, James E.B. "Out of the Body: Mark Rothko's Painting." In *The Body Imaged: The Human Form and Visual Culture since the Renaissance*, edited by Kathleen Adler and Marcia Pointon. Cambridge: Cambridge University Press, 1993.

"British Painting." *The Studio*, 1931, 150–51.

Broome, Karl. "Kids Love Ink: London's Youth Tattoo Milieux." PhD diss., Goldsmiths College, University of London, 2006.

Browning, Christopher. *The Origins of the Final Solution: The Evolution of Nazi Jewish Policy, September 1939 to March 1942.* Lincoln: University of Nebraska Press, 2004.

Brumberg, Joan Jacobs. *The Body Project: An Intimate History of American Girls.* New York: Random House, 1997.

Bruno, Giuliana. "M Is for Mapping: Art, Apparel, Architecture Is for Peter Greenaway." In *Atlas of Emotion: Journeys in Art, Architecture, and Film.* New York: Verso, 2002.

Bryder, Linda. *Below the Magic Mountain: A Social History of Tuberculosis in Twentieth-Century Britain*. Oxford: Clarendon, 1988.

Buhrich, N. "The Association of Erotic Piercing with Homosexuality, Sadomasochism, Bondage, Festishes and Tattoos." *Archives of Sexual Behavior* 12 (1983): 161–71.

Burke, Timothy. *Lifebouy Men and Lux Women: Commodification, Consumption, and Cleanliness in Modern Zimbabwe*. Durham, NC: Duke University Press, 1996.

Burleigh, Michael. *The Third Reich: A New History*. London: Pan, 2000.

Burleigh, Michael, and Wolfgang Wippermann. *The Racial State: Germany 1933–1945*. Cambridge: Cambridge University Press, 1991.

Busfield, Joan. "Mental Illness." In Cooter and Pickstone, *Companion to Medicine*, 633–51.

Butler, Judith. *Bodies That Matter: On the Discursive Limits of "Sex."* New York: Routledge, 1993.

Butler, Judith. "Foucault and the Paradox of Bodily Inscriptions." *Journal of Philosophy* 86 (1989): 601–7.

Butler, Judith. *Gender Trouble: Feminism and the Subversion of Identity*. London: Routledge, 1990.

Bynum, William F. "Policing Hearts of Darkness: Aspects of the International Sanitary Conferences." *History of Philosophy and Life Sciences* 15 (1993): 421–34.

Bynum, William F. *Science and the Practice of Medicine in the Nineteenth Century*. Cambridge: Cambridge University Press, 1994.

Cahn, Susan K. *Coming on Strong: Gender and Sexuality in Twentieth-Century Women's Sport*. New York: Free Press, 1994.

Canguilhem, Georges. *The Normal and the Pathological*. Translated by Caroline Fawcett. 1943. New York: Zone Books, 1993.

Cantor, David, ed. *Cancer in the Twentieth Century*. Baltimore: Johns Hopkins University Press, 2008.

Cantor, David. "The Diseased Body." In Cooter and Pickstone, *Companion to Medicine*, 347–66.

Čapek, Karel. *R.U.R.* Translated by Claudia Novack. London: Penguin, 2004.

Carden-Coyne, Ana. "American Guts and Military Manhood." In *Cultures of the Abdomen: Diet, Digestion and Fat in the Modern World*, ed. Ana Carden-Coyne and Christopher E. Forth, 71–85. New York: Palgrave, 2005.

Carden-Coyne, Ana. "From Pieces to Whole: The Sexualisation of Muscles in Post War Bodybuilding." In Forth and Crozier, *Body Parts*.

Carden-Coyne, Ana, and Christopher E. Forth, eds. *Cultures of the Abdomen: Diet, Digestion and Fat in the Modern World*. New York: Palgrave, 2005.

Cariani, P. "Emergence and Artificial Life." In *Artificial Life II*, edited by C. G. Langton, C. Taylor, J. D. Farmer, and S. Rasmussen, 775–97. Redwood, CA: Addison-Wesley, 1992.

Cartwright, Lisa, and Brian Goldfarb. "On the Subject of Neural and Sensory Prostheses." In Smith and Morra, *The Prosthetic Impulse*, 125–54.

Casper, Monica J. *The Making of the Unborn Patient: A Social Anatomy of Fetal Surgery*. New Brunswick, NJ: Rutgers University Press, 1998.

Castleman, Michael. "Porn-Star Secrets: Going Naked in Front of the Camera Necessitates Lots of Hair-Removal Tricks." Salon.com, 2000. http://dir.salon.com/sex/feature/2000/09/06/hair_removal/index.html.

Charlesworth, B. "Fisher, Medawar, Hamilton and the Evolution of Aging." *Genetics* 156 (2000): 927–31.

Chase, Marilyn. "To Avoid Surgery, Eat This Camera." *Wall Street Journal*, August 15, 2002. Available at www.cpmc.org/advanced/endoscopy/wsj_article_aug2002.html.

Chipp, Herschel B., ed. *Theories of Modern Art*. Berkeley: University of California Press, 1968.

Cixous, Hélène. "The Laugh of the Medusa." *Signs* 4 (1976): 875–93.

Cixous, Hélène. *Portrait of Jacques Derrida as a Young Saint*. New York: Columbia University Press, 2004.

Clair, Jean. *Identity and Alterity: Figures of the Body 1895–1995*. Venice Bienale. Venice: Marsilio, 1995.

Clarke, Julie. "The Sacrificial Body of Orlan." *Body and Society* 5 (1999): 185–207.

Classen, Constance. "McLuhan in the Rainforest: The Sensory Worlds of Oral Cultures." In *The Empire of the Senses*, edited by David Howes, 147–63. New York: Berg, 2005.

Cocks, Harry. "Saucy Stories: Pornography, Sexology and the Marketing of Sexual Knowledge in Britain, c. 1918–1970." *Social History* 29 (2004): 465–84.

Cohen, Lawrence. "Where It Hurts: Indian Material for an Ethics of Organ Transplantation." *Daedelus* 128 (1999): 135–65.

Colb, Sherry F. "What Is a Mother? The California 'Egg Donor' Case Gets It Wrong." *FindLaw Legal News and Commentary*. 2004. http://writ.news.findlaw.com/colb/20040519.html.

Cole, Anna, and Anna Haebich. "Corporal Punishments and Corporeal Publics: A Cross-Cultural Historical Approach to the Techniques and Cosmologies of Body Modification." Paper presented at Body Modification, Mark II, Macquarie University, Sydney, April 21–23, 2005.

Cole, Thomas R. *The Journey of Life: A Cultural History of Aging in America*. Cambridge: Cambridge University Press, 1992.

Collins, Harry. "The TEA Set: Tacit Knowledge and Scientific Networks." *Science Studies* 4 (1974): 165–86.

Comiskey, Carolyn Ward. "Cosmetic Surgery in Paris in 1926: The Case of the Amputated Leg." *Journal of Women's History* 16 (2004): 30–54.

Comiskey, Carolyn Ward. "'I Will Kill Myself . . . If I Have to Keep My Fat Calves!': Legs and Cosmetic Surgery in Paris in 1926." In Forth and Crozier, *Body Parts*, 247–64.

"Conference of Medical Royal Colleges, 'Diagnosis of Brain Death.'" *British Medical Journal* 273 (1976): 1187–88.

Connell, R. W. *Masculinities*. 2nd ed. Berkeley: University of California Press, 1995.

"Contemporary Figure Painters." *The Studio*, 1925, 1–18.

Cook, Hera. *The Long Sexual Revolution*. New York: Oxford University Press, 2004.

Cooke, Rachel. "Bold, Brave, Beautiful." *The Observer*, September 18, 2005.

Cooter, Roger. "The Dead Body." In Cooter and Pickstone, *Companion to Medicine*, 469–86.

Cooter, Roger, and Bill Luckin. "Accidents in History: An Introduction." In *Accidents in History: Injuries, Fatalities and Social Relations*, edited by Roger Cooter and Bill Luckin, 1–16. Amsterdam and Atlanta: Rodopi, 1997.

Cooter, Roger, and John Pickstone, eds. *Companion to Medicine in the Twentieth Century*. London: Routledge, 2000.

Cooter, Roger, and Steve Sturdy. "Of War, Medicine and Modernity: Introduction." In *War, Medicine and Modernity*, edited by Roger Cooter, Mark Harrison, and Steve Sturdy, 1–21. Phoenix Mill, UK: Sutton, 1998.

Cork, Richard. "British Sculpture in the Late Twentieth Century." In *Breaking the Mould: British Art of the 1980s and 1990s: The Weltkunst Collection*. London: Lund Humphries, 1997.

Cottingham, Laura. "The Feminist Continuum: Art after 1970." In *The Power of Feminist Art: The American Movement of the 1970s, History and Impact*, edited by Norma Broude and Mary D. Garrard, 279–82. New York: Abrams, 1994.

Cowling, Elizabeth, and Jennifer Mundy, eds. *On Classic Ground: Picasso, Léger, de Chirico and the New Classicism, 1910–1930*. London: Tate Gallery, 1990.

Craig, Maxine Leeds. *Ain't I a Beauty Queen? Black Women, Beauty, and the Politics of Race*. New York: Oxford University Press, 2002.

Crapanzano, Vincent. "Rite of Return: Circumcision in Morocco." In *Psychoanalytic Study of Society*, edited by Warner Meunsterberger and L. Bryce Boyer, 9:15–36. New York: Library of Psychological Anthropology, 1981.

Crew, David F., ed. *Nazism and German Society, 1933–1945*. London: Routledge, 1994.

Crimp, Douglas. "Cultural Analysis/Cultural Activism." In *Melancholia and Moralism: Essays on AIDS and Queer Politics*, 27–41. Cambridge, MA: MIT Press, 2002.

Cross, Stephen J., and Randall Albury. "Walter B. Cannon, L. J. Henderson, and the Organic." *Osiris* 3 (1987): 165–92.

Crossley, Nick. "Mapping Reflexive Body Techniques: On Body Modification and Maintenance." *Body and Society* 11 (2005): 1–35.

Crown Agents for the Colonies. *Hints on the Preservation of Health in Tropical Africa*. 1938. London: The Crown Agents for the Colonies, 1943.

Crozier, Anna. "Sensationalising Africa: British Medical Impressions of Sub-Saharan Africa 1890–1939." *Journal of Imperial and Commonwealth History* 35 (2007): 393–415.

Crozier, Anna. "What Was Tropical about Tropical Neurasthenia? The Utility of the Diagnosis in the Management of British East Africa." *Journal of the History of Medicine and Allied Sciences* 64 (2009): 518–48.

Crozier, Ivan. "All the Parts Were Perfectly Natural." In Forth and Crozier, *Body Parts*, 65–84.

Crozier, Ivan. "British Medical Writing before Havelock Ellis: The Missing Story." *Journal for the History of Medicine and Allied Sciences* 63 (2008): 65–102.

Crozier, Ivan. "Introduction: Havelock Ellis, John Addington Symonds and the Construction of *Sexual Inversion*." In *Sexual Inversion: A Critical Edition*, edited by Ivan Crozier, 1–86. Basingstoke, UK, and New York: Palgrave Macmillan, 2008.

Crozier, Ivan. "The Medical Construction of Homosexuality and Its Relation to the Law in Nineteenth-Century England." *Medical History* 45 (2001): 61–82.

Crozier, Ivan. "William Acton and the History of Sexuality: The Professional and Medical Contexts." *Journal of Victorian Culture* 5 (2000): 1–27.

Csordas, Thomas. *The Sacred Self: A Cultural Phenomenology of Charismatic Healing*. Berkeley: University of California Press, 1994.

Cunningham, Andrew. *The Anatomical Tradition: The Resurrection of the Anatomical Projects of the Ancients*. Aldershot, UK: Scolar Press, 1997.

Darby, Robert. *A Surgical Temptation: The Demonisation of the Foreskin in Victorian Medicine*. Chicago: Chicago University Press, 2005.

Davidson, Arnold. *The Emergence of Sexuality*. Cambridge, MA: Harvard University Press, 2001.

Davidson, Roger, and Lesley Hall, eds. *Sex, Sin and Suffering: Venereal Disease and European Society since 1870*. London: Routledge, 2001.
Davis, Kathy, "'A Dubious Equality': Men, Women and Cosmetic Surgery." *Body and Society* 8, no. 1 (2002): 58–60.
Davis, Kathy. *Reshaping the Female Body: The Dilemma of Cosmetic Surgery*. London: Routledge, 1995.
de Francia, Peter. *Fernand Léger*. New Haven, CT, and London: Yale University Press, 1983.
de Grazia, Victoria. *How Fascism Ruled Women: Italy, 1922–1945*. Berkeley: University of California Press, 1992.
De Michelis, Elizabeth. *A History of Modern Yoga: Patañjali and Western Esotericism*. London: Continuum, 2004.
Del Mar, Norman. *Richard Strauss: A Critical Commentary on His Life and Works*. 3 vols. London: Barrie & Rockliff, 1962–1972.
Delaporte, Sophie. *Gueules cassées de la grande guerre*. Paris: Agnès Viénot, 1996.
Deleuze, Gilles, and Felix Guattari. *Capitalism and Schizophrenia: Anti-Oedipus*. 1972. Minneapolis: University of Minnesota Press, 1983.
Deleuze, Gilles, and Felix Guattari. *A Thousand Plateaus: Capitalism and Schizophrenia*, trans. Brian Massumi. 1980. Minneapolis: University of Minnesota Press, 1987.
DeMello, Margo. *Bodies of Inscription: A Cultural History of the Modern Tattoo Community*. Durham, NC: Duke University Press, 2000.
Derrida, Jacques. "Circonfession." In *Jacques Derrida, Collaboration with Geoffrey Bennington*. Chicago: University of Chicago Press, 1991.
Derrida, Jacques. *Given Time* [*Donner le temps*]. Chicago: University of Chicago Press, 1991.
Derrida, Jacques. *Given Death* [*Donner le mort*]. Chicago: University of Chicago Press, 1995.
Deter, R. L., G. T. Caldwell, and A. I. Folsom. "A Clinical and Pathological Study of the Posterior Female Urethra." *Journal of Urology* 55 (1946): 651–21.
Donald, Ian. *Practical Obstetric Problems*. London: Lloyd-Luke, 1969.
Doesburg, Theo van. "Meditations at the Frontier," *De Avondpost*, 1915. Reprinted in Harrison and Wood, *Art in Theory*.
Donne, John. "Divine Poems X." In *The Complete English Poems*, edited by C. A. Patrides. London: Everyman, 1991.
Don't Leave Me This Way: Art in the Age of AIDS. National Gallery of Australia, Thames and Hudson, Melbourne.
Dorling, Daniel. *Death in Britain: How Local Mortality Rates Have Changed: 1950s to 1990s*. York, UK: Joseph Rowntree Foundation, 1997.
Douglas, Mary. *Purity and Danger: An Analysis of the Concepts of Pollution and Taboo*. London: Ark Paperbacks, 1966.
Downing, Lisa. "The Measure of Sexual Dysfunction: A Plea for Theoretical Limitlessness." In "Regions of Sexuality," edited by Iain Morland and Wendy O'Brien. Special issue, *Transformations: Region, Culture, Society* 8 (2004), http://transformations.cqu.edu.au/journal/issue_08/article_02.shtml#return2 (accessed July 2008).
Dreger, Alice Domurat. *Hermaphrodites and the Medical Invention of Sex*. Cambridge, MA: Harvard University Press, 1998.
Dreger, Alice Domurat. *One of Us: Conjoined Twins and the Future of the Normal*. Cambridge, MA: Harvard University Press, 2004.

Drummond, Murray J.N. "Men's Bodies: Listening to the Voices of Young Gay Men." *Men and Masculinities* 7, no. 3 (2005): 270–90.

du Boulay, Shirley. *Cicely Saunders: Founder of the Modern Hospice Movement*. London: Hodder and Stoughton, 1984.

Duden, Barbara. *Disembodying Women: Perspectives on Pregnancy and the Unborn*. Cambridge, MA: Harvard University Press, 1993.

Duden, Barbara. *The Woman beneath the Skin: A Doctor's Patients in Eighteenth-Century Germany*. Translated by Thomas Dunlap. Cambridge, MA: Harvard University Press, 1991.

Dumit, Joseph. *Drugs for Life*. Durham, NC: Duke University Press, forthcoming.

Durbach, Nadja. *Bodily Matters: The Anti-vaccination Movement in England, 1853–1907*. Durham, NC, and London: Duke University Press, 2005.

Durkheim, Émile. *The Elementary Forms of Religious Life*. Translated by Carol Cosman. 1912. Oxford: Oxford University Press, 2001.

Dutton, Kenneth R. *The Perfectible Body: The Western Ideal of Male Physical Development*. New York: Continuum, 1995.

Dwyer, Leslie, and Degung Santikarma. "When the World Turned to Chaos: 1965 and Its Aftermath in Bali, Indonesia." In *The Spector of Genocide: Mass Murder in Historical Perspective*, edited by Robert Gellately and Ben Kiernan, 289–306. Cambridge: Cambridge University Press, 2003.

Dye, Lee. "How Snakes See Two Ways: How Snake Eyes Could Lead to Smarter Missiles and Stop Cancer," ABCnews.com, January 9, [2008?], http://abcnews.go.com/Technology/story?id=98115&page=1.

Dyer, Richard. *White*. London: Routledge, 1997.

Eberle, Matthias. *World War One and the Weimar Artists: Dix, Grosz, Beckmann, Schlemmer*. New Haven, CT, and London: Yale University Press, 1985.

Echenberg, Myron. *Black Death, White Medicine: Bubonic Plague and the Politics of Public Health in Colonial Senegal, 1914–1945*. Portsmouth, NH: Heinemann; Oxford: James Currey; Cape Town: David Philips, 2002.

Edwards, Paul N. *The Closed World: Computers and the Politics of Discourse in Cold War America*. Cambridge, MA: MIT Press, 1996.

El Saadawi, Nawal. *The Hidden Face of Eve*. London: Zed Books, 1980.

Elias, Norbert. *The History of Manners: The Civilizing Process*. Vol. 1. New York: Pantheon Books, 1978.

Eller, Jonathan, and William F. Touponce. *Ray Bradbury: The Life of Fiction*. Kent, OH: Kent State University, 2004.

Ellis, Havelock. "Birth Control and Eugenics." In *The Philosophy of Conflict and Other Essays in War-Time*, 128–41. London: Constable, 1919.

Ellis, Havelock. *Evolution of Modesty*. Philadelphia: F. A. Davis, 1900.

Ellis, Havelock. *Psychology of Sex*. London: Heineman, 1933.

Ellis, Havelock. *Sex in Relation to Society*. Philadelphia: F. A. Davis, 1910.

Ellis, Havelock. *Studies in the Psychology of Sex*. Philadelphia: F. A. Davis, 1935.

Epstein, Steven. *Impure Science: AIDS, Activism, and the Politics of Knowledge*. Berkeley: University of California Press, 1996.

Ettinger, Robert C.W., and R. Michael Perry. *The Prospect of Immortality*. Edited by Charles Tandy. 1965. Palo Alto, CA: Ria University Press, 2005.

Ettorre, Elizabeth. "Reproductive Genetics, Gender and the Body: 'Please Doctor, May I Have a Normal Baby?'" *Sociology* 34 (2000): 403–20.

Ettorre, Elizabeth. *Reproductive Genetics, Gender and the Body*. London: Routledge, 2002.

Eyler, John M. *Victorian Social Medicine: The Ideas and Methods of William Farr*. Baltimore: Johns Hopkins University Press, 1979.

Fanon, Frantz. *Black Skin, White Masks*. Translated by Charles Lam Markmann. New York: Grove, 1967.

Fanon, Frantz. *The Wretched of the Earth*. New York: Grove, 1963.

Farley, John. *Bilharzia: A History of Imperial Tropical Medicine*. Cambridge: Cambridge University Press, 1991.

Farnell, Ross. "In Dialogue with 'Posthuman' Bodies: Interview with Stelarc." *Body and Society* 5 (1000): 129–47.

Featherstone, Mike. "The Body in Consumer Culture." In *The Body: Social Process and Cultural Theory*, edited by Mike Featherstone, Mike Hepworth, and Bryan S. Turner, 170–96. London: Sage, 1991.

Featherstone, Mike, ed. *Body Modification*. London: Sage, 2000.

Featherstone, Mike, and M. Hepworth. "The Mask of Ageing." In *The Body: Social Process and Cultural Theory*, edited by Mike Featherstone, Mike Hepworth, and Bryan S. Turner, 371–89. London: Sage, 1991.

Featherstone, Mike, and A. Wernick. Introduction to *Images of Ageing: Cultural Representations of Later Life*, edited by Mike Featherstone and A. Wernick, 1–19. London: Routledge, 1995.

Feld, Steven. "Places Sensed, Senses Placed." In *Empire of the Senses*, edited by David Howes, 179–91. New York: Berg, 2005.

Feld, Steven. *Sound and Sentiment*. Philadelphia: University of Pennsylvania Press, 1982.

Ferguson, H. "Body Piercing." *British Medical Journal* 319 (1999): 1627–29.

Figlio, Karl. "The Historiography of Scientific Medicine: An Invitation to the Human Sciences." *Comparative Studies in Society and History* 19 (1977): 262–86.

Finnegan, Frances. *Poverty and Prostitution*. Cambridge: Cambridge University Press, 1979.

Fischer, Michael M.J. "Cultural Critique with a Hammer, Gouge, and Woodblock: Art and Medicine in the Age of Social Retraumatization." In Fischer, *Emergent Forms of Life and the Anthropological Voice*, 90–144. Durham, NC: Duke University Press, 2003.

Fischer, Michael M.J. *Emergent Forms of Life and the Anthropological Voice*. Durham, NC: Duke University Press, 2003.

Fischer, Michael M.J. "Legal Postulates in Flux: Justice, Wit and Hierarchy in Iran." In *Law and Politics in the Middle East*, edited by D. Dwyer, 115–42. New York: J. F. Bergin, 1989.

Fischer, Michael M.J. *Mute Dreams, Blind Owls, and Dispersed Knowledges*. Durham, NC: Duke University Press, 2004.

Fischer, Michael M.J. "Technoscientific Intrastructures and Emergent Forms of Life: A Commentary." *American Anthropologist* 107 (2005): 55–61.

Fischer, Michael M.J. "With a Hammer, a Gouge, and a Woodblock: The Work of Art and Medicine in the Age of Social Retraumatization—The Texas Woodcut Art of Dr. Eric Avery." In *Para-Sites: A Casebook against Cynical Reason*, edited by George E. Marcus. Late Editions, 7: Cultural Studies for the End of the Century. Chicago: Chicago University Press, 2000.

Fisher, Kate. *Birth Control, Sex and Marriage in Britain, 1918–1960*. Oxford: Oxford University Press, 2006.

Fitzpatrick, Sheila. "Ascribing Class: The Constitution of Social Identity in Soviet Russia." *Journal of Modern History* 65 (1993): 745–70.

Fitzpatrick, Sheila. *The Commissariat of Enlightenment, Soviet Organisation of Education and the Arts under Lunacharsky*. Cambridge: Cambridge University Press, 1970.

Fitzpatrick, Sheila. *Everyday Stalinism: Ordinary Life in Extraordinary Times, Soviet Russia in the 1930s*. New York: Oxford University Press, 1999.

Fitzpatrick, Sheila. "Signals from Below: Soviet Letters of Denunciation of the 1930s." In *Accusatory Practices: Denunciation in Modern European History, 1789–1989*, edited by Sheila Fitzpatrick and Robert Gellately, 85–120. Chicago: Chicago University Press, 1997.

Folsom, Alfred, and Harold O'Brien. "The Female Obstructing Prostate." *Journal of the American Medical Association* 121 (1943): 573–80.

Forrest, David. " 'We're Here, We're Queer, and We're Not Going Shopping': Changing Gay Male Identities in Contemporary Britain." In *Dislocating Masculinity: Comparative Ethnographies*, edited by Andrea Cornwall and Nancy Lindisfarne, 97–110. London: Routledge, 1994.

Forth, Christopher E., and Ivan Crozier, eds. *Body Parts: Critical Explorations in Corporeality*. Lanham, MD: Lexington Books, 2005.

Forth, Christopher E., and Ivan Crozier. "Introduction: Parts, Wholes and People." In Forth and Crozier, *Body Parts*, 1–16.

Foucault, Michel. "Body/Power." In *Power/Knowledge: Selected Interviews and Other Writings, 1972–1977*, edited by Colin Gordon, 55–62. Brighton, UK: Harvester, 1980.

Foucault, Michel. *Discipline and Punish: The Birth of the Prison*. Translated by Alan Sheridan. New York: Vintage, 1977.

Foucault, Michel. *History of Sexuality I: The Will to Knowledge*. Translated by Robert Hurley. London: Penguin, 1978.

Foucault, Michel. "Nietzsche, Genealogy, History." Translated by D. F. Bouchard. In *Language, Counter-Memory, Practice*, edited by S. Simon and D. F. Bouchard, 139–64. Ithaca, NY: Cornell University Press, 1977. Reprinted in *The Foucault Reader*, edited by Paul Rabinow, 76–100 (London: Penguin, 1984).

Foucault, Michel. "The Simplest of Pleasures." In *Foucault Live: Collected Interviews, 1961–1984*, edited by S. Lotringer, 295–97. New York: SEMIOTEXT(E), 1996. Also available at http://www.thefoucauldian.co.uk/simple.pdf.

Fox, Renée C., and Judith P. Swazey. *The Courage to Fail: A Social View of Organ Transplants and Dialysis*. Chicago: University of Chicago Press, 1974.

Fox, Renée C., and Judith P. Swazey. *Spare Parts: Organ Replacement in American Society*. New York and Oxford: Oxford University Press, 1992.

Freud, Sigmund. *Three Essays on the Theory of Sexuality*. In *Standard Edition*, translated and edited by James Strachey, 125–245. London: Hogarth, 1953.

Freud, Sigmund. *Das Unbehagen in der Kultur* [*Civilisation and Its Discontents*]. Vienna: Psychoanalytischer Verlag, 1930.

Frevert, Ute. *Women in German History: From Bourgeois Emancipation to Sexual Liberation*. Translated by Stuart McKinnon Evans. Oxford: Berg, 1989.

Frost, Laura. *Sex Drives: Fantasies of Fascism in Literary Modernism*. Ithaca, NY: Cornell University Press, 2002.

Frueh, Joanna. "The Body through Women's Eyes." In *The Power of Feminist Art: The American Movement of the 1970s, History and Impact*, edited by Norma Broude and Mary D. Garrard, 190–207. New York: Abrams, 1994.

Fussell, Sam. *Muscle: Confessions of an Unlikely Bodybuilder*. London: Sphere Books, 1991.

Gabo, Naum. "The Constructive Idea in Art." In *Circle: International Survey of Constructive Art*, edited by J. L. Martin, B. Nicholson, and N. Gabo, 1–10. London: Praeger, 1937.

Gaiman, Neil. *Death: The High Cost of Living*. N.p.: DC Comics Vertigo, 1993.

Galison, Peter. "Die Ontologie des Feindes: Norbert Wiener und die Vision der Kybernetik." In *Räume des Wissens, Repräsentation, Codierung, Spur*, edited by Hans-Jörg Rheinberger, Michael Hagner, and Bettina Wahrig-Schmidt, 281–324. Berlin: Akademie Verlag, 1997.

Ganssmüller, Christian. *Die Erbgesundheitspolitik des Dritten Reiches*. Cologne: Böhlau, 1987.

Garb, Tamar. *Bodies of Modernity: Figure and Flesh in Fin-de-Siècle France*. London: Thames and Hudson, 1998.

Gaston, Sawnie R. "The Role of Leadership in the Quality of Fracture Care." *Bulletin for the American College of Surgeons* 60 (November 1975): 16–23.

Gatens, Moira. *Imaginary Bodies: Ethics, Power and Corporeality*. London and New York: Routledge, 1996.

Gates, Henry Louis. "A Liberalism of Heart and Spine." *New York Times*, March 27, 1994.

Geary, James. *The Body Electric: An Anatomy of the New Bionic Senses*. London: Weidenfeld and Nicholson, 2002.

Geldzahler, Henry. *Pop Art 1955–70*. New York: Museum of Modern Art, in conjunction with the International Cultural Corporation of Australia, 1985. An exhibition catalog.

Gell, Alfred. *Wrapping in Images: Tattooing in Polynesia*. Oxford: Oxford University Press, 1993.

Gellatelly, Robert. "Denunciations in Twentieth-Century Germany." In *Accusatory Practices: Denunciation in Modern European History, 1789–1989*, edited by Sheila Fitzpatrick and Robert Gellately, 185–221. Chicago: University of Chicago Press, 1997.

Gershon, Michael. *The Second Brain*. New York: Harper Collins, 1998.

Gibson, William. "Introduction to 'the Body.'" In Smith, *Stelarc*.

Gibson, William. *Neuromancer*. New York: Berkeley Publishing Group, 1984.

Giddens, Anthony. *Modernity and Self-Identity: Self and Society in the Late Modern Age*. Cambridge, UK: Polity, 1991.

Giddens, Anthony. *The Transformation of Intimacy: Sexuality, Love and Eroticism in Modern Societies*. Cambridge, UK: Polity, 1992.

Giddins, Gary. *A Pocketful of Dreams: The Early Years, 1903–1940*. New York: Little, Brown, 2001.

Gill, Walt, and Lucy Gilson. "Reforming the Health Sector in Developing Countries: The Central Role of Policy Analysis." *Health Policy and Planning* 9 (2004): 353–70.

Gilman, Charlotte Perkins. *The Living of Charlotte Perkins Gilman: An Autobiography*. 1935. New York: Arno, 1972.

Gilman, Sander L. *Making the Body Beautiful: A Cultural History of Aesthetic Surgery*. Princeton, NJ: Princeton University Press, 1999.

Gleason, Abbot, Peter Kenez, and Richard Stites, eds. *Bolshevik Culture: Experiment and Order in the Russian Revolution*. Bloomington: Indiana University Press, 1985.

Golan, Romy. *Modernity and Nostalgia: Art and Politics between the Wars*. New Haven, CT, and London: Yale University Press, 1995.

Goldberg, Carey. "Ultra-Tiny Knitting Thread Helps Restore Brain Function." *Boston Globe*, March 20, 2006.

Golden, Janet L. *Message in a Bottle: The Making of Fetal Alcohol Syndrome*. Cambridge, MA: Harvard University Press, 2005.

Golden, Thelma. "My Brother." In *Black Male: Representations of Black Masculinity in Contemporary American Art*, ed. Thelma Golden. New York: Whitney Museum of Art and Abrams, 1994.

Golding, John, and Christopher Green, eds. *Léger and Purist Paris*. London: Tate Gallery, 1970. An exhibition catalog.

Goldman, Wendy Z. *Women, the State, and Revolution: Soviet Family Policy and Social Life, 1917–1936*. Cambridge: Cambridge University Press, 1993.

Goldschmidt, Walter. *The Sebei: A Study in Adaptation*. New York: Holt, Rinehart & Winston, 1986.

Good, Byron. *Medicine, Rationality and Experience*. Cambridge: Cambridge University Press, 1994.

Goodall, Jane. "An Order of Pure Decision: Un-Natural Selection in the Work of Stelarc and Orlan." *Body and Society* 5 (1999): 149–70.

Goodall, Jane. "The Will to Evolve." In Smith, *Stelarc*, 1–32.

Gordon, David. "A Sword of Empire? Medicine and Colonialism at King William's Town, Xhosaland, 1856–91." In *Medicine and the Colonial Identity*, edited by Mary P. Sutphen and Bridie Andrews, 41–60. London: Routledge, 2003.

Gordon, H. L. "The Mental Capacity of the African." *Journal of the African Society* 33 (1934): 226–42.

Gorringe, Hugo, and Irene Rafanell. "The Embodiment of Caste: Oppression, Protest and Change." *Sociology* 41 (2007): 97–114.

Gradmann, Christoph. "'It Seemed About Time to Try One of Those Modern Medicines': Animal and Human Experimentation in the Chemotherapy of Sleeping Sickness 1905–1908." In *Twentieth Century Ethics of Human Subjects Research: Historical Perspectives on Values, Practices, and Regulations*, edited by Giovanni Maio and Volker Roelcke, 83–97. Stuttgart, Germany: Steiner, 2004.

Gräfenberg, Ernest. "The Role of Urethra in Female Orgasm." *International Journal of Sexology* 3 (1950): 145–48.

Greenfield, Douglas. "Constipation and Utopia." Paper presented at the MLA Annual Convention, San Diego, December 2003.

Greenwood, David. *Antimicrobial Drugs*. Oxford: Oxford University Press, 2008.

Greer, Germaine. "Lady, Love Your Cunt." In *The Mad Woman's Underclothes: Essays and Occasional Writings 1968–1985*, 74–77. London: Picador, 1986.

Griffin, John Howard. *Black Like Me*. Boston: Houghton Mifflin, 1961.

Griffin-Bonazzi, Elizabeth. "Griffin, John Howard." The Handbook of Texas Online. http://www.tshaonline.org/handbook/online/articles/GG/fgr99.html.

Griffiths, Paul, Karola Stotz, and Rob Knight, "How Scientists Conceptualise Genes: An Empirical Study." *Studies in History and Philosophy of Biological and Biomedical Sciences* 35 (2004): 647–73.

Gross, Otto. *Von geschlechtlicher Not zur sozialen Katastrophe*. Hamburg: Edition Nautilus, 2000.

Grossman, David. Interview in "Reflections on the Body Politic." By Mark Sorkin. *The Nation*, July 11, 2005. http://www.thenation.com/doc/20050711/grossman.

Grossmann, Anita. *Reforming Sex: The German Movement for Birth Control and Abortion Reform, 1920–1950*. New York: Oxford University Press, 1995.

Grosz, Elizabeth. *Space, Time and Perversion: Essays on the Politics of Bodies*. New York: Routledge; Sydney: Allen and Unwin, 1995.

Grosz, Elizabeth. *Volatile Bodies: Towards a Corporeal Feminism*. Bloomington: Indiana University Press, 1994.

Gullette, Margaret Morganroth. *Aged by Culture*. Chicago: Chicago University Press, 2004.

Günsberg, Maggie. *Italian Cinema: Gender and Genre*. New York: Palgrave Macmillan, 2005.

Gutting, Gary. *Michel Foucault's Archaeology of Scientific Reason*. Cambridge: Cambridge University Press, 1989.

Hacking, Ian. "Making Up People." In *Reconstructing Individuality: Autonomy, Individuality and the Self in Western Thought*, edited by Thomas C. Heller, Morton Sosna, and David E. Wellbery, 222–36. Stanford, CA: Stanford University Press, 1986.

Haiken, Elizabeth. "The Making of the Modern Face: Cosmetic Surgery." *Social Research* 67 (2000): 81–97.

Haiken, Elizabeth. "Plastic Surgery and American Beauty at 1921." *Bulletin for the History of Medicine* 68 (1994): 429–53.

Haiken, Elizabeth. *Venus Envy: A History of Cosmetic Surgery*. Baltimore: Johns Hopkins University Press, 1997.

Haley, Bruce. *The Healthy Body and Victorian Culture*. Cambridge, MA: Harvard University Press, 1978.

Halfin, I., and J. Hellbeck. "Rethinking the Stalinist Subject: Stephan Kotkin's 'Magnetic Mountain' and the State of Soviet Historical Studies." *Jahrbücher für Geschichte Osteuropas* 44 (1996): 456–63.

Hall, Lesley A. "Hauling Down the Double Standard: Feminism, Social Purity and Sexual Science in Late Nineteenth-Century Britain." *Gender and History* 16 (2004): 36–56.

Hall, Lesley A. *Hidden Anxieties: Male Sexuality, 1900–1950*. Cambridge, UK: Polity, 1991.

Hall, Lesley A. *Sex, Gender and Social Change in Britain since 1880*. London: MacMillan, 2000.

Hallowell, Nina. "Reconstructing the Body or Reconstructing the Woman? Perceptions of Prophylactic Mastectomy for Hereditary Breast Cancer Risk." In *Ideologies of Breast Cancer: Feminist Perspectives*, edited by L. Potts, 153–80. London: Macmillan, 2000.

Hallowell, Nina, C. Foster, R. Eeles, A. Ardern-Jones, and M. Watson. "Accommodating Risk: Women's Responses to BRCA1/2 Genetic Testing Following a Cancer Diagnosis." *Social Science and Medicine* 59 (2004): 553–65.

Halperin, David. *How to Do the History of Homosexuality*. Chicago: Chicago University Press, 2002.

Haraway, Donna. *Modest_Witness@Second_Millennium.FemaleMan©Meets _OncoMouse: Feminism and Technoscience*. New York: Routledge, 1997.

Haraway, Donna. *Simians, Cyborgs, and Women: The Reinvention of Nature*. London: Routledge, 1991.

Harrison, Charles, and Paul Wood, eds. *Art in Theory, 1900–1990: An Anthology of Changing Ideas*. Oxford: Blackwell, 1995.

Harrison, Mark. " 'The Tender Frame of Man': Disease, Climate, and Racial Difference in India and the West Indies, 1760–1860." *Bulletin of the History of Medicine* 70 (1996): 68–93.

Hass, Wolf. *Komm, Süsser Tod*. Vienna: Rowohlt, 1998.

Hau, Michael. *The Cult of Health and Beauty in Germany: A Social History, 1890–1930*. Chicago: Chicago University Press, 2003.

Hausman, Bernice L. *Changing Sex: Transsexualism, Technology, and the Idea of Gender*. Durham, NC, and London: Duke University Press, 1995.

Hellbeck, Jochen. "Fashioning the Stalinist Soul: The Diary of Stephan Podlubnyi (1931–1939)." *Jahrbücher für Geschichte Osteuropas* 44, no. 3 (1996): 344–73.

Helmreich, Stefan. *Silicon Second Nature: Culturing Artificial Life in a Digital World*. Berkeley: University of California, 1998.

Herzig, Rebecca. "Removing Roots: 'North American Hiroshima Maidens' and the X-ray." *Technology and Culture* 40 (1999): 723–45.

Hetherington, Penelope. "Female Circumcision and the Writing of History." *Indian Ocean Studies Review* 5, no. 3 (1992): 1–13.

Hetherington, Penelope. "The Politics of the Clitoris: Contaminated Speech, Feminism and Female Circumcision." *African Studies Review and Newsletter* 19, no. 1 (1997): 4–9.

Heuser, Stephen. "A Case of Mind over Matter: Mind-Reading Devices Show Promise." *Boston Globe*, April 2, 2006.

Higgs, Edward. *Life, Death and Statistics: Civil Registration, Censuses and the Work of the General Register Office, 1836–1952*. Hatfield, UK: Local Population Studies, 2004.

Hilberg, Raul. *Die Vernichtung der europäischen Juden*. Rev. ed. Frankfurt am Main: Fischer, 1990.

Hill, Matthew. "Ukraine Babies in Stem Cell Probe." BBC News, December 12, 2006. http://news.bbc.co.uk/1/hi/world/europe/6171083.stm.

Hirschauer, Stefan. "The Manufacture of Bodies in Surgery." *Social Studies of Science* 21 (1991): 279–319.

Hirschauer, Stefan. "Performing Sexes and Genders in Medical Practices." In *Differences in Medicines: Unraveling Practices, Techniques, and Bodies*, edited by Marc Berg and Annemarie Mol, 13–27. Durham, NC, and London: Duke University Press, 1998.

Hirschauer, Stefan. *Die soziale Konstruktion der Transsexualität*. Frankfurt am Main: Suhrkamp, 1993.

Hite, Shere. *The Hite Report on Female Sexuality*. New York: MacMillan, 1976.

Hobbes, Thomas. *The Elements of Law Natural and Politic*. Edited by F. Toennies and M. M. Goldsmith. London: Simpkin, Marshall, 1889.

Hoberman, John M. *Darwin's Athletes: How Sport Has Damaged Black America and Preserved the Myth of Race*. New York: Houghton Mifflin, 1997.

Hoberman, John M. *Mortal Engines: The Science of Performance and the Dehumanization of Sport*. New York: Free Press, 1992.

Hoberman, John M. *Testosterone Dreams: Rejuvenation, Aphrodisia, Doping*. Berkeley: University of California Press, 2005.

Hoche, Alfred E. *Krieg und Seelenleben*. Leipzig, Germany: Speyer & Koerner, 1915.

Hogle, Linda F. "Enhancement Technologies and the Body." *Annual Review of Anthropology* 34 (2005): 695–716.

Hogle, Linda F. *Recovering the Nation's Body: Cultural Memory, Medicine, and the Politics of Redemption.* New Brunswick, NJ, and London: Rutgers University Press, 1999.

Holliday, Ruth, and Jacqueline Sanchez Taylor. "Aesthetic Surgery as False Beauty." *Feminist Theory* 7 (2006): 179–95.

Holquist, Peter. "To Count, to Extract, to Exterminate: Statistics and Population Politics in Late Imperial and Soviet Russia." In *A State of Nations: Empire and Nation-Making in the Age of Lenin and Stalin*, edited by T. Martin and R. Suny, 111–44. Oxford: Oxford University Press, 2001.

Hope, Christine. "Caucasian Female Body Hair and American Culture." *Journal of American Culture* 5 (1982): 93–99.

Hornibrook, Ettie. *Restoration Exercises for Women.* London: William Heinemann, 1931.

Houlbrook, Matt. *Queer London: Perils and Pleasures in the Sexual Metropolis, 1918–1957.* Chicago: Chicago University Press, 2005.

Hoyt, David. "Fin-de-Siècle Ethnographic Discourse in Western Europe." *History of Science* 39 (2001): 331–54.

Huffman, John. "Clinical Significance of the Paraurethral Ducts and Glands." *Archives of Surgery* 62 (1951): 615–26.

Huffman, John. "The Detailed Anatomy of the Paraurethral Ducts in the Adult Human Female." *American Journal of Obstetrics and Gynecology* 55 (1948): 86–101.

Huffman, John. "The Development of the Periurethral Glands in the Human Female." *American Journal of Obstetrics and Gynecology* 46 (1946): 773–85.

Hughes, James J. "The Future of Death: Cryonics and the Telos of Liberal Individualism." *Journal of Evolution and Technology* 6 (2001): 8–22.

Hulten, Pontus. "The Blind Lottery of Reputation or the Duchamp Effect." In *Marcel Duchamp*, 13–19. London: Thames and Hudson, 1993.

Hundert, E. J. "The European Enlightenment and the History of the Self." In *Rewriting the Self: Histories from the Renaissance to the Present*, edited by Roy Porter, 72–83. London: Routledge, 1997.

Hung, Wu. "Between Past, Present and Future: A Brief History of Contemporary Chinese Photography." In *Between Past, Present and Future: New Photography and Video from China*, edited by Wu Hung and Christopher Phillips. Göttingen, Germany: Steidl, 2004. Published in conjunction with an exhibit shown at the Smart Museum of Art at the University of Chicago and the International Center of Photography in New York.

Huppauf, Bernd. "Languemarck, Verdun and the Myth of a New Man in Germany after the First World War." *War and Society* 6, no. 2 (1988): 70–103.

Hurwitz, Emanuel. *Otto Gross, Paradies Sucher zwischen Jung und Freud, Leben und Werk.* Zurich: Suhrkamp, 1979.

Hutcherson, Hilda. "5 Myths of Anal Sex Uncovered." iVillage. http://love.ivillage.com/lnssex/sextaboos/0,,jdgc,00.html.

Hyde, Alan. *Bodies of Law.* Princeton, NJ: Princeton University Press, 1997.

Ince, Kate. *Orlan: Millenial Female.* Oxford: Berg, 2000.

Irigary, Luce. *The Sex Which Is Not One.* Translated by Catherine Porter and Carolyn Burke. Ithaca, NY: Cornell University Press, 1985.

Jackson, Jean. *"Camp Pain": Talking with Chronic Pain Patients*. Philadelphia: University of Pennsylvania Press, 2000.

Jackson, Jean. *Women in Pain*. Philadelphia: University of Pennsylvania Press, 1994.

Jane, Candid. "Our Search for Beauty: Men Are the Guilty Ones." *Australian Women's Weekly*, July 8, 1933, 7.

Jarvis, Christine. *The Male Body at War: American Masculinity during World War II*. Dekalb: University of Northern Illinois Press, 2004.

Jay, Mike. *Blue Tide: Search of Soma*. London: Autonomedia, 2002.

Jeanneret, Edouard [Le Corbusier], and Amédée Ozenfant. *Après le Cubisme*. Paris: Édition des Commentaires, 1918.

Jeffords, Susan. *The Remasculinization of America: Gender and the Vietnam War*. Bloomington: Indiana University Press, 1989.

Jennet, Bryan. "Brain Death and the Vegetative State." In *Oxford Textbook of Medicine*, edited by D. J. Weatherall et al. (Oxford: Oxford University Press, 1966).

Jennett, Bryan. "Brain Death and the Vegetative State." In *Encyclopedia of Life Sciences*, 3:410–13. Chichester, UK: Wiley, 2002.

Jervis, John. *Exploring the Modern*. Oxford: Blackwell, 1998.

Jobs, Richard I. "Tarzan under Attack: Youth, Comics, and Cultural Reconstruction in Postwar France." *French Historical Studies* 26, no. 4 (2003): 705–13.

Johnson, Eric A. *Nazi Terror: The Gestapo, Jews, and Ordinary Germans*. New York: Basic Books, 1999.

Jones, Amelia. *Body Art: Performing the Subject*. Minneapolis: University of Minnesota Press, 1998.

Jones, Amelia. *Irrational Modernism: A Neurasthenic History of New York Dada*. Cambridge, MA: MIT Press, 2004.

Jones, Amelia. "Stelarc's Technological Transcendence/Stelarc's Wet Body: The Insistent Return of the Body." In Smith, *Stelarc*, 87–124.

Jones, Amelia, and Andrew Stephenson, eds. *Performing the Body, Performing the Text*. London and New York: Routledge, 1999.

Jones, James. *Kinsey: A Public/Private Life*. New York: W. W. Norton, 1997.

Joralemon, Donald. "Organ Wars: The Battle for Body Parts." *Medical Anthropology Quarterly*, n.s., 9 (1995): 335–56.

Jordan, John W. "The Rhetorical Limits of the 'Plastic Body.'" *Quarterly Journal of Speech* 90 (2004): 327–58.

Jordanova, Ludmilla. *Sexual Visions: Images of Gender in Science and Medicine between the Eighteenth and Twentieth Centuries*. New York: Harvester Wheatsheaf, 1989.

Joyce, Patrick. *The Rule of Freedom: Liberalism and the Modern City*. London and New York: Verso, 2003.

Jumeau-Lafond, Jean-David. *Carlos Schwabe: Symboliste et Visionnaire*. Paris: ACR Édition, 1994.

Juno, Andrea, and V. Vale, eds. *Angry Women*. San Francisco: Re/Search, 1991.

Jupp, Peter C. *From Dust to Ashes: Cremation and the British Way of Death*. Basingstoke, UK: Palgrave Macmillan, 2006.

Kalipeni, Ezekiel, ed. *HIV and AIDS in Africa: Beyond Epidemiology*. Malden, MA: Blackwell, 2004.

Kantorowicz, Ernst. *The King's Two Bodies: A Study in Medieval Political Theology*. Princeton, NJ: Princeton University Press, 1957.

Katz, Robert S. "Influenza 1918–1919: A Study in Mortality." *Bulletin of the History of Medicine* 48 (1974): 416–22.

Kaw, Eugenia. "Medicalization of Racial Features: Asian-American Women and Cosmetic Surgery." In *The Politics of Women's Bodies: Sexuality, Appearance, and Behavior*, edited by Rose Weitz, 167–83. New York and Oxford: Oxford University Press, 1998.

Kay, Lily. *Who Wrote the Book of Life: A History of the Genetic Code*. Stanford, CA: Stanford University Press, 1999.

Kellaher, L., and D. Prendergast. "Resistance, Renewal or Reinvention: The Removal of Ashes from Crematoria." *Pharos International* 70 (2004): 10–13.

Kelves, Daniel. *In the Name of Eugenics: Genetics and the Uses of Human Heredity*. Cambridge, MA: Harvard University Press, 1995.

Kemp, Martin, and Marina Wallace. *Spectacular Bodies: The Art and Science of the Human Body from Leonardo to Now*. Berkeley: University of California Press, 2000.

Kenez, Peter. *Cinema and Soviet Society, 1917–1953*. London: I. B. Tauris, 2001.

Kennedy, Dane. "Diagnosing the Colonial Dilemma: Tropical Neurasthenia and the Alienated Briton." In *Decentering Empire: Britain, India, and the Transcolonial World*, by Dane Kennedy and Durba Ghosh, 157–81. Hyderabad, India: Orient Longman, 2006.

Kennedy, Dane. *Islands of White: Settler Society and Culture in Kenya and Southern Rhodesia, 1890–1939*. Durham, NC: Duke University Press, 1987.

Kennedy, Dane. "The Perils of the Midday Sun: Climatic Anxieties in the Colonial Tropics." In *Imperialism and the Natural World*, edited by John M. MacKenzie. Manchester, UK: Manchester University Press, 1990.

Kern, Edith. *Existential Thought and Fictional Technique: Kierkegaard, Sartre, Beckett*. New Haven, CT: Yale University Press, 1970.

Kern, Stephen. *The Culture of Time and Space, 1880–1918*. Cambridge, MA: Harvard University Press, 1983.

Kero, A., U. Högberg, L. Jacobsson, and A. Lalos. "Legal Abortion: A Painful Necessity." *Social Science and Medicine* 53 (2001): 1481–90.

Kershaw, Ian. *Hitler, 1889–1936: Hubris*. London: Penguin, 1999.

Kinsey, Alfred, Wardell Pomeroy, C. E. Martin, and P. H. Gebhart. *Sexual Behaviour in the Human Female*. New York: W. B. Saunders, 1953.

Kinsey, Alfred, Wardell Pomeroy, C. E. Martin, and P. H. Gebhart. *Sexual Behaviour in the Human Male*. New York: W. B. Saunders, 1948.

Kirby, Vicki. "On the Cutting Edge: Feminism and Clitoridectomy." *Australian Feminist Studies* 5 (1987): 35–55.

Kirby, Vicki. *Telling Flesh: The Substance of the Corporeal*. London: Routledge, 1997.

Klee, Ernst. *Euthanasie im NS Staat*. Frankfurt am Main: Fischer, 1985.

Klein, Alan M. *Little Big Men: Bodybuilding Subculture and Gender Construction*. Albany: State University of New York Press, 1993.

Knorr-Cetina, Karin, and Aaron Cicourel, eds. *Advances in Social Theory and Methodology: Toward an Integration of Micro- and Macrosociologies*. London: Routledge and Kegan Paul, 1981.

Koch, Gertrud. "Cosmos in Film: On the Concept of Space in Walter Benjamin's 'Work of Art' Essay." In *Walter Benjamin's Philosophy*, edited by Andrew Benjamin and Peter Osborne, 205–15. London and New York: Routledge, 1994.

Koch, Thomas. *Zwangssterilisation im Dritten Reich: Das Beispiel der Uni-Frauenklinik Göttingen*. Frankfurt am Main: Mabuse, 1994.

Kocher, Emil Theodor. "Concerning Pathological Manifestations in Low-Grade Thyroid Diseases, Nobel Lecture, December 11, 1909." In *Nobel Lectures: Physiology or Medicine, 1901–1921*, edited by the Nobel Foundation, 330–83. Amsterdam: Elsevier, 1967.

Koonz, Claudia. *Mothers in the Fatherland*. New York: St. Martin's, 1987.

Koplow, David A. *Smallpox: The Fight to Eradicate a Global Scourge*. Berkeley: University of California Press, 2003.

Kotkin, Stephen. *Magnetic Mountain: Stalinism as a Civilisation*. Berkeley: University of California Press, 1995.

Kouba, Leonard J., and Judith Muasher. "Female Circumcision in Africa: An Overview." *African Studies Review* 28, no. 1 (1985): 95–110.

Kristeva, Julia. *Powers of Horror: An Essay on Abjection*. Translated by Leon S. Roudiez. New York: Columbia University Press, 1982.

Krüger, Arnd. "Breeding, Rearing and Preparing the Aryan Body." *International Journal of the History of Sport* 16 (1999): 42–68.

Krüger, Arnd, and Dietrich Ramba. "Sparta or Athens? The Reception of Greek Antiquity in Nazi Germany." In *The Olympic Games Through the Ages: Greek Antiquity and Its Impact on Modern Sport*, edited by Roland Renson and Manfred Lammer, 345–56. Athens, Greece: Hellenic Sports Research Institute, 1991.

Kübler-Ross, Elisabeth. *On Death and Dying*. New York: Simon and Schuster, 1969.

Kühl, Stefan. *Die Internationale der Rassisten: Aufstieg und Niedergang der internationalen Bewegung für Eugenik und Rassenhygiene im 20. Jahrhundert*. Frankfurt am Main: Campus, 1997.

Kuhn, T. S. "The Function of Measurement in the Modern Physical Sciences." In *The Essential Tension*. Chicago: Chicago University Press, 1977.

Kurzweil, Ray. *The Age of Intelligent Machines*. Cambridge, MA: MIT Press, 1990.

Kusch, Martin. *Psychological Knowledge*. London: Routledge, 1998.

Kveim Lie, Anne. "Stories of Origin and the Norwegian Radesyge." *Social History of Medicine* 20 (2007): 563–79.

Labre, Magdala Peixoto. "The Brazilian Wax: New Hairlessness Norm for Women?" *Journal of Communication Inquiry* 26 (2002): 113–32.

Lacan, Jacques. *The Seminar of Jacques Lacan II: The Ego in Freud's Theory and in the Technique of Psychoanalysis, 1954–1955*. Edited by Jacques-Alain Miller. Translated by Sylvana Tomaselli. New York: W. W. Norton, 1988.

La Capra, Dominique. "Trauma, Absence, Loss." *Critical Inquiry* 25 (1999): 696–727.

Lacoue-Labarthe, Philippe. *Typography: Mimesis, Philosophy, Politics*. Stanford, CA: Stanford University Press, 1998.

Landecker, Hannah. *Culturing Life: How Cells Became Technologies*. Cambridge, MA: Harvard University Press, 2007.

Landecker, Hannah. "New Times for Biology: Nerve Cultures and the Advent of Cellular Life in Vitro." *Studies in History and Philosophy of Biological and Biomedical Sciences* 33 (2002): 667–94.

Lapper, Alison (with Guy Feldman). *My Life in My Hands*. New York: Simon and Schuster, 2005.

Laqueur, Thomas. "Amor Veneris, Vel Dulcedo Appeletur." In *Zone 3: Fragments for a History of the Human Body*, edited by Michel Feher, Ramona Naddaff, and Nadia Tazi, 90–131. New York: Zone Books, 1989.

Laqueur, Thomas. "Diary." *London Review of Books*, September 7, 2006. http://www.lrb.co.uk/v28/n17/laqu01_.html (Accessed August 2008).
Laqueur, Thomas. *Making Sex: The Body and Gender from the Greeks to Freud*. Cambridge, MA: Harvard University Press, 1990.
Laqueur, Thomas. *Solitary Sex: A Cultural History of Masturbation*. New York: Zone Books, 2003.
Lawrence, Christopher. "Democratic, Devine and Heroic: The History and Historiography of Surgery." In *Medical History and Surgical Practice: Studies in the History of Surgery*, edited by Christopher Lawrence, 1–47. London and New York: Routledge, 1992.
Lawrence, Christopher, and George Weiss, eds. *Greater than the Parts: Holism in Biomedicine, 1920–1950*. New York and Oxford: Oxford University Press, 1998.
Le Rider, Jacques. "Hans Gross, criminologue, et son fils Otto Gross, 'délinquant sexuel' et psychoanalyste." In *Kulturwissenschaft im Vielvölkerstaat: Zur Geschichte der Ethnologie und verwandter Gebiete in Österreich ca. 1780–1918*, edited by Britta Rupp-Eisenreich und Justin Stegl, 229–40. Vienna: Böhlau, 1995.
Lears, Jackson. *Fables of Abundance: A Cultural History of Advertising in America*. New York: Basic Books, 1994.
Lears, Jackson. *No Place of Grace: Antimodernism and the Transformation of American Culture, 1880–1920*. New York: Pantheon Books, 1981.
Léger, Fernand. "Art and the People." *Arts de France*, 1946. Reprinted in *Fernand Léger, 1881–1955*, edited by Robert T. Buck, Edward F. Fry, and Charlotta Kotik, 146–47. New York: Abbeville, 1982.
Léger, Fernand. "The New Realism Goes On." *Art Front* 3, no. 1 (February 1937): 493–96.
Léger, Fernand. "A New Realism—the Object." In Chipp, *Theories of Modern Art*, 277–80.
Legman, Gershoy. *Love and Death: A Study in Censorship*. 1949. New York: Hacker Art Books, 1963.
Leit, Richard A., Harrison G. Pope, Jr., and James J. Gray. "Cultural Expectations of Muscularity in Men: The Evolution of Playgirl Centerfolds." *International Journal of Eating Disorders* 29 (2001): 90–93.
Lerner, Paul. *Hysterical Men: War, Psychiatry, and the Politics of Trauma in Germany, 1890–1930*. Ithaca, NY: Cornell University Press, 2003.
Lightfoot-Klein, Hanny. *Prisoners of Ritual: An Odyssey into Female Genital Circumcision in Africa*. New York: Haworth, 1989.
Lilienthal, Georg. *Der "Lebensborn e.V."* Frankfurt am Main: Fischer, 2003.
Livingstone, Julie. *Debility and the Moral Imagination in Botswana*. Bloomington: Indiana University Press, 2005.
Lock, Margaret. "Death in Technological Time: Locating the End of Meaningful Life." *Medical Anthropology Quarterly*, n.s., 10 (1996): 575–600.
Lock, Margaret. *Twice Dead: Organ Transplants and the Reinvention of Death*. Berkeley: University of California Press, 2002.
Los Angeles County Museum of Art. *Degenerate Art: The Fate of the Avant-Garde in Nazi Germany*. New York: Abrams, 1991.
Loudon, Irvine. *Death in Childbirth: An International Study of Maternal Care and Maternal Mortality, 1800–1950*. Oxford: Clarendon, 1992.
Loudon, Irvine. "On Maternal and Infant Mortality." *Social History of Medicine* 4 (1991): 29–73.

Lovgren, Stefan. "HIV Originated with Monkeys, Not Chimps, Study Finds." *National Geographic*, June 12, 2003. http://news.nationalgeographic.com/news/2003/06/0612_030612_hivvirusjump.html.

Lupton, Deborah. "The Social Construction of Medicine and the Body." In *The Handbook of Social Studies in Health and Medicine*, edited by Gary L. Albrecht, Ray Fitzpatrick and Susan C. Scrimshaw, 50–63. London: Sage, 2000.

Lyons, Maryinez. *The Colonial Disease: A Social History of Sleeping Sickness in Northern Zaire, 1900–1940*. Cambridge: Cambridge University Press, 1992.

Lyotard, Jean-François. *Libidinal Economy*. 1974. London: Athlone, 1993.

Macaulay, Thomas B. "Horatius." In *Lord Macaulay's Essays and Lays of Ancient Rome*. London: Longmans, 1905.

Mackintosh, Maureen, and Meri Koivusalo, eds. *Commercialization of Healthcare: Global and Local Dynamics and Policy Responses*. Basingstoke, UK: Palgrave Macmillan, 2005.

Macleod, Roy, and Milton Lewis, eds. *Disease, Medicine and Empire: Perspectives on Western Medicine and the Experience of European Expansion*. New York: Routledge, 1988.

Maglen, Krista. "A World Apart: Geography, Australian Quarantine, and the Mother Country." *Journal of the History of Medicine and Allied Sciences* 60 (2005): 196–217.

Maines, Rachel P. *The Technology of Orgasm: "Hysteria," the Vibrator, and Women's Sexual Satisfaction*. Baltimore: Johns Hopkins University Press, 1999.

Malevich, Kasimir. "The Question of Imitative Art" (Smolensk, 1920). Reprinted in part in Harrison and Wood, *Art in Theory*.

Mallmann, Klaus-Michael, and Gerhard Paul. "Omniscient, Omnipotent, Omnipresent? Gestapo, Society and Resistance." In Crew, *Nazism and German Society*, 166–96.

Malvern, Sue. *Modern Art, Britain and the Great War: Witnessing, Testimony and Remembrance*. New Haven, CT, and London: Yale University Press, 2004.

Mangan, J. A., and James Walvin. *Manliness and Morality: Middle-Class Masculinity in Britain and America, 1800–1940*. New York: St. Martin's, 1987.

Mani, Lata. "Contentious Traditions: The Debate on Sati in Colonial India." In *Recasting Women: Essays in Indian Colonial History*, edited by K. Sangari and V. Sudesh, 88–126. New Brunswick, NJ: Rutgers University Press, 1990.

Mann, Steve. *Cyborg: Digital Destiny and Human Possibility in the Age of the Wearable Computer*. New York: Random House Doubleday, 2001.

Marinetti, F. T. "Multiplied Man and the Reign of the Machine." In *Marinetti: Selected Writings*, translated by R. W. Flint and Arthur A. Coppotelli, 90–93. New York: Farrar, Straus and Giroux, 1971.

Marshall, Barbara L., and Stephen Katz. "Forever Functional: Sexual Fitness and the Ageing Male Body." *Body and Society* 8, no. 4 (2002): 43–70.

Martin, Aryn. "Can't Anybody Count? Counting as an Epistemic Theme in the History of Human Chromosomes." *Social Studies of Science* 34 (2004): 923–48.

Martin, Emily. *Flexible Bodies: Tracking Immunity in American Culture—From the Days of Polio to the Age of AIDS*. Boston: Beacon, 1994.

Martin, J. L., B. Nicholson, and N. Gabo, eds. *Circle: International Survey of Constructive Art*. London: Praeger, 1937.

Martin, Richard. "Mind Control: Matthew Nagle is Paralyzed. He's Also a Pioneer in the New Science of Brain Implants." *Wired* 13 (2005), http://www.wired.com/wired/archive/13.03/brain.html?pg=2&topic=brain&topic_set.

Massumi, Brian. "The Evolutionary Alchemy of Reason." In Smith, *Stelarc*, 125–92.
Masters, William, and Virginia Johnston. *Human Sexual Response*. Boston: Little, Brown, 1966.
Mauss, Marcel. "Techniques of the Body." *Economy and Society* 2 (1973): 70–88.
Mayor, Susan. "Authors Reject Interpretation Linking Autism and MMR Vaccine." *BMJ* 328 (2004): 602.
McCrae, M. "The Scottish Roots of the National Health Service." PhD diss., University of Edinburgh, 2001.
McCulloch, Jock. *Colonial Psychiatry and the African Mind*. Cambridge: Cambridge University Press, 1995.
McKeown, Thomas. *The Modern Rise of Population*. London: Edward Arnold, 1976.
McLaren, Angus. *Twentieth-Century Sexuality*. Oxford: Blackwells, 1999.
McNay, M. B., and J. E. Fleming. "Forty Years of Obstetric Ultrasound 1957–1997: From A-Scope to Three Dimensions." *Ultrasound in Medicine and Biology* 25 (1999): 3–56.
Meggit, M. J. *Desert People: A Study of the Walbiri Aborigines of Central Australia*. Sydney: Angus and Robertson, 1971.
Mejia, Michael. *Forgetfulness*. Tuscaloosa: University of Alabama Press, 2005.
Mensendieck, Bess M. *"It's Up to You."* New York: Mensendieck System, 1931.
Merewether, Charles. "The Unspeakable Condition of Figuration." In *Body*, 155–56. Sydney, Australia: Art Gallery of New South Wales and Bookman, 1997.
Merleau-Ponty, Maurice. "The Body as Expression and Speech." In *The Phenomenology of Perception*, 202–34. London: Routledge, 2002.
Merleau-Ponty, Maurice. *The Phenomenology of Perception*. New York: Humanities Press, 1962.
Meyerowitz, Joanne. *How Sex Changed: A History of Transsexuality in the United States*. Cambridge, MA: Harvard University Press, 2002.
Mialet, Hélène. "Do Angels Have Bodies? Two Stories about Subjectivity in Science: The Cases of William X and Mister H." *Social Studies of Science* 29 (1999): 551–81.
Mialet, Hélène, "Do Angels Have Bodies: Two Stories about Subjectivity in Science, The Cases of William X and Mr. H." In *The Philosophy of Expertise*, edited by E. Selinger and R. P. Crease, 246–79. New York: Columbia University Press, 2006.
Miglietti, Francesca Alfano. *Extreme Bodies: The Use and Abuse of the Body in Art*. Milan, Italy: Skira, 2003.
Miller, James. *The Passion of Michel Foucault*. New York: Simon and Schuster, 1993.
Millner, V. S., B. H. Eichold, T. H. Sharpe, and S. Lynn. "First Glimpse of the Functional Benefits of Clitoral Hood Piercings." *American Journal of Obstetrics and Gynecology* 193 (2005): 675–76.
Mills, James H. "Body as Target, Violence as Treatment: Psychiatric Regimes in Colonial and Post-colonial India." In Mills and Sen, *Confronting the Body*, 80–101.
Mills, James H. *Madness, Cannabis and Colonialism: The "Native Only" Lunatic Asylums of British India, 1857 to 1900*. Basingstoke, UK: Palgrave, 2000.
Mills, James H., and Satadru Sen, eds. *Confronting the Body: The Politics of Physicality in Colonial and Post-colonial India*. London: Anthem, 2004.
Minden, Michael, and Holger Bachmann, eds. *Fritz Lang's* Metropolis: *Cinematic Visions of Technology and Fear*. Rochester, NY: Camden House, 2000.
Ministry of Health. *Report on an Investigation into Maternal Mortality*. Cmd. 5422. London: HMSO, 1937.

Möhring, Maren. "Working Out the Body's Boundaries: Physiological, Aesthetic, and Psychic Dimensions of the Skin in German Nudism, 1890–1930." In Forth and Crozier, *Body Parts*, 229–46.

Mol, Annemarie. *The Body Multiple: Ontology in Medical Practice*. Durham, NC: Duke University Press, 2002.

Mol, Annemarie. "Lived Reality and the Multiplicity of Norms: A Critical Tribute to George Canguilhem." *Economy and Society* 27 (1998): 274–84.

Mondrian, Piet. "Neo-Plasticism: The General Principle of Plastic Equivalence" (January 1921). In Harrison and Wood, *Art in Theory.*

Mort, Frank. *Dangerous Sexualities: Medico-moral Politics in England since 1830.* 1987. London: Routledge, 2000.

Morton, Donald, ed. *The Material Queer: A Lesbigay Cultural Studies Reader*. Boulder, CO: Westview, 1996.

Mosse, George. "Beauty without Sensuality: The Exhibition *Entartete Kunst*." In Los Angeles County Museum of Art, *Degenerate Art*, 25ff.

Motluk, Alison. "Seeing with Your Ears." *New York Times Magazine*, December 11, 2005. http://www.nytimes.com/2005/12/11/magazine/11ideas_section3-14.html?scp=19&sq=voice&st=nyt.

Mulkay, Michael. *The Embryo Research Debate: Science and the Politics of Reproduction*. Cambridge, UK: Cambridge University Press, 1997.

Mulvey, Laura. "Visual Pleasure and Narrative Cinema." *Screen* 16, no. 3 (1975).

Murphy, Timothy F. *Gay Science: The Ethics of Sexual Orientation Research*. New York: Columbia University Press, 1997.

Murray, Jocelyn. "The CMS and the Female Circumcision Issue in Kenya, 1929–1932." *Journal of Religion in Africa* 8 (1976): 92–104.

Neuhaus, Jessamyn. "The Joy of Sex Instruction: Women and Cooking in Marital Sex Manuals, 1920–1963." In *Kitchen Culture in America: Popular Representations of Food, Gender, and Race*, edited by Sherrie A. Inness, 95–118. Philadelphia: University of Pennsylvania Press, 2001.

New York Times. "Coma Victim Gives Birth." March 19, 1996. http://query.nytimes.com/gst/fullpage.html?res=9D05E2DC1739F93AA25750C0A960958260 (accessed May 2006).

New York Times. "The Red-Faced Reds." October 23, 1964.

New York Times. "Sex Test Disqualifies Athlete." September 16, 1967.

New York Times. "US Track Team Leads in Moscow." July 21, 1963.

Nicolson, Malcolm. "Ian Donald, Diagnostician and Moralist." Web site of the Royal College of Physicians of Edinburgh. 2004. http://www.rcpe.ac.uk/library/donald/donald1.html.

Nnaemeka, Obioma, ed. *Female Circumcision and the Politics of Knowledge: African Women in Imperialist Discourses*. Westport, CT, and London: Praeger, 2005. Nye, Robert A. *Crime, Madness and Politics in Modern France: The Medical Concept of National Decline*. Princeton, NJ: Princeton University Press, 1984.

Nye, Robert A. "The Evolution of the Concept of Medicalization in the Late Twentieth Century." *Journal of History of the Behavioral Sciences* 39 (2003): 115–29.

O'Keefe, Paul. "Art, Action and the Machine." In *Dynamism: The Art of Modern Life before the Great War*, edited by Penelope Curtis. Liverpool: Tate Gallery, 1991. An exhibition catalog.

Oaks, Laury. "Antiabortion Positions and Young Women's Life Plans in Contemporary Ireland." *Social Science and Medicine* 56 (2003): 1973–86.

Obeyesekere, Gananath. *Medusa's Hair: An Essay on Personal Symbols and Religious Experience*. Chicago: University of Chicago Press, 1981.

Oosterhuis, Harry. *Step-Children of Nature*. Chicago: Chicago University Press, 1999.

ORLAN. "Carnal Art Manifesto." http://www.dundee.ac.uk/transcript/volume2/issue2_2/orlan/orlan.htm.

ORLAN. "What's All This Body Art?" *Flash Art* 26, no. 168 (January–February 1993): 50–53.

Oudshoorn, Nelly. *Beyond the Natural Body: An Archaeology of Sex Hormones*. London: Routledge, 1994.

Oudshoorn, Nelly. *The Male Pill*. Durham, NC: Duke University Press, 2003.

Ouedraogo, Arouna. "Food and the Purification of Society: Dr. Paul Carton and Vegetarianism in Interwar France." *Social History of Medicine* 14 (2001): 223–45.

Outram, Dorinda. *The Body and the French Revolution: Sex, Class and Political Culture*. New Haven, CT: Yale University Press, 1989.

Owen, Wilfred. "Dulce et Decorum est." In *The War Poems of Wilfred Owen*, edited by Jon Stallworthy. London: Chatto and Windus, 1994.

Pearl, Raymond. *The Biology of Death*. Philadelphia: Lippincott, 1922.

Pearson, Karl. "The Chances of Death." In *Chances of Death and Other Studies of Evolution*. 2 vols. London: Edward Arnold, 1897.

Pederson, Susan. "National Bodies, Unspeakable Acts; the Sexual Politics of Colonial Policy Making." *Journal of Modern History* 63 (1991): 647–80.

Peiss, Kathy. *Hope in a Jar: The Making of America's Beauty Culture*. New York: Metropolitan Books, 1998.

Pendergast, Tom. *Creating the Modern Man: American Magazines and Consumer Culture, 1900–1950*. Columbia: University of Missouri Press, 2000.

Peukert, Detlev J.K. "The Genesis of the 'Final Solution' from the Spirit of Science." In Crew, *Nazism and German Society*, 274–99.

Peukert, Detlev J.K. *Inside Nazi Germany: Conformity and Opposition in Everyday Life*. New Haven, CT: Yale University Press 1989.

Pick, Daniel. *Faces of Degeneration*. Cambridge: Cambridge University Press, 1989.

Pinkus, Karen. *Bodily Regimes: Italian Advertising under Fascism*. Minneapolis: University of Minnesota Press, 1995.

Pitts, Victoria. *In the Flesh: The Cultural Politics of Body Modification*. New York: Palgrave/MacMillan, 2003.

Pitts-Taylor, Victoria. *Surgery Junkies: Wellness and Pathology in Cosmetic Culture*. New Brunswick, NJ: Rutgers University Press, 2007.

Pohl, Frederic, and Hans Moravec. "Souls in Silicon." In *The American Body in Context: An Anthology*, ed. Jessica R. Johnston, 49–59. 1993. Wilmington, DE: Scholarly Resources, 2001.

Polanyi, Michael. *Personal Knowledge: Towards a Post-Critical Philosophy*. London: Routledge, 1958.

Pomfret, David. "The 'City of Evil' and the 'Great Outdoors': The Modern Health Movement and the Urban Young, 1918–1940." *Urban History* 28 (2001): 405–27.

Poole, Roger. "The Unknown Kierkegaard: Twentieth-Century Reception." In *The Cambridge Companion to Kierkegaard*, edited by Alistair Hannay and Gordin D. Marino, 48–75. Cambridge: Cambridge University Press, 1998.

Pope, Harrison G., Jr., Katharine A. Phillips, and Roberto Olivardia. *The Adonis Complex: The Secret Crisis of Male Body Obsession*. New York: Free Press, 2000.

Porter, Dorothy. "The Healthy Body." In Cooter and Pickstone, *Companion to Medicine*, 201–16.

Porter, Roy. *The Greatest Benefit of Mankind: A Medical History of Humanity.* New York: W. W. Norton, 1999.

Porter, Roy, and Lesley Hall. *The Facts of Life: The Creation of Sexual Knowledge in Britain, 1650–1950.* New Haven, CT: Yale University Press, 1995.

Porteus, Stanley. *The Psychology of a Primitive People: A Study of the Australian Aborigines.* Sydney: Books for Libraries Press, 1931.

Powell, Earle, III, and James Wood. Foreword. In *Degenerate Art: The Fate of the Avant-Garde in Nazi Germany*, edited by the Los Angeles County Museum of Art. New York: Abrams, 1991.

Pratchett, Terry. *The Colour of Magic.* London: Smythe, 1989.

Presner, Todd Samuel. "'Clear Heads, Solid Stomachs, and Hard Muscles': Max Nordau and the Aesthetics of Jewish Regeneration." *Modernism/Modernity* 10, no. 2 (2003): 269–96.

Pressman, Jack D. *Last Resort: Psychosurgery and the Limits of Medicine.* Cambridge: Cambridge University Press, 1998.

Preston, S. H. *Mortality Patterns in National Populations.* New York: Academic Press, 1976.

Pultz, John. *The Body and the Lens: Photography 1893 to the Present.* New York: Abrams, 1995.

Rabinbach, Anson. *The Human Motor: Energy, Fatigue, and the Origins of Modernity.* Berkeley: University of California Press, 1990.

Rabinow, Paul. "Severing the Ties: Fragmentation and Dignity in Late Modernity." *Knowledge and Society: The Anthropology of Science and Technology* 9 (1992): 125–92.

Rafanell, Irene. "Durkheim and the Performative Model: Reconfiguring Social Objectivity." In *Sociological Objects: The Reconfiguration of Social Theory*, edited by Geoff Cooper, Andrew King and Ruth Retti, 59–76. London: Ashgate, 2009.

Rafanell, Irene. "The Sexed and Gendered Body as a Social Institution: A Critical Reconstruction of Two Social Constructionists' Models." PhD diss., University of Edinburgh, 2003.

Ramsey, Paul. *The Patient as Person: Explorations in Medical Ethics.* 2nd ed. New Haven, CT: Yale University Press, 2002.

Rangel de Almeida, João. "The International Sanitary Conferences (1851–1866): A Social Locale for the Negotiation of Knowledge." Master's thesis, University of Edinburgh, 2006.

Ranger, Terence, and Paul Slack, eds. *Epidemics and Ideas: Essays on the Historical Perception of Pestilence.* Cambridge: Cambridge University Press, 1992.

Rapley, Cath. "Pubic Relations." *The Observer*, April 14, 2002. http://observer.guardian.co.uk/life/story/0,,683936,00.html.

Rapp, Dean. "The Early Discovery of Freud by the British General Educated Public, 1912–1919." *Social History of Medicine* 3 (1990): 217–43.

Read, Herbert. *Art Now: An Introduction to the Theory of Modern Painting and Sculpture.* 1933. London: Faber and Faber, 1960.

Read, Herbert. *A Concise History of Modern Sculpture.* London: Thames and Hudson, 1971.

Reeves, E. A. *Hints to Travellers.* 11th ed. 2 vols. London: Royal Geographical Society, 1935–1938.

Reiser, Stanley J. "The Science of Diagnosis: Diagnostic Technology." In *Companion Encyclopedia of the History of Medicine*, edited by William F. Bynum and Roy Porter, 826–51. London: Routledge, 1993.

Rejali, Darius. *Torture and Modernity: Self, Society and State in Modern Iran*. Boulder, CO: Westview, 1994.

Report on the Work of the Horn Scientific Expedition to Central Australia. Melville, Mullen and Slade, 1896. Fascimile Bundaberg, Australia: Corkwood, 1994.

Resnick, Mitchell. *Turtles, Termites, and Traffic Jams: Explorations in Massively Parallel Microworlds*. Cambridge, MA: MIT Press, 1994.

Reynolds Whyte, Susan, and Benedicte Ingstad, eds. *Disability and Culture*. Berkeley: University of California Press, 1995.

Rheinberger, Hans-Jörg. "Molekulare Medizin als Paradigma? Gentechnologie im Blick von Wissenschaftstheorie und medizinischer Ethik." In *Meilensteine der Medizin*, edited by Heinz Schott, 555–61. Dortmund, Germany: Harenberg, 1996.

Rhodes, Rosemary. "Death and Dying." In *Encyclopedia of Life Sciences*, 5:343–50. Chichester, UK: Wiley, 2002.

Riefenstahl, Leni, dir. *Triumph of the Will*. 1935. DVD, Detroit: Synapse Films, 2001.

Roberts, J. "US Rape Case Leaves Ethical Uproar." *British Medical Journal* 312 (1996): 329.

Roberts, Mary Louise. *Civilisation without Sexes: Reconstructing Gender in Postwar France, 1917–1927*. Chicago and London: University of Chicago Press, 1994.

Roosth, Sophia. "Sonic Eukaryotes: Sonocytology, Cytoplasmic Milieu, and the *Temps Intèrieur*." Unpublished manuscript, 2006.

Rose, Nikolas. "The Neurochemical Self and Its Anomalies." In *Risk and Morality*, edited by Richard V. Ericson and Aaron Doyle, 407–37. Toronto: University of Toronto Press, 2003.

Rosenberg, Charles. "The Bitter Fruit: Heredity, Disease and Social Thought in Nineteenth Century America." *Perspectives in American History* 8 (1974): 189–235.

Rosenberg, Charles. "What Is Disease? In Memory of Owsei Temkin." *Bulletin of the History of Medicine* 77 (2003): 491–505.

Rosenthal, Norman. "The Blood Must Continue to Flow." In *Young British Artists from the Saatchi Collection*. London: Royal Academy of the Arts, 1997.

Roth, Rachel. *Making Women Pay: The Hidden Costs of Fetal Rights*. Ithaca, NY: Cornell University Press, 2000.

Rowling, J. K. *Harry Potter and the Deathly Hallows*. London: Bloomsbury, 2007.

Rowling, J. K. *Harry Potter and the Philosopher's Stone*. London: Bloomsbury, 1997.

Rusche, Georg, and Otto Kirchheimer. *Punishment and Social Structure*. New York: Columbia University Press, 1939.

Rutherford, Joseph F. *Millions Now Living Will Never Die*. Brooklyn, NY: International Bible Students Association, 1920.

Ryan, Charles J. *Health Preservation in West Africa*. London: John Bale, Sons & Danielsson, 1914.

Ryan, Susan. "New FX Series Has Families Trading Races." *Boston Globe*, March 4, 2006.

Sacks, Oliver. "The Mind's Eye: What the Blind See." In *The Empire of the Senses*, edited by David Howes, 25–42. New York: Berg, 2005.

Saletan, William. "Ass Backwards: The Media's Silence about Rampant Anal Sex." Slate.com, September 20, 2005. http://www.slate.com/?id=2126643&nav=tap2/.

Sanal, Aslihan. "Flesh Yours, Bones Mine: The Making of the Biomedical Body in Turkey." PhD diss., Massachusetts Institute of Technology, 2005.

Sanders, Clinton. *Customizing the Body: The Art and Culture of Tattooing*. Philadelphia: Temple University Press, 1989.

Sarno, John E. *The Divided Mind: The Epidemic of Mindbody Disorders*. New York: Reagan Books, 2006.

Sartre, Jean-Paul. *No Exit and Three Other Plays*. New York: Vintage, 1989.

Sauerteig, Lutz L.D.H., and Roger Davidson, eds. *Shaping Sexual Knowledge: A Cultural History of Sex Education in Twentieth-Century Europe and North America*. London: Routledge, 2009.

Scary, Elaine. *The Body in Pain: The Making and Unmaking of the World*. Oxford: Oxford University Press, 1985.

Scheper-Hughes, Nancy. "Bodies for Sale—Whole or in Parts." In Scheper-Hughes and Wacquant, *Commodifying Bodies*, 1–8.

Scheper-Hughes, Nancy, and Loic Wacquant, eds. *Commodifying Bodies*. London: Sage, 2002.

Schiebinger, Londa. *Nature's Body: Gender in the Making of Modern Science*. Boston: Beacon, 1993.

Schlich, Thomas. "The Emergence of Modern Surgery." In *Medicine Transformed: Health, Disease and Society in Europe, 1800–1939*, edited by Deborah Brunton, 61–91. Manchester, UK: Manchester University Press, 2004.

Schlich, Thomas. *Die Erfindung der Organtransplantation: Erfolg und Scheitern des chirurgischen Organersatzes (1880–1930)*. Frankfurt am Main: Campus, 1998.

Schlich, Thomas. "Die Konstruktion der notwendigen Krankheitsursache: Wie die Medizin Krankheit beherrschen will." In *Anatomien medizinischen Wissens*, edited by Cornelius Borck, 201–29. Frankfurt am Main: Fischer, 1996.

Schlich, Thomas. "Körper und Person: Kultur, Chirurgie und persönliche Identität." *Zeitschrift für medizinische Ethik* 48 (2002): 237–45.

Schlich, Thomas. "Surgery, Science and Modernity: Operating Rooms and Laboratories as Spaces of Control," *History of Science* 45 (2007): 231–56.

Schlich, Thomas. *Surgery, Science and Industry: A Revolution in Fracture Care, 1950–1990s*. Basingstoke, UK: Palgrave, 2002.

Schlich, Thomas. "Tod, Geschichte, Kultur." In *Hirntod: Zur Kulturgeschichte der Todesfeststellung*, edited by Thomas Schlich and Claudia Wiesemann, 9–42. Frankfurt am Main: Suhrkamp, 2001.

Schlich, Thomas. *The Origins of Organ Transplantation: Surgery and Laboratory Science, 1880s–1930s*. Rochester, NY: The University of Rochester Press, 2010.

Schlich, Thomas. "Trauma Surgery and Traffic Policy in Germany in the 1930s: A Case-Study in the Co-evolution of Modern Surgery and Society." *Bulletin for the History of Medicine* 80 (2006): 73–94.

Schneemann, Carolee. *Imaging Her Erotics: Essays, Interviews, Projects*. Cambridge, MA, and London: MIT Press, 2002.

Schultz, Tim, and Robert Feldman. "New Moves in Sex Surgery." *Cleo* 77 (March 1979): 50–59.

Schyfter, Pablo. "Tackling the 'Body Inescapable' in Sport: Body-Artifact Kinesthetics, Embodied Skill, and the Community of Practice in Lacrosse Masculinity." *Body and Society* 14, no. 3 (2008): 81–103.

Sengoopta, Chandak. "Glandular Politics: Experimental Biology, Clinical Medicine, and Homosexual Emancipation in Fin-de-Siècle Central Europe." *Isis* 89 (1998): 445–73.

Serlin, David. "Broadcasting the Body: Performing Live Surgery on Television and the Internet since 1945." In *Imagining Illness: Public Health and Visual Culture*, edited by David Serlin. Minneapolis: University of Minnesota Press, 2007.
Seuphor, Michel. "In Defence of an Architecture." In *The Tradition of Constructivism*, edited by Stephen Bann. London: Thames and Hudson, 1974.
Sharma, J. A., A. Angelucci, and M. Sur. "Induction of Visual Orientation Modules in Auditory Cortex." *Nature* 404 (2000): 841–47.
Sharp, Lesley A. "The Commodification of the Body and Its Parts." *Annual Revue of Anthropology* 29 (2000): 287–328.
Shilling, Chris. *The Body and Social Theory*. London: Sage, 1993.
Shorter, Edward. *A History of Psychiatry*. New York: Basil Wiley, 1997.
Shtier, Rachel. *Striptease: The Untold Story of the Girlie Show*. New York: Oxford University Press, 2004.
Shutts, David. *Lobotomy: Resort to the Knife*. New York: Van Nostrand Reinhold, 1982.
Siegelbaum, Lewis H. *Stakhanovism and the Politics of Productivity in the USSR, 1935–1941*. Cambridge: Cambridge University Press, 1988.
Sigerist, Henry E. *American Medicine*. New York: W. W. Norton, 1934.
Silver, Kenneth. *Esprit de Corps: The Art of the Parisian Avant-Garde and the First World War, 1914–1925*. Princeton, NJ: Princeton University Press, 1989.
Simmel, Georg. "Die Grossstädte und das Geistesleben." *Die Grossstädte: Vorträge und Aufsätze zur Städteausstellung; Jahrbuch der Gehe-Stiftung zu Dresden* 9 (1902–1903): 185–206.
Skeggs, Bev. *Formations of Class and Gender: Becoming*. London: Sage, 1997.
Skene, Alex. "The Anatomy and Pathology of Two Important Glands of the Female Urethra." *American Journal of Obstetrics Diseases of the Woman and Child* 13 (1880): 265–70.
Slattery, Dennis Patrick. *The Wounded Body*. Albany: State University of New York Press, 2000.
Smith, Benjamin Richard. "Adjusting the Quotidian: Ashtanga Yoga as Everyday Practice." 2004. http://wwwmcc.murdoch.edu.au/cfel/docs/Smith_FV.pdf.
Smith, Bonnie. *Changing Lives: Women in European History since 1700*. Toronto: D. C. Heath, 1989.
Smith, Dale C. "Appendicitis, Appendectomy, and the Surgeon." *Bulletin for the History of Medicine* 70 (1996): 414–44.
Smith, Erin A. *Hard-Boiled: Working-Class Readers and Pulp Magazines*. Philadelphia: Temple University Press, 2000.
Smith, F. B. "Labouchere's Amendment to the Criminal Law Amendment Act." *Historical Studies* 17 (1976): 16–73.
Smith, Marquard, ed. *Stelarc, the Monograph*. Cambridge, MA: MIT Press, 2005.
Smith, Marquard, and Joanne Morra, eds. *The Prosthetic Impulse: From a Posthuman to a Biocultural Future*. Cambridge, MA: MIT Press, 2006.
Sontag, Susan. *AIDS and Its Metaphors*. New York: Farrar, Straus, Giroux, 1989.
Sontag, Susan. *Illness as Metaphor*. New York: Vintage Books, 1978.
Sontag, Susan. *Illness as Metaphor and Aids and Its Metaphors*. New York: St. Martin's, 2001.
Sorkin, Aaron. "What Kind of Day Has It Been?" *The West Wing*. Season 1, episode 22, broadcast on May 17, 2000, by NBC.

Spinrad, Norman. "The Neuromantics." *Isaac Asimov's Science Fiction Magazine*, May 1986, 186.

Spongberg, Mary. *Feminizing Venereal Disease*. New York: New York University Press, 1997.

Spronk, Rachel. "Ambiguous Pleasures: Sexuality and New Self-Definitions in Nairobi." PhD diss., University of Amsterdam, 2006.

Stanworth, Michelle. "Reproductive Technologies and the Deconstruction of Motherhood." In *Reproductive Technologies: Gender, Motherhood and Medicine*, edited by Michelle Stanworth, 10–35. Minneapolis: University of Minnesota Press, 1988.

Starr, Douglas. *Blood: An Epic History of Medicine and Commerce*. New York: Alfred A. Knopf, 1998.

Steele, Valerie. *The Corset: A Cultural History*. New Haven, CT: Yale University Press, 2001.

Steele, Valerie. *Fashion and Eroticism: Ideals of Feminine Beauty from the Victorian Era to the Jazz Age*. New York: Oxford University Press, 1985.

Stelarc and Marquard Smith. "Animating Bodies, Mobilizing Technologies: Stelarc in Conversation." In Smith, *Stelarc*, 215–42.

Stelarc and Ross Farnell, "In Dialogue with 'Posthuman' Bodies: Interview with Stelarc." In Featherstone, *Body Modification*, 129–47.

Stephens, John, and Rickard Christophers. "The Malaria Infection of Native Children." In *Reports to the Malaria Committee*, ed. Royal Society, 3rd ser., 4–14. London: Royal Society, 1900.

Stephens, John, and Rickard Christophers. "The Native as a Prime Agent in the Malaria Infection of Europeans." In *Further Reports to the Malaria Committee*, ed. Royal Society, 3–19. London: Royal Society, 1900.

Stephens, John, and Rickard Christophers. "The Segregation of Europeans." In *Reports to the Malaria Committee*, ed. Royal Society, 3rd ser., 21–24. London: Royal Society, 1900.

Stephenson, Neal. *The Diamond Age*. New York: Bantam, 1995.

Stewart, Mary Lynn. *For Health and Beauty: Physical Culture for Frenchwomen, 1880s–1930s*. Baltimore: Johns Hopkins University Press, 2001.

Stirn, A. "Body Piercing: Medical Consequences and Psychological Motivations." *Lancet* 361 (2003): 1205–15.

Stites, Richard. *Russian Popular Culture: Entertainment and Society since 1900*. Cambridge: Cambridge University Press, 1992.

Stites, Richard. *The Women's Liberation Movement in Russia: Feminism, Nihilism, and Bolshevism, 1860–1930*. Princeton, NJ: Princeton University Press, 1978.

Stone, James L. "Dr. Gottlieb Burckhardt—The Pioneer of Psychosurgery." *Journal of the History of the Neurosciences* 10 (2001): 79–92.

Stopes, Marie. *Married Love*. London: A. C. Fifield, 1918. Reprinted in an edition by Ross McKibbin (Oxford: Oxford University Press, 2004).

Stopes, Marie. *Radiant Motherhood: A Book for Those Who Are Creating the Future*. London: G. P. Putnam and Sons, 1926.

Strong, Carson. "Consent to Sperm Retrieval and Insemination after Death or Persistent Vegetative State." *Journal of Law and Health* 14 (1999): 243.

Sullivan, Nikki. *Tattooed Bodies: Subjectivity, Textuality, Ethics and Pleasure*. Westport, CT: Praeger, 2001.

Suny, Ronald Grigor. *The Soviet Experiment: Russia, the USSR, and the Successor States*. Oxford: Oxford University Press, 1998.

Surya, Michel. *Georges Bataille: An Intellectual Biography*. Translated by E. T. Krzysztof Fijalkowski and Michael Richardson. New York: Verso, 1992.

Susman, Warren I. " 'Personality' and the Making of Twentieth-Century Culture." In *Culture as History: The Transformation of American Society in the Twentieth Century*, 271–85. New York: Pantheon, 1984.

"Suspicion Surrounds Flo-Jo's Death." BBC News: Sport, September 23, 1998. http://news.bbc.co.uk/1/hi/sport/177433.stm.

Synnott, Anthony. *The Body Social: Symbolism, Self and Society*. London: Routledge, 1993.

Taithe, Bertrand. "The Rise and Fall of European Syphilization: The Debates on Human Experimentations and Vaccination of Syphilis, c. 1845–1870." In *Sexual Cultures in Europe, 1700–1995*, edited by Franz Eder, Lesley Hall, and Gert Hekma, 2:34–57. Manchester, UK: Manchester University Press, 1998.

Taschen, Angelika, ed. *Aesthetic Surgery*. Cologne: Taschen, 2005.

Tatlin, Vladimir, T. Shapiro, I. Meyerzon, and Pavel Vinogradov. "The Work Ahead of Us" (December 31, 1920), Moscow. In *The Tradition of Constructivism*, edited by Stephen Bann. Documents of 20th-Century Art Series. London: Thames and Hudson, 1974.

Taussig, Michael. *Mimesis and Alterity: A Particular History of the Senses*. New York: Routledge, 1993.

Temkin, Owsei. "The Role of Surgery in the Rise of Modern Medical Thought." *Bulletin for the History of Medicine* 25 (1951): 248–59.

Tepper, Shelley, Jaishree Jagirdar, Desmond Heath, and Stephen Geller. "Homology between the Female Paraurethral (Skene's) Glands and the Prostate." *Archives of Pathology and Laboratory Medicine* 108 (1984): 423–25.

Terry, Jennifer. *An American Obsession: Science, Medicine, and Homosexuality in Modern Society*. Chicago: University of Chicago Press, 2000.

Theweleit, Klaus. *Male Fantasies*. Vol. 2, *Male Bodies: Psychoanalyzing the White Terror*. Minneapolis: University of Minnesota Press, 1989.

Thomas, Dylan. "Do Not Go Gentle into That Good Night." In *Collected Poems 1934–1952*, 116. London: Dent, 1952.

Thompson, Charis. *Making Parents: The Ontological Choreography of Reproductive Technologies*. 2005. Cambridge, MA: MIT Press, 2007.

Tichi, Cecelia. *Shifting Gears: Technology, Literature, and Culture in Modernist America*. Chapel Hill: University of North Carolina Press, 1987.

Times. "Kiev Ban on Top Woman Sprinter." September 15, 1967.

Times, August 11, 1983.

Timmermann, Carsten. "To Treat or Not to Treat: Drug Research and the Changing Nature of Essential Hypertension." In *The Risks of Medical Innovation: Risk Perception and Assessment in Historical Context*, edited by Thomas Schlich and Ulrich Tröhler, 133–47. London and New York: Routledge, 2005.

Torgourvich, Mariana. "Piercing." In *Late Imperial Culture*, edited by Román de la Capa, E. Ann Kaplan, and Michael Sprinker. London: Verso, 1995.

Toubia, Nahid. "Female Circumcision as a Public Health Issue." *New England Journal of Medicine* 331, no. 11 (1994): 712–16.

Treicher, Paula. "AIDS and HIV Infection in the Third World: A First World Chronicle." In *Remaking History*, edited by Barbara Kruger and Phil Mariani, 31–86. Seattle, WA: Bay Press, 1989.

Trollope, Anthony. *Barchester Towers*. Harmondsworth, UK: Penguin, 1987.

Tuck, Michael. "Venereal Disease, Sexuality and Society in Uganda." In Davidson and Hall, *Sex, Sin and Suffering*, 191–203.

Turda, Marius, and P. J. Weindling, eds. *Blood and Homeland: Eugenics and Racial Nationalism in Central and Southeast Europe, 1900–1940*. Budapest, Hungary: Central European University Press, 2007.

Turner, Terence. "The Social Skin." In *Not Work Alone: A Cross-Cultural View of Activities Superfluous to Survival*, edited by Jeremy Cherfas and Roger Lewin, 112–39. Beverly Hills, CA: Sage, 1980.

Turner, Victor. "Bodily Marks." In *The Encyclopedia of Religion*, edited by Mircea Eliade, 2:269–75. New York: Macmillan Free Press, 1978.

Turner, Victor. *Dramas, Fields and Metaphors: Symbolic Action in Human Society*. Cornell University Press, 1974.

Turner, Victor. *Drums of Affliction: A Study of Religious Processes among the Ndembu*. Oxford: Clarendon, 1968.

Turner, Victor. *Forest of Symbols: Aspects of Ndembu Ritual*. Ithaca, NY: Cornell University Press, 1967.

Turner, Victor. *The Ritual Process: Structure and Anti-Structure*. Chicago: Aldine, 1969.

Turner, William. "Teaching of Moses Maimonides," *The Catholic Encyclopedia*. Vol. 9. New York: Robert Appleton, 1910. Available at http://www.newadvent.org/cathen/09540b.htm.

Turney, J.D.N., and Brian Balmer. "The Genetic Body." In Cooter and Pickstone, *Companion to Medicine*, 399–415.

Ungerleider, Steven. *Faust's Gold: Inside the East German Doping Machine*. New York: St. Martin's, 2001.

Vale, V. *Modern Primitives: An Investigation of Contemporary Adornment and Ritual*. San Francisco: Research Publications, 1989.

van der Will, Wilfried. "The Body and the Body Politic as Symptom and Metaphor in the Transition of German Culture to National Socialism." In *The Nazification of Art: Art, Design, Music, Architecture and Film in the Third Reich*, edited by Brandon Taylor and Wilfried van der Will. Winchester, UK: Winchester School of Art, 1990.

Vaughan, Megan. *Curing Their Ills: Colonial Power and African Illness*. Cambridge, UK: Polity, 1991.

Veatch, Robert. "The Impending Collapse of the Whole-Brain Definition of Death." *Hastings Center Report* 23 (1993): 18–24.

Vigarello, Georges. *Concepts of Cleanliness: Changing Attitudes in France since the Middle Ages*. Translated by Jean Birrell. Cambridge: Cambridge University Press, 1988.

von Geldern, James, and Richard Stites, eds. *Mass Culture in Soviet Russia*. Bloomington: Indiana University Press, 1995.

von Hagens, Gunther. *Body Worlds: The Anatomical Exhibition of Real Human Bodies*. Heidelberg, Germany: Institute for Plastination, 2004. An exhibition catalog.

Voracek, Martin, and Maryanne L. Fisher. "Shapely Centrefolds? Temporal Change in Body Measures: Trend Analysis." *British Medical Journal* 325 (2002): 1447–48.

Vyleta, Daniel M. "City of the Devil, Bulgakovian Moscow and the Search for the Stalinist Subject." *Rethinking History* 4, no. 1 (2000): 37–53.

Vyleta, Daniel M. "The Cultural History of Crime." In *Blackwells Companion to Nineteenth Century Europe*, edited by Stephan Berger, chap. 27. London: Blackwell, 2005.

Wachsmann, Nikolaus. *Hitler's Prisons: Legal Terror in Nazi Germany*. New Haven, CT: Yale University Press, 2004.

Wakefield, A. J., S. H. Murch, A. Anthony, J. Linnell, D. M. Casson, M. Malik, M. Berelowitz, et al. "Ileal-Lymphoid-Nodular Hyperplasia, Non-Specific Colitis, and Pervasive Developmental Disorder in Children." *Lancet* 351 (1998): 637–41.

Walkowitz, Judith. *Prostitution and Victorian Society: Women, Class and the State*. Cambridge: Cambridge University Press, 1980.

Wallechinksy, David. *The Complete Book of the Olympics*. London: Penguin Books, 1984.

Waller, John. "'The Illusion of an Explanation': The Concept of Hereditary Disease, 1770–1870." *Journal of the History of Medicine and Allied Sciences* 57 (2002): 410–48.

Walley, Christine. "Searching for Voices: Feminism, Anthropology, and the Global Debate over Female Genital Operations." *Cultural Anthropology* 12, no. 3 (1997): 405–38.

"Walter." *My Secret Life*. http://www.my-secret-life.com/index.php.

Warner, W. Lloyd. *The Family of God: A Symbolic Study of Christian Life in America*. New Haven, CT: Yale University Press, 1961.

Warthin, Aldred Scott. *The Physician of the Dance of Death: A Historical Study of the Evolution of the Dance of Death Mythus in Art*. New York: Hoeber, 1931.

Warwick, Andy. "Exercising the Student Body: Mathematics and Athleticism in Victorian Cambridge." In *Science Incarnate*, edited by Christopher Lawrence and Steven Shapin, 288–326. Chicago: Chicago University Press, 1998.

Watkins, Elizabeth Siegel. *On the Pill: A Social History of Oral Contraceptives, 1950–1970*. Baltimore: Johns Hopkins University Press, 1998.

Watson, Christine. "Touching the Land: Towards an Aesthetic of Balgo Contemporary Painting." In *From the Land: Dialogues with the Kluge-Ruhe Collection of Australian Aboriginal Art*, edited by H. Morphy and M. Smith-Bowles, 163–92. Charlottesville: University of Virginia Press, 1999.

Weatherall, D. J. "Introduction." *Journal of Internal Medicine* 247 (2000): 3–5.

Webster, Andrew. *Health Technology and Society: A Sociological Critique*. Basingstoke, UK: Palgrave Macmillan, 2007.

Weindling, Paul. *Health, Race and German Politics between National Unification and Nazism, 1870–1945*. Cambridge: Cambridge University Press, 1989.

Weingart, Peter, Jürgen Kroll, and Kurt Bayertz. *Rasse, Blut und Gene: Geschichte der Eugenik und Rassenhygiene in Deutschland*. Frankfurt am Main: Suhrkamp, 1988.

Weismann, August. *On Heredity*. Oxford: Clarendon, 1891.

Weiss, Gail. *Body Images: Embodiment as Intercorporeality*. New York: Routledge, 1999.

Wertham, Fredric. *Seduction of the Innocent*. New York: Rinehart, 1953.

Westlake, Michael. *Imaginary Women*. Winnipeg, MB: Paladin Books, 1989.

Wetzell, Richard. *Inventing the Criminal: A History of German Criminology, 1880–1945*. Chapel Hill: University of North Carolina Press, 2000.

Whipple, Beverley. "Beyond the G Spot: New Research on Human Female Sexual Anatomy and Physiology." *Scandinavian Journal of Sexology* 3 (2000): 35–42.

Whipple, Beverley, and B. R. Komisaruk. "Beyond the G Spot: Recent Research on Female Sexuality." *Medical Aspects of Human Sexuality* 1 (1998): 19–23.

White, Kevin. *The First Sexual Revolution: The Emergence of Male Heterosexuality in Modern America*. New York: New York University Press, 1993.

Whitehead, Barbara. *Sweet Death, Come Softly*. London: Headline, 1993.

Whitelaw, Mitchell. *Metacreation*. Cambridge, MA: MIT Press, 2004.

Wilkinson, Richard. *Unhealthy Societies: The Afflictions of Inequality*. London: Routledge, 1996.

Willmott, F. E. "Body Piercing: Lifestyle Indicator or Fashion Accessory?" *International Journal of STD and AIDS* 12 (2001): 358–60.

Willoquet-Maricondi, Paula. "Fleshing the Text: Greenaway's *Pillow Book* and the Erasure of the Body." *Postmodern Culture* 9 (1999): 52–82. http://muse.jhu.edu/journals/postmodern_culture/v009/9.2willoquet.html.

Wilmoth, John R., and Shiro Horiuchi. "Rectangularization Revisited: Variability of Age at Death within Human Populations." *Demography* 36 (1999): 475–95.

Wilson, Elizabeth. "The Brain in the Gut." In *Psychosomatic: Feminism and the Neurological Body*, chap. 2. Durham, NC: Duke University Press, 2004.

Witkop, Bernhard. "Paul Ehrlich and His Magic Bullets—Revisited." *Proceedings of the American Philosophical Society* 143 (1999): 540–57.

Witz, A. "Whose Body Matters? Feminist Sociology and the Corporeal Turn in Sociology and Feminism." *Body and Society* 6 (2000): 1–24.

Wolf, Naomi. *The Beauty Myth*. New York: Double Day, 1991.

"Woman Has First Face Transplant." BBC News, November 30, 2005. http://news.bbc.co.uk/1/hi/health/4484728.stm

Woods, R. "The Measurement of Historical Trends in Fetal Mortality in England and Wales." *Population Studies* 59 (2005): 147–62.

Worboys, Michael. *Spreading Germs: Disease Theories and Medical Practice in Britain, 1865–1900*. New York: Cambridge University Press, 2000.

Yalom, Marilyn. *A History of the Breast*. New York: Knopf, 1997.

Zamyatin, Yevgeny. *We*. Translated by Clarence Brown. London: Penguin, 1993.

Zimmermann, Michael. *Rassenutopie und Genozid: Die nationalsozialistische "Lösung der Zigeunerfrage."* Hamburg, Germany: Wallstein, 1996.

Žižek, Slavoj. *The Sublime Object of Ideology*. London: Verso, 1989.

Žižek, Slavoj. *Organs without Bodies: On Deleuze and Consequences*. New York: Routledge, 2004.

Zweiniger-Bargielowska, Ina. "The Body and Consumer Culture." In *Women in Twentieth-Century Britain*, edited by Ina Zweiniger-Bargielowska, 183–97. London: Pearson, 2001.

CONTRIBUTORS

Ana Carden-Coyne is codirector of the Centre for the Cultural History of War, University of Manchester, United Kingdom.

Anna Cole is a visiting fellow at the Department of Anthropology at Goldsmiths College, University of London, United Kingdom.

Anna Crozier is a lecturer at the Centre for Medical History at the University of Exeter, United Kingdom.

Ivan Crozier is a Senior Lecturer in the Science Studies Unit at the University of Edinburgh, United Kingdom.

Michael M. J. Fischer is professor of anthropology and science and technology studies at the Massachusetts Institute of Technology, United States.

Christopher E. Forth is professor of history and Howard Chair of Humanities and Western Civilization at the University of Kansas, United States.

Anna Haebich is a professor and a director of the Centre for Public Culture and Ideas at Griffith University, Australia.

Malcolm Nicolson is the director of the Centre for the History of Medicine, University of Glasgow, United Kingdom.

Dan O'Connor is a Greenwall Foundation postdoctoral fellow with joint appointments in the Berman Institute of Bioethics and the Institute of the History of Medicine at Johns Hopkins University, United States.

Thomas Schlich is professor and Canada Research Chair in History of Medicine at the Department of Social Studies of Medicine at McGill University, Canada.

D. M. Vyleta is a novelist and a historian. He lives in the USA.

INDEX

Abbas, Ali Ismail, 148–9
Abeylegesse, Elvan, 114
abjection, 3, 18
abortion, 120
acne, 129
acupuncture, 13
Adams, Parveen, 190–1
Adler, Alfred, 134
aesthetics, 2, 6, 79, 127–45, 167
Africa, 100, 104
African Americans, 9
AIDS, 10, 11, 16, 32, 33, 35, 51, 57, 93, 102, 104, 105, 167
Albury, W. R., 30
alcohol, 12
Aldebaran, Sara, 6
amputation, 21
anabolic steroids, 114
anaesthesia, 72
anal sex, 49, 50, 53
Anderson, Pamela, 119
Anderson, Warwick, 100
anorexia nervosa, 135
anthropology, 48
anti-abortion lobby, 37
antisepsis, 72
Antonioni, Michelangelo, 54
apartheid, 8
Apollinaire, Guillaume, 203
Aristotle, 221
Armstrong, Neil, 122
Arnold, David, 99
art/medicine action, 13–18
Artaud, Antonin, 181, 187, 190
artificial
  penis, 82
  vagina, 82
Ashley, April, 117
Asimov, Isaac, 197
aspirin, 143
assisted suicide, 121
atherosclerosis, 95
Atlas, Charles, 138
Austin, J. L., 47
Australian Aboriginals, 8, 151
Australian Women's Weekly, 136
Avery, Eric, 13–18, 52, 167
avian influenza *see* H5N1

Bach, Johann Sebastian, 26
Back, Les, 162
Bacon, Kevin, 121
Baden-Powell, Robert, 5
Balzac, Honoré de, 221
Banks, Iain M., 247
Barker, Francis, 168
Barker, Joanna, 117
Barnes, Barry, 45–7
Barthes, Roland, 67
Bataille, Georges, 171
Bauman, Zygmunt, 144
BDSM, 57
Beatles, The, 51
beauty, 6, 129
  contests, 133
Beccaria, Cesare, 223
Belgium, 27
Bell, Clive, 203

Beloved, 176
Benjamin, Walter, 166, 168, 191
Bertolucci, Bernardo, 50
BIDD, 117–19
BIG SICK LIVER, 14–15
Billings, Dwight, 82
bioart, 179
biomedical bodies, 12
biopolitics, 198
biosurgical art, 182–95
biotechnology, 168
Birbaumer, Nils, 195
Bird-Rose Deborah, 152
binary opposition, 148
Birkin, Jane, 54
birth control *see* contraception
"Blackwash," 9
blood, 164
  transfusion, 86
Boccioni, Umberto, 203
Boddy, Janice, 159
Bodies Without Organs, 187
body, 1, 18
  building, 135–9
  and identity, 18–21, 117
  image, 132–5
  marked, 147–65
  marking, 150
  marks, 148
  modifications of, 48, 162 *see also* branding; cicatrization; suspension; tattoo; tattooing
  reappropriation of, 164
  and sexuality, 155–9
  working-class, 228, 240
Body Identity Disintegration Disorder (BIDD), 117–18
Body Modification Ezine (BME), 179
Botham, Ian, 9
Botox, 6
bottom, 20
Bradbury, Ray, 197, 198
brain, 30
brain death, 74, 89, 90, 120
BrainGate, 186
branding, 58, 148, 162
Braque, Georges, 203
Brazilian, 55, 55–7
breasts, 132–4
  enhancement of, 53
British Board of Film Classification, 65
British Medical Association, 27
  BMA Ethics Commission, 27
British Medical Journal, 60
Broome, Karl, 161–2
Brown, Helen Gurley, 51
bulimia, 135
Buñel, Luis, 170, 172, 202
Burgess, Anthony, 244–5
Butler, Judith, 7, 10, 148

cadavers, 124
Caidin, Martin, 118
cancer, 93
cannabis, 13
Canter, David, 94
Čapek, Karel, 229–31
Caplan, Abe, 186
Caribbeans, 8
casualty, 125
casual sex, 99
Catholicism, 107
Catts, Oran, 187, 195
  see also Zurr, Ionatt
censorship, 65, 123, 124
chaining, 20
"Chelsea girl," 57, 61, 62
childbirth, 6, 21, 39
child mortality, 26
Children's Fund, 102
cholera, 95
Christianity, 26
Christian morality, 49
cicatrization, 148, 152–4, 178
cinema, 6, 44
circumcision
  female, 155–9, 176
  male, 175–6
civilization, 155
Cixous, Hélène, 177
class, 2, 5
Classen, Constance, 198
classicism, 203
clinical science, 94
clitoridectomy, 156
clitoris, 63, 157, 158
A Clockwork Orange (film and novel), 244–5
clothing, 6
Clynes, Manfred, 194
Cobb, Portia, 211
Cold War, 111, 113
Colombo, Realdo, 63
colonialism, 9, 99–102
colonization, 8, 9
Communism, 113, 114, 228, 233, 238–41
condoms, 11, 51, 105

contagion, 95–9
Contagious Diseases Acts, 96
contraception, 6, 12, 48, 51, 68
convulsive therapy, 76
Cook, James, 160
corpse, 18
cosmetics, 129–30, 166
  male use, 130–1
cosmetic surgery, 73, 78–80, 128, 131, 134, 144
Courbet, Gustave, 54, 55
Crapanzano, Vincent, 169
cremation, 31–2
cricket, 9
criminology, 160, 226
Crosby, Bing, 27
Cruise, Tom, 121
cryonics, 33
cultural history, 1
culture, 10, 21, 45
cunnilingus, 49, 56, 57
cybernetics, 89
cyberpower, 179
cyberspace, 246
cyborgs, 61, 118, 119, 194
cytheria, 65

Dadaism, 201
Dali, Salvador, 170
Daly, Bob, 112
Daly, Mary, 156
Daly, Richard, 112
Dawkins, Richard, 244
Dawson, David, 33
DDT, 102
Dean, James, 120
death, 23–36
decolonization, 102
degeneration, 226
Deleuze, Gilles, 160, 187
  and Guattari, Félix, 196
de Ligt, Bart, 206
DeMello, Margo, 161
democracy, 242
Dénatalité, 204
depression, 76
Derrida, Jacques, 175
Desbonnet, Edmond, 138
diet, 5, 12, 99, 93, 118
difference, 3
Dignity in Dying, 27
discipline, 5
disciplined bodies, 3–4
disease, 72, 93–108, 149, 198
divine eye, 169–74
DNA, 87, 88
Dobson, Betty, 68
Dobson, Frank, 204
docile bodies, 4, 43
dogging, 68
Donne, John, 24, 33
Dostoyevsky, Fyodor, 221242
Douglas, Mary, 18
Dreger, Alice, 75, 80
drosophilia, 173
drugs, 198
  use of, 20, 198
Duchamp, Marcel, 201
Duden, Barbara, 116, 119
Durbach, Nadja, 98
Durkheim, Emile, 224, 226

ECDC, 97
education, 2
Ehrlich, Paul, 104
electroencephalograph, 29
Elias, Norbert, 1
Ellis, Havelock, 49, 63, 141
Eliot, George, 221
England, 9
Enlightenment, 10, 222
E.R., 125
estrogen, 129
ethnicity, 2, 8
eugenics, 234–35
European Athletics Championships, 112
European Centre for Disease Prevention and Control, 96
euthanasia, 27, 120–1
Ewald, Manfred, 113
exercise, 4
existentialism, 33
exoticism, 8
Exposition Nationale de la Maternité et de l'Enfance, 204
Extreme Makeover, 2
Eyton, Audrey, 118

face, 128–31
facial hair, 6
Fairbanks, Douglas, 140
Fanon, Frantz, 102, 166
Farr, William, 23
Fascism, 171, 242
fashion, 6, 19
fat, 5, 6, 99, 134
Fat Liberation Manifesto, 6
fellatio, 49, 58

female
body, 162–4
  ejaculation, 62–6
  feminism, 6, 10, 44
fetish, 57
fetus, 37, 40, 106
Figlio, Karl, 75
Fleming, Jane, 110
Fletcher, Horace, 228
Flint, Larry, 55
food, 5
Ford, Henry, 143
Fordism, 202, 228
Forth, Christopher, 5
Foucault, Michel, 1, 2, 3–4, 10, 48, 68, 160, 165, 168
Fox, Renée, 85, 86, 184
Freeman, Walter, 77
Freespirit, Judy, 6
Freikörperkultur see naturism
Freud, Sigmund, 44, 62, 141, 157, 186, 187, 233
Fresnaye, Roger de la, 204
Friedrich, Ernst, 127
Fry, Roger, 203
Furness, Steve, 172
Fussell, Sam, 139
futurism, 228

G-spot, 62, 64, 65
Gabo, Naum, 207
Gattaca, 168, 195
Gaudier-Brezska, Henri, 203
gay
  culture, 160
  men, 104
  rights, 243
Gell, Alfred, 164
gender, 5, 81
gendered bodies, 6–8
genetic
  counselling, 106
  disease, 93, 106–7
genital piercings, 57, 67, 68
genomics, 107
germs, 96
Gibson, William, 179, 191, 245
Giddens, Anthony, 50, 127, 160
GI Joe, 139
Gill, Eric, 204
Gilman, Charlotte Perkins, 26
Gilman, Sander, 118
Gimzewski, Jim, 183
Glasgow, 26
Gleizes, Albert, 204
Go Ask Alice, 121
Golan, Romy, 204
Goldschmidt, Walter, 159
Gómez-Peña, Guillermo, 179
Gough, Steve, 67
Goya, Francesco, 54
Gräffenberg, Ernest, 63
Greenaway, Peter, 177
Grieg, Tony, 9
Griffin, John Howard, 173, 174–8
Griffith Joyner, Florence, 109, 110
Gris, Juan, 203
Gross, Otto, 233
Grosz, Elizabeth, 148, 155
Gulbenkian Foundation, 217
Gulick, Luther, 143
gushing, 64

H5N1 (avian influenza), 12, 167
Habermas, Jürgen, 196, 198
Hadlee, Richard, 9
Haiken, Elizabeth, 79, 87, 118
hairlessness, 56–8
Hamilton, Richard, 210
Haraway, Donna, 118, 173, 217
Harbou, Thea von, 232
Harvard Medical School, 29, 120
Hass, Wolf, 26
Hausman, Bernice, 80, 81
Hawking, Steven, 183–4
Hays Code, 123–5
health, 4, 21
  practices, 5
healthy bodies, 4–6
heat, 8
Hemmings, David, 54
hepatitis C virus (HCV), 13–18, 52
herbalism, 13
heterosexual, 68
Hetherington, Penelope, 157–8
Hills, Gillian, 54
Hinton, S.E., 121
historiography, 1
Hite, Shere, 44, 62
Hitler, Adolf, 234
HIV, 10, 12, 13–18, 16, 51, 52, 57, 167
Hobbes, Thomas, 223
Hoberman, John, 9, 113, 141
Holding, Michael, 9
home birthing, 13

homosexual
  body, 68
  rights, 44
homosexuality, 16, 47, 51, 68
Horatius, 27
Hornby, Wilfred, 168
Hornibrook, Fred, 205
Houston Police Department, 52
Huffman, John, 63
Hughes, Howard, 123
Human Genome Project, 244
humiliation, 57
Hustler, 55
hygiene, 55, 57, 93, 96

Ice Cube, 174
identity, 19, 43
immunology, 87
individual, 5
individualism, 5, 242
influenza, 24, 35
Internet, 11, 44, 66, 70
intravenous injections, 105
in vitro fertilization, 11, 107

jail, 20
jazz, 201
Jews, 131, 135, 235
Johnston, Virginia, 62, 63, 64
  see also Masters, William
Jonson, Ben, 114
Jorgensen, Christine, 117
Jorlemon, Donald, 86
Joseph, Jacques, 131

Kafka, Franz, 160
Kahnweiler, Daniel, 203
Kant, Immanuel, 223
Kaufmann, Denis Abramovich, 173
Kelly, Paula, 54
Kelogg, John Harvey, 228
Kennedy, John Fitzgerald, 123, 124
Kevorkian, Jack, 120
Kierkegaard, Søren, 32
Kikuyu, 157
Kinsey, Alfred, 44, 138
Kirby, Vivki, 148,157
Klein, Melissa, 179
Kline, Nathan, 194
Kłobukowska, Ewa, 112
Knight, Laura, 204
Kocher, Theodor, 72, 83
Kratochvilova, Djamila, 109
Kristeva, Julia, 18, 19, 117, 200
Krupskania, Nadezdha, 237
Kübler-Ross, Elisabeth, 36
Kubrik, Stanley, 244
kuruwarri see cicatrization
Kutjungka people, 153

Lacan, Jacques, 191
lactation, 149
Lancet, 61, 98
land, 153
Lang, Fritz, 232
Lapper, Alison, 150
Laqueur, Thomas, 65
largesse, 6
Latin America, 31
law, 20–1
Lawrence, Christopher, 72, 75
Le Corbusier, 205
Léger, Fernand, 203, 204, 205, 206, 207
Legman, Gershoy, 138
legs, 133, 144
Lemoine-Luccioni, Eugenie, 188
Lenon, V. I., 237
Lerner, Paul, 227
Lewis, Carl, 114
lifestyle, 4
Lightfoot, Henry, 159
Lillee, Dennis, 9
Lindeland, Liv, 55
liposuction, 6
LIVER DIE, 14–16
lobotomy, 76–7
Loeb, Gerald, 185
Lyotard, Jean-François, 182, 196

M*A*S*H, 125
Macauly, Thomas, 27
Macfarlane Burnet, Frank, 87
machines, 228–33
McLuhan, Marshall, 188, 191
magazine, 6, 44, 66
Malevich, Kasimir, 206
Mann, Steve, 172
Man Ray, 201
Marinetti, Filippo Tommaso, 140
marriage, 6
Martin, Emily, 87
Marvel comics, 116
Marx, Karl, 221, 224, 226
Marxism, 1, 237–41
Massumi, Brian, 193

Masters, William, 62, 63, 64
  see also Johnston, Virginia
masturbation, 62, 63, 65
maternal mortality, 39
medical
  history, 1
  innovations, 6
  technology, 29
Medical Research Council, 217
medicine, 2, 4, 10
Meet Joe Black, 32
Meijer, Peter, 172
Mejia, Michael, 169
Mensendieck, Bess, 132
menstruation, 149
Merleau-Ponty, Maurice, 198
methicilin-resistant *Staphylococus aureus* (MRSA), 12
Metropolis (film), 224, 232–3
Mexico, 31
Mialet, Hélène, 181, 183, 184
Michelangelo, 167
midwifery, 13
migration, 101
Mises, Ludwig von, 242
Miss America, 134
Miyake, Issey, 188
modernism, 2, 6, 207
Modigliani, Armado, 54
Mol, Annemarie, 95
Mondrian, Piet, 207
Moniz, Egals, 76
morality, 12
Morris, Jan, 117
Morrison, Toni, 176
mosquito, 102
motherhood, 133
mothers
  genetic, 36
  gestational, 36
muscularity, 135–9

Nagle, Matthew, 186
nanotechnology, 194, 246
natural
  bodies, 44
  kinds, 21, 45–7
nature, 44
naturism, 5, 66, 67
Nazism, 222, 234–7
  Neo-Nazism, 243
Neshat, Shirin, 170, 171
Netherlands, 27
neurasthenia, 101
New Man, 202
New Woman, 133
Ngong Hills (Kenya), 8
NHS, 98, 117
Nichols, Kelly, 55
Nietszche, Frederich, 160, 173, 233
nipples, 58
Nixon, Richard, 123
Nobel Prize, 72
Noël, Suzanne, 131
Notting Hill Carnival, 9
nudity, 66–7, 133

obesity, 244
Obeyesekere, Gananath, 169
obstetrics, 40
O'Connor, Flannery, 168, 175
Olympic Games, 109–11, 113, 125, 140
Orbach, Susie, 6
organic chic, 5
organ transplants, 12, 30
orgasm, 44, 62, 65, 66
  clitoral, 62
  vaginal, 62
  vaginal v clitoral, 157
ORLAN, 19, 183, 187–92
Orwell, George, 230
Owen, Wilford, 27
Ozenfant, Amédée, 204, 205

paedophilia, 48
pain, 21, 152, 164
  relief of, 12
Pasolini, Pier Paolo, 50
Pavlov, Ivan, 242
Pearl, Raymond, 24
Pearson, Karl, 23, 24
penis enhancement, 53
Penthouse, 55
performance, 3, 142
performativity, 7, 47, 62, 70, 127
persistent vegetative state, 30–1
Pevsner, Antoine, 207
pharmacogenomics, 108
Phoenix, Isus, 67
photography, 2
Picasso, Pablo, 203, 204
piercings, 19, 20, 151, 162
  Prince Albert, 58, 62
  Princess Diana, 58
the pill, 10, 51
The Pillow Book, 177
Piss Christ, 214
Pitts, Victoria, 162, 168

plague, 100
plastic surgery, 6, 7
Playboy, 54–5, 135
Playgirl, 139
pleasure, 2, 20, 21
Polanyi, Michael, 3
pornography, 11, 50, 54, 55, 65, 68–9
Porter, Dorothy, 114
Porteus, Stanley, 155
post-colonialism, 8
Powell, Asafa, 114
power, 5, 6, 47, 68, 108
practice, 21, 46
Pratchett, Terry, 32
pregnancy, 36–7, 41
Press, Tamara, 111, 112
Pressman, Jack, 77
Prince Albert piercing *see* piercings
Princess Diana piercing *see* piercings
Proctor, Dod, 204
prohibition, 20–1
prostate, 63
prostitution, 54, 96
Prozac, 143
psychiatry, 76, 93, 226–7
psychoanalysis, 73, 223
psychosurgery, 75
psychotropic drugs, 73, 77
pubic
  hair, 6, 48, 53–8, 67–8
  shaving, 54, 55
public health, 52, 105
punishment, 21

Quinn, Marc, 150

Rabanne, Paco, 188
Rabinow, Paul, 86
race, 8, 128–9, 234
racial
  bodies, 8–10
  science, 234–5
Rambo, 139
Ramsay, Paul, 85
Rant, Justice, 20
rape, 164
Read, Herbert, 204
regulation, 13
religion, 11, 12, 31, 33, 71, 106, 168, 169
representation, 3
resistance, 108
Rhodes, Rosemary, 29
Richards, Renée, 117
ritual, 179
Roberts, William, 204
robots, 229–30
Rodlubnyi, Stephan, 242
Roe v. Wade, 119, 120
Romanticism, 221
Rose, Nikolas, 143
Rout, Ettie, 205
Rowling, J. K., 32
Royal Geographical Society, 101
R.U.R., 229–31
Russell, Jane, 123
Rutheford, Joseph, 33

sadomasochism, 20, 178, 179
safe sex, 10, 68
St. Elsewhere, 125
sanctioning, 2, 45, 70
Sandow, Eugen, 138
Sanger, Margaret, 205
Santa Muerte, 32
Sarno, John, 186
Sartre, Jean-Paul, 33
Saunders, Cicely, 36
scarification, 151–5, 162, 178
Schlemmer, Oskar, 206
Schottmuller, Oda, 208
Schwabe, Carlos, 32
Schwarzenegger, Arnold, 139
schyzophrenia, 76
scouting, 5
scrotum (shaved), 56, 57
secularization, 70
self, 18
  expression of, 10
  referentiality of, 46–7
sex, 21, 43–70
  education, 11, 44, 105
sexology, 43, 44, 140–1
sex reassignment surgery, 74, 80–2
sexual
  emancipation, 142
  intercourse, 11
  performance, 49
  piercings, 58–62
  pleasure, 6, 10, 43, 49
  reproduction, 11
sexual bodies, 10–13, 43–70
sexual body modification, 43, 53–66
sexuality, 129, 140, 162
sexually transmitted infections (STI), 48, 50, 57, 68, 104–6
Shimla (India), 8
shock therapy, 76

Sigerist, Henry E., 73
silicone implantations, 134
Simmel, Georg, 224, 226
situated bodies, 4
Skene, Alex, 63
Skene's glands, 63
skin, 129
Skinheads, 243
sleeping sickness, 100
slenderness, 132–5
Smallpox, 12, 98
smoking, 244
snakes, 172
Social Darwinism, 234
social kinds, 44–7
Socialism, 237–42
sociology, 243
Sorine, Sandy, 204
South Africans, 8
South Asians, 8
Soviet Union, 237–41
Spanner trial, 20
spatiality, 95, 108
speed limits, 74, 75
Spiderman, 116
sport, 19
Stakhanov, Aleksei, 240
Stakhanovite, 143, 240
Stalin, Joseph, 143, 238
Stalinism, 222, 224–41
Star Wars, 139
Steinach, Eugen, 141
Steinem, Gloria, 156
Stelarc, 19, 183, 187, 188, 191–5
stem cells, 107
  research on, 12
Stephenson, Neal, 246
stethoscope, 29
stillbirth, 40, 41
Stopes, Marie, 49, 205
Strathern, Andrew, 168
Strathern, Marilyn, 168
Strauss, Richard, 41
subjectivity, 7
Sudden Acute Respiratory Syndrome
    (SARS), 16, 96
suicide, 26
suicide bombers, 166
Suleiman, Elia, 170
Sundahl, Deborah, 65
Superman, 116
surgery, 71–92
surrealism, 201
suspension, 58, 162
The Swan, 7
Swazey, Judith, 85–6, 184
swinging, 68
syphilis, 10, 93

Tarzan, 138
Tatlin, Vladimir, 203
tattooed body, 160–4
tattooing, 148, 162
tattoos, 19, 20, 53, 58, 160–4, 168,
    174–8
  and class, 161
  and gender, 162–4
Taussig, Michael, 198
Taylor, Frederic Winslow, 142
Taylorism, 202, 205, 228, 231, 240
technological fix, 71
teenage body, 121–2
Ten Years Younger, 2, 7
test-tube babies *see* in vitro
    fertilization
Thaipusam, 151, 154
thanatology, 36
Third Ear, 195
Thomas, Dylan, 33
Tiergarten (Berlin), 67
Tolstoy, Leo, 221
topless sunbathing, 133
Toubia, Nahid, 159
training, 5, 21, 66
transexuality, 74, 117
transplantation, 74, 83–7
trauma, 92–83
Trollope, Anthony, 23
tropical medecine, 99–104
tuberculosis, 93
Turner, Terence, 168
Turner, Victor, 168, 175, 177, 178

ultrasound, 37, 41
Un Chien Andalou, 170
UNICEF, 102
United Nations, 102
Urban, Thomas, 82
urination, 63–4
U.S. Center for Disease Control, 96
utopia, 206

vaccination, 98
  MMR, 98, 108
Valentino, Rudolf, 140
van Damme, Jean-Claude, 119
van Doesburg, Teo, 206, 207
van de Velde, Theodoor, 49, 50, 141

vasectomy, 141
vegetarianism, 5
Verhoeven, Paul, 119
vertical clitoral hood piercing, 58–60, 61
  Warlpiri people, 152–4
Viagra, 141
vibrator, 65
Visine, 56
Voluntary Euthanasia Society, 27
von Hagen, Gunther, 183

Walley, Christine, 156
Walter, 54
war, 1, 166
Warlpiri, 152, 153
Watson, Christine, 153
Watt, James, 77
waxing, 56, 57
Weismann, August, 24
Weiss, Gail, 67
Wertham, Fredric, 134
West Indies, 9
whipping, 20, 148
Whipple, Beverly, 64
White, Karin, 141
whiteness, 8
Wiendling, Paul, 11
Wiener, Norbert, 89
Wilks, Michael, 27
Wolpaw, Jonathan, 186
women, 2
women's bodies, 2, 149
Wonderwoman, 116
World Health Organization, 102
World League for Sexual Reform, 44
World War I, 6, 24, 78, 84, 100, 117, 127, 166, 201, 202, 222, 227
World War II, 6, 76

X Men, 116
X-ray, 129

Yarralin People, 152
Yeats, W.B., 141
yoga, 5
  nude, 67

Zamyatin, Yevgeny, 230
Zaretsky, Adam, 183
Žižek, Slavoj, 196, 241
Zurr, Ionatt, 187, 195
  see also Catts, Oran